BEYOND INITIAL RESPONSE
Using the National Incident Management System's Incident Command System

by

Tim Deal — Michael de Bettencourt — Vickie Deal

Gary Merrick — Chuck Mills

AuthorHouse™
1663 Liberty Drive
Bloomington, IN 47403
www.authorhouse.com
Phone: 1-800-839-8640

© 2010 Tim Deal, Vickie Deal, Michael de Bettencourt, Gary Merrick, and Chuck Mills. All rights reserved.

No part of this book may be reproduced, stored in a retrieval system, or transmitted by any means without the written permission of the author.

First published by AuthorHouse 1/28/2010

ISBN: 978-1-4389-8861-0 (sc)

Library of Congress Control Number: 2009912653

Printed in the United States of America
Bloomington, Indiana

This book is printed on acid-free paper.

— This Book Is Dedicated —

TO

THOSE

WHO

Save Lives

Protect Property

Protect Environment

Book Review published in the International Association of Emergency Managers Newsletter:

By Daryl Lee Spiewak, CEM, TEM, Emergency Programs Manager, Brazos River Authority, Waco, Texas

"Overall, I found *Beyond Initial Response: Using the National Incident Management System's Incident Command System* to be a highly readable reference valuable for training, exercises and actual response. I recommend that all first responders and emergency managers seeking to better understand and implement ICS add this book to their library. I have it in mine and it is already wearing out from use."

CONTENTS

Figures x

Second Edition xiii

How to Use This Book.. xv

Acknowledgments xvii

Foreword: Bringing Order Out of Chaos xix

1. **Incident Command System Overview**.. 1-1
   - Common Terminology 1-2
   - Span-of-Control 1-14
   - Modular Organization 1-15
   - Unified Command Structure 1-33
   - Incident Action Plan 1-34

2. **Incident Command System Planning Process** 2-1
   - The Operational Period 2-3
   - The Operational Planning "P" 2-4
   - The ICS Planning Process 2-5
   - The Operational Period and Operational Planning "P" 2-15
   - Determining the Meeting Schedule 2-17

3. **Incident Commander (IC)** 3-1
   - Section One: Initial Response 3-3
   - Section Two: Document Your Initial Response 3-6
   - Section Three: Transfer of Command 3-11
   - Section Four: Incident Commander and the ICS Planning Process 3-15

4. **Unified Command (UC)** 4-1
   - Makeup of the Unified Command 4-2
   - Unified Command Responsibilities and Expectations 4-4
   - Unified Command and the ICS Planning Process 4-5

5. **Command Staff**... 5-1
   - Safety Officer 5-2
   - Safety Officer's Role in the ICS Planning Process 5-6

- Liaison Officer 5-24
- The Liaison Officer's Role in the ICS Planning Process 5-29
- Public Information Officer 5-31
- The Public Information Officer's Role in the ICS Planning Process.................. 5-39
- Intelligence/Investigations Function 5-41
- The Intelligence/Investigations Officer's Role in the ICS Planning Process........ 5-43

6. **OPERATIONS SECTION CHIEF (OSC)** 6-1

- The OSC Role in the ICS Planning Process 6-5
- Daily Self-Evaluation of the Operations Section 6-25

7. **PLANNING SECTION CHIEF (PSC)** 7-1

- Establishing the Units of the Planning Section 7-6
- Managing the Planning Team 7-19
- Resources Unit Leader 7-23
- Situation Unit Leader 7-57
- Documentation Unit Leader 7-69
- Demobilization Unit Leader 7-75

8. **LOGISTICS SECTION CHIEF (LSC)** 8-1

- Support Branch Organization 8-6
- Service Branch Organization 8-9
- LSC Role in the ICS Planning Process 8-12

9. **FINANCE / ADMINISTRATION SECTION CHIEF (FSC)** 9-1

- FSC Role in the ICS Planning Process 9-8

10. **AREA COMMAND** 10-1

- Determination to Activate an Area Command 10-2
- Area Command Operating Cycle 10-3
- Area Command Coordination 10-17
- Area Command Organization 10-18

11. **MULTI-AGENCY COORDINATION** 11-1

- On-scene Incident Command and the EOC 11-3
- On-scene Incident Command, Area Command, and the EOC 11-6
- On-scene Incident Command, Area Command, EOC, and JFO 11-8

12. EMERGING COMMUNICATIONS 12-1
- What is New Media? 12-2
- Why New Media is Important to you the Incident Commander 12-5
- Actual uses of New Media during Response Operations................................... 12-6

CONCLUSION

APPENDIXES

APPENDIX A: SAMPLE INCIDENT ACTION PLAN (IAP)

APPENDIX B: POSITION-SPECIFIC OPERATIONAL PLANNING "P"s

APPENDIX C: MANAGING RISKS USING THE ICS PLANNING PROCESS

APPENDIX D: EXAMPLE ICS-214 UNIT LOG

APPENDIX: E JOINT INFORMATION CENTER MEDIA ANALYSIS WORKSHEET

APPENDIX F: EXAMPLE INCIDENT BRIEFING FORM, ICS-201

APPENDIX G: JOINT INFORMATION CENTER QUERY RECORD

APPENDIX H: EXAMPLE OPERATIONS BRIEFING CHECKLIST

APPENDIX I: PLANNING SECTION CHIEF SUPPORT KIT CHECKLIST

APPENDIX J: EXAMPLE DEMOBILIZATION PLAN "YAZ NORTHERN INCIDENT"

APPENDIX K: MEETING RULES AND AGENDAS

APPENDIX L: INSTRUCTIONS TO THE FIELD OBSERVERS

APPENDIX M: BEST BRIEFING PRACTICES

APPENDIX N: FACILITIES NEEDS ASSESSMENT WORKSHEET

APPENDIX O: INCIDENT COMMAND POST CHECK-OFF SHEET

APPENDIX P: INCIDENT COMMAND POST MOVING PLAN "MERIDIAN FLOOD INCIDENT"

APPENDIX Q: AREA COMMAND OPERATIONAL CYCLE (COMMAND ACTIVITIES)

APPENDIX R: DELEGATION OF AUTHORITY/ DELINEATION OF RESPONSIBILITIES

APPENDIX S: AREA COMMAND OPERATING GUIDE

ABOUT THE AUTHORS

FIGURES

A Large Operations Organization	1-26
Area Command Coordination	10-17
Area Command Operating Cycle	10-3
Area Command Operational Cycle (Command Activities)	Q-1
Area Command Organization	10-18
Assignment List (ICS-204)	5-22
Assignment List (ICS-204)	7-52
Back of the Rose-color T-card	7-33
Back of the White-color T-card	7-35
Check-in List ICS-211	7-28
Command and General Staff	3-14
Command and General Staff Meeting	K-4
Command Staff	1-17
Components of the Incident Action Plan	1-36
Demobilization Checkout Form	7-82
Demobilization Resource Tracking Table	7-83
Documentation Unit Staffing Chart	7-71
Evolution of an Incident Action Plan	2-14
Example Area Command Operating Guide Contents	10-15
Example ICS-214 Unit Log	D-1
Example Incident Briefing Form, ICS-201	F-1
Expanding Response Organization	1-24
Facilities Needs Assessment Worksheet	N-1
Facility Symbols	1-13
Finance/Administration Section	9-5
Finance/Administration Section Organization	1-32
Front of the Rose-color T-card	7-32
Front of the White-color T-card	7-34
General Staff	1-19
Helibase	1-10
Helicopter Typing Matrix	1-7
ICS Map Symbols	7-62
ICS Organization Chart	1-1
ICS Organization Chart	7-24
Identification of Response Strategies and Tactics	5-9
Incident Action Plan	7-50
Incident Commander/Unified Command Operational Planning "P"	B-2
Incident Command Post Check-off Sheet	O-1
Incident Map	1-14

Initial Response Organization	1-23
Initial Unified Command Meeting Agenda	K-2
Job Safety Analysis	C-13
Joint Information Center Media Analysis Worksheet	E-1
Joint Information Center Query Record	G-1
Level of Risk Matrix	C-4
Logistics Section Chief Operational Planning "P"	B-8
Logistics Section Organization	1-30
Management Activities in the Incident Command System	1-16
Meeting Rules and Agendas	K-1
On-scene Incident Command, Area Command, the EOC, and the Joint Field Office	11-8
Operational Periods	2-16
Operational Planning "P"	1-35
Operational Planning Worksheet (ICS-215)	5-11
Operations Briefing	K-7
Operations Organization Continued Growth	1-25
Operations Section Chief Operational Planning "P"	B-4
Operations Section Organization	1-20
Organizational Elements	1-3
Organization Assignment List (ICS-203)	7-54
Organization Chart ICS-207	7-56
Planning Meeting	K-6
Planning Section Chief Operational Planning "P"	B-5
Planning Section Daily Work Schedule	7-21
Planning Section Organization	1-28
Position Titles	1-4
Principal Components of the Incident Action Plan	2-12
Relationship of the On-scene Command, Area Command and the EOC	11-7
Resource Assignment Flowchart During an Operational Period	7-41
Resources Unit Leader Operational Planning "P"	B-6
Resources Unit Organization	7-27
Resources Unit Staffing Guide (per 12-hour period)	7-26
Safety Officer Operational Planning "P"	B-3
Sample Incident Action Plan	A-2
Seven T-Card Colors	7-31
Situation Unit Leader Operational Planning "P"	B-7
Span-of-Control	1-15
Strike Team	1-8
Support Position Titles	1-4
Tactics Meeting	K-5
Task Force	1-8
Unified Area Command	10-19
Unified Command	1-34

Unified Command Develop/Update Objectives Meeting K-3
Upper Left Corner of the ICS-215A 5-16
Upper Left Section of the ICS AC-215 10-9
Upper Left Side of the Operational Planning Worksheet 5-12
Upper Right Corner of the ICS-215A 5-17
Upper Right Corner of the Operational Planning Worksheet 5-13
Upper Right Section of the ICS AC-215 10-11

SECOND EDITION

It has been a little more than three years since we first published *Beyond Initial Response*. Since that time the book has been put through its paces. Thousands of responders ranging from those who are highly skilled to those who are just entering the response discipline have used the book.

We have made several changes in the second edition of *Beyond Initial Response*. The changes are based on using the book in numerous Incident Command System courses and exercises. The major changes that we made included three new chapters: Area Command, Multi-agency Coordination, and Emerging Communications. In addition to the new chapters, we significantly expanded our discussion of the Safety Officer.

The Area Command chapter is written from the perspective of the Area Commander and is the most comprehensive written explanation of Area Command that we have seen published. The Multi-agency Coordination chapter is a brief overview of how multi-agency coordination works from the perspective of the Incident Commander, Area Commander, Emergency Operations Center, and Joint Field Office. It was a difficult chapter to write as the subject of multi-agency coordination can easily be a stand-alone book. Our goal was to provide some understanding of how all of the response organizations interact together to provide an efficient and effective response. The chapter on Emerging Communications introduces the world of New Media (a term to indicate the emergence of digital, computer, network, and communications technologies) and its potential challenges and benefits to incident management teams.

For the Safety Officer chapter we expanded the text to include a thorough discussion on how Safety Officers conduct a hazard analysis using the ICS process and forms such as the ICS-215a, Incident Action Plan Safety Analysis Worksheet. There is also a detailed discussion in the Appendix on managing risk by using the Incident Command System planning process.

You will also find some new "tools" in the appendixes of the book. These include: Managing Risk, Best Briefing Practices, Sample Instructions to the Field Observer, Delegation of Authority, OSC Operations Briefing Outline, and a sample Area Command Operating Guide.

Beyond Initial Response, second edition, is loaded with many examples of ICS forms that we have filled in so you can see how the forms are used. The struggle we had as authors was trying to determine whose forms to use. Many agencies have developed their own versions. The Department of Homeland Security has worked to create all-hazard forms, the wildland firefighting community has versions of ICS forms that they have been using, and the US Coast Guard has a version that they implemented. What we decided was to select the forms that best enabled us to demonstrate how the form is used. Every form used in this book is available from these agencies, and we have listed the sources for the various forms in the section of the book titled Conclusion.

We hope that you find Beyond Initial Response, second edition, a useful tool in implementing the Incident Command System.

Chuck Mills

Chuck Mills

President

Emergency Management Services International, Inc.

The views expressed in this book do not necessarily represent the views of the U.S. Government, the Department of Homeland Security, or the Federal Emergency Management Agency.

HOW TO USE THIS BOOK

Beyond Initial Response is designed to be used both as a reference and as a response tool. You can choose to read the whole book from cover to cover or select one of the Incident Command System position-specific sections. The concepts of the Incident Command System are so intertwined that they can be difficult to peel apart and present in a building block fashion. In organizing this book, we struggled with whether to discuss the ICS Planning Process first or each of the key ICS positions on the Incident Management Team. We opted to discuss the ICS Planning Process first and then key positions on an Incident Management Team and each of their roles in the ICS Planning Process. We recommend that if you are going to read only about a particular position that you consider reading Chapter 2, Incident Command System Planning Process, as well.

Position Codes

Throughout this book, you will see the use of position codes. For example, the position code for the Resources Unit Leader is RESL. The codes that we have used in this book come from the National Wildfire Coordinating Group (NWCG). Some of the position codes on the NWCG list contain a number after them such as the code for Safety Officer. For example, the Safety Officer position code can be written as SOF1 or SOF2. The number indicates the level of qualification. The highest ICS-trained and -experienced Safety Officers would have a position code of SOF1. We do not place any numbers at the end of the codes in the book, so you'll see only SOF for the Safety Officer. Position codes are also another example of common terminology so you want to get comfortable using them.

Position codes used in this book:

(AOBD) Air Operations Branch Director
(BCMG) Base/Camp Manager
(COML) Communications Unit Leader
(COST) Cost Unit Leader
(DIVS) Division/Group Supervisors
(DMOB) Demobilization Unit Leader
(DOCL) Documentation Unit Leader
(FACL) Facilities Unit Leader
(FDUL) Food Unit Leader
(FOBS) Field Observers
(FSC) Finance/Administration Section Chief
(GSUL) Ground Support Unit Leader
(HEB) Helibase Manager
(IC) Incident Commander
(INTL) Intelligence Officer

(LOFR) Liaison Officer
(LSC) Logistics Section Chief
(MEDL) Medical Unit Leader
(OSC) Operations Section Chief
(PIO) Public Information Officer
(PROC) Procurement Unit Leader
(PSC) Planning Section Chief
(RESL) Resources Unit Leader
(SITL) Situation Unit Leader
(SOF) Safety Officer
(STAM) Staging Area Manager
(SPUL) Supply Unit Leader
(THSP) Technical Specialists
(TIME) Time Unit Leader

ACKNOWLEDGMENTS

The genesis for the original edition of this book went back to the mid-1990s. It was a time when organizations other than the wildland firefighting agencies, which developed the Incident Command System (ICS), began to explore the possibility of using the ICS for response to all-hazard incidents. The US Coast Guard was one of those nonwildland firefighting agencies, and several of the authors worked for the individuals who brought ICS to the US Coast Guard. Among those insightful emergency responders, James Spitzer and Larry Hereth were instrumental in our indoctrination and that of the Coast Guard. The Coast Guard was also very fortunate to have an exceptional mentor from the California Department of Forestry and Fire Protection, Deputy Chief Ralph Alworth. Chief Alworth led the development of a course that provided some basic ICS skills to help the Coast Guard "jump-start" its implementation of ICS. To Ralph we owe a debt of gratitude—your lessons were well-received. Thank you. Our appreciation also extends to California Department of Fish and Game, Deputy Administrator, Office of Spill Prevention and Response Lisa Curtis for her support in those early days of learning.

Writing can be a painful business and doubt can occasionally cast a shadow over the writers who are struggling to take a proven concept that has a library of material already written about it and wonder if what they're creating will help responders across all response disciplines. Fortunately for the five of us, we have excellent colleagues and professionals who steadied us and kept us moving forward. There is always a risk in listing names, but we think the risks are worth it. To Joe Couch, Archie Gresham, Ed Doyle, Soo Klein, Kelly Roberts, Lia de Bettencourt, Danny Cruz, Roger Laferriere, Karen Jones, Jill Bessetti, Dave Ormes, Kristy Plourde, Bill Whitson, Becky Jones, and Bob Ward thank you all for your input and recommendations … it made a difference.

Lastly, we cannot fail to recognize the exceptional support provided by Pat Mills. Her patience and meticulous attention to detail in providing us many of the visual aids you will find in this book were absolutely crucial in making this book such a valuable response tool. Pat takes Chuck's words and puts them into pictures that we can all quickly understand and use in emergency response. A picture truly is worth a thousand words, in which case, Pat authored many of the words included within. Thank you, Pat.

The second edition of our book reflects our continued growth and understanding of different areas where responders can use more guidance. As with the first edition of the book, we were able to tap into the expertise and knowledge of many professionals who gave their time in reading the revised manuscript. We thank all of them for their generosity and support. Specifically, the authors want to acknowledge Pete Bakersky, one of the country's legends in urban search and rescue; Lanney Holmes and Dean Matthews from the Federal Emergency Management Agency; Ron Cantin Vice-President of Emergency Management Services International; and Greg Greenhoe and Buck Latapie who both have over three decades of experience in wildland firefighting and are highly recognized professionals in the use of the Incident Command System.

Once again, we reached out to Pat Hadley-Miller for copyediting, and as with the first edition of *Beyond Initial Response,* she was excellent. Thank you, Pat.

The authors want to thank Mike Deal for the new book cover and all of the help that he provided in developing the many figures that you will see throughout the book.

To Frank Shelley from the US Coast Guard you always provided unblemished feedback, and we appreciate all you did for us.

A special thanks to Garry Briese and Jeannette Sutton who authored Chapter 12, Emerging Communications.

Garry has tremendous experience in emergency management with more than 36-years in the fire service, private industry, and the federal government. He served as the Executive Director of the International Association of Fire Chiefs (IAFC), Regional Administrator for the Federal Emergency Management Agency, Vice President, Emergency Management & Homeland Security for ICF International, and many other high visibility emergency management related positions. Garry is a well known author and lecturer on leadership and on the future challenges for the fire and emergency services community and has co-authored two first responder emergency medical textbooks as well as an innovative textbook for the basic training of firefighters. He is the co-founder of The Center for New Media & Resiliency. Garry has a BA from University of South Florida and a MPA from Nova Southern University.

Jeannette is a Disaster Sociologist at the Trauma, Health and Hazards Center, University of Colorado at Colorado Springs. Her primary focus of research is the uses of emerging technologies and social media in crisis and disaster situations. Jeannette has authored numerous academic publications and regularly consults with public sector organizations from the federal to local level. She is the co-founder of The Center for New Media & Resiliency. Jeannette has a Ph.D. in sociology from the University of Colorado at Boulder.

Finally, we want to acknowledge you, our readers, who are building your ICS skills and are out there putting them into practice every day in the service of our country.

FOREWORD

— Bringing Order Out of Chaos —

There is nothing more important to the successful outcome of an incident than a well-organized and functioning Incident Command System (ICS), fully prepared to be integrated into the larger National Incident Management System (NIMS).

But, why are the ICS and NIMS so important? Simply because *"No plan survives first contact with the enemy."*

This is a thought often attributed to many different military leaders, from Dwight Eisenhower to Napoleon to George Patton. But the observation, according to the Quote Verifier, *"actually originated with Helmuth von Moltke in the mid-nineteenth century ... but his version was not so succinct.... therefore no plan of operations extends with any certainty beyond the first contact with the main hostile force. In a process that's routine in the world of quotation, von Moltke's actual words were condensed into a pithier comment over time, then placed in more-familiar mouths."*

It makes no matter if the basic "plan" is for military, federal, or civilian use. While having a plan is almost always better (failing to plan is planning to fail), in truth, in many ways, it often doesn't matter if a specific plan exists for the specific incident since each incident has unique aspects that need to be accommodated and addressed.

If we agree in concept with the basic ideas behind *"No plan survives first contact with the enemy,"* then what really does matter are the skills and abilities of those in command responsible positions, from first arriving local responders, to regional and state mutual aid, to federal and military assistance, and their collective abilities to bring order out of chaos.

Interagency, multi-agency, multi-disciplinary, multi-governmental, multi-jurisdictional, are the words that describe today's operating environment. The incidents we are facing today continue to grow more and more complex, requiring a structured and commonly understood leadership and management structure, which is ICS and NIMS. No single agency, local, state or federal, has the expertise, authority, and resources to manage these increasingly complex situations.

This is exactly why the ICS and NIMS were developed ... a common structure, common language, common and effective communications, common operating picture, and a common planning process are all key elements of ICS and NIMS. *Beyond Initial Response* presents many aspects of ICS and NIMS in an easy-to-understand and organized approach.

But, being book-smart must be tempered and enhanced with real-world experience. A wise mentor of mine once told me, when I was facing a professional challenge, that "*This experience is like when you go to school to learn, and the pain you feel is the tuition you are paying for learning these lessons.*"

How decisions are made affects the quality of the decisions and the words and ideas on these pages are written by "been there, done that, learned the lesson" made the corrections real-world practitioners. The authors have met challenges and made mistakes and turned those challenges and mistakes into lessons and corrective actions, then used their lessons to provide you with the most comprehensive approach ever written for both ICS and NIMS.

Right now, this is just another book. But, you can make this book into a personal resource for you…. read and highlight or underline key words and phrases, tab and dog-ear the pages, write notes to yourself on the pages, ask yourself questions about the content, compare these words to other books and articles about ICS and NIMS.

Make the personal effort to transfer the "information" in this book into "knowledge" you can use to make yourself, and your team, into better and safer leaders. After all, that's what it is really all about.

Garry Briese —

FEMA Region 8 Administrator (2008–2009)

Vice-President, Emergency Management & Homeland Security, ICF International (2007–2008)

Executive Director, International Association of Fire Chiefs (1985–2007)

Executive Director, Florida College of Emergency Physicians (1979–1985)

CHAPTER 1
INCIDENT COMMAND SYSTEM OVERVIEW

The Incident Command System (ICS) is like math or a sport: if you do not practice it, you become rusty. We put this Incident Command System Overview chapter in this book in the event that you have not used ICS in a while and need a refresher on some of the ICS principles, terminology, processes, or organizational structure that make ICS the powerful response tool that it is. If you're reading one of the ICS position-specific chapters and come across something you are unsure of, or just don't remember, the answer may lie below in this overview. For example, span-of-control may be mentioned in Chapter 6, Operations Section Chief, and you do not remember the ICS rule of thumb for span-of-control. In this chapter, you'll find the answer. The overview does not cover everything about ICS, but we have hit the highlights. The rest of the book has much more detail.

The ICS organizational chart in *Figure 1.1* shows the most familiar aspects of the Incident Command System. Much of this chapter is devoted to dissecting the organizational chart so that you can see the different functional responsibilities within the ICS system, have a good grasp on the common terminology, and have an understanding of the major responsibilities of those positions.

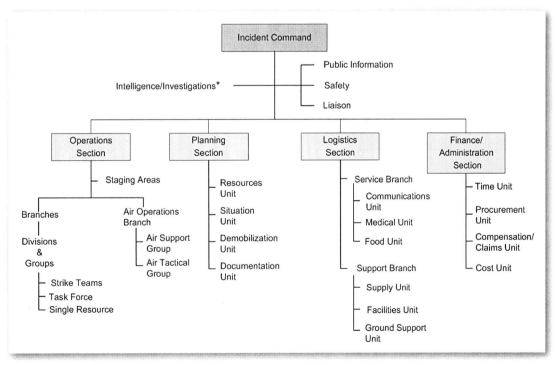

*Figure 1.1. The fully developed ICS organization. *Although Intelligence/Investigations is shown as a Command Staff position, the Intelligence/Investigations function can be assigned to several places within the incident command team.*

Specifically, this chapter focuses on some of the basic management characteristics of the ICS. The ones we will touch on are:

- Common Terminology
- Span-of-control
- Modular Organization
- Unified Command Structure
- Incident Action Plans

Common Terminology

The best place to start a discussion on the Incident Command System is with terminology. During a response to an incident, you do not have the luxury of trying to learn a new language, and with multi-agency response operations, if you do not have a common language, confusion will follow. One of the primary reasons ICS came about was to provide the response community with common terminology to ensure that response agencies could come together and rapidly form into a focused and coordinated team. Common terminology is applied to the following:

- Organizational elements
- Position titles
- Supporting position titles
- Resources
- Facilities

We're going to take a look at each of these in the following pages. You're learning the basics of a tested and proven system where a common language across response agencies and organizations is critical to an efficient response. Your ability to understand the language of ICS is vital when working in a multi-agency response operation.

Organizational Elements

In the ICS there's a consistent pattern for designating each level of the organization. *Figure 1.2* is a list of the organizational elements that start at the top with the Incident Command and go down to the Unit level. Think of it as a hierarchy. The one exception is the Command Staff. Although the Command Staff positions are shown above the General Staff, they are not actually in the chain of command. Not every incident will have all of these organizational elements established. The incident will dictate the management structure.

Organizational Elements

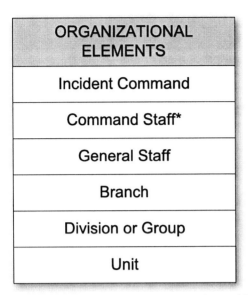

*Figure 1.2. The organizational elements within the ICS that may be established to manage an incident. *The elements are in hierarchical order with the exception of the Command Staff, which is not in the chain of command.*

Position Titles

Each one of the organizational elements listed in *Figure 1.2* has its own unique position title. The use of position titles serves three different purposes. First, titles provide a common standard for multi-agency use at an incident. For example, if one agency uses the title Section Chief, another uses the title Section Director, and another Section Manager – things can get confusing! Position titles provide standardization.

A second important purpose for using common position titles is that the use of these distinctive titles allows for the positions to be filled by persons most qualified in ICS independent of the rank or position they hold in their daily jobs. For example, if a mid-grade manager is the best qualified to fill an ICS position because they have the appropriate experience, training, and qualifications then they are assigned to that position.

The third purpose for common position titles is that they provide standardization for ordering more personnel resources. If more Division Supervisors are needed, anyone who is dispatched to the incident as a Division Supervisor is expected to have that qualification regardless of what agency or organization they come from.

Figure 1.3 is a list of ICS position titles that are linked to the various organizational elements that you saw in *Figure 1.2*. For example, if you're in a Command Staff position in the ICS organization you would be an Officer (e.g., Safety Officer).

Position Titles

ORGANIZATIONAL ELEMENTS	POSITION TITLES
Incident Command	Commander
Command Staff	Officer
General Staff	Chief
Branch	Director
Division or Group	Supervisor
Unit	Leader

Figure 1.3. The position titles are shown next to their organizational element.

Support Position Titles

We have one more set of position titles to cover before we're ready to move on to resources. These are the titles for other positions that can be called in to support the core management team. *Figure 1.4* shows the organizational elements, position titles, and the supporting titles. There's a very important distinction that you need to understand between the supporting titles of deputy and assistant. A deputy is fully qualified and capable of performing the duties of his or her superior. As you can see in *Figure 1.4*, deputies can be assigned to the Incident Commander, General Staff, or Branch Directors. For example, a Deputy Operations Section Chief is just as qualified as the Operations Section Chief they are supporting. Assistants, on the other hand, are not as qualified as the person performing the primary function, but they have the technical capability and qualifications to provide important support to the primary position. Assistants can work for the Command Staff and Unit Leaders, for example, an Assistant Safety Officer can work for the Safety Officer.

Support Position Titles

ORGANIZATIONAL ELEMENTS	POSITION TITLES	SUPPORT POSITION TITLES
Incident Command	Commander	Deputy
Command Staff	Officer	Assistant
General Staff	Chief	Deputy
Branch	Director	Deputy
Division or Group	Supervisor	not applicable
Unit	Leader	Assistant

Figure 1.4. The ICS support position titles.

A Word on Deputies

Deputies can be extremely important to a response organization, but you want to ensure that you use them correctly. Below is some guidance:

Rules for Deputies:

- Only the Incident Commander, General Staff, and Branch Directors have deputies
- You can have more than one deputy
- Deputies must be as qualified as the ICS position that they work for (e.g., Deputy Operations Section Chief should be as qualified as the Operations Section Chief)

Use of Deputies:

- Can provide relief to the primary position holder
- Can support the primary position during high operational tempo
- Can be assigned to perform specific tasks that require their level of knowledge and experience

Resources Terminology

The ability to manage resources in an emergency is absolutely critical to an effective and efficient response. We dedicate many pages in this book to resource management for that very reason. For now, we're only going to touch on the basics, to provide some of the common terminology that you will hear regarding resources on an incident.

Under the Incident Command System, resources consist of all personnel and major tactical equipment available or potentially available for assignment to incident operations and for which a status is maintained. Equipment resources will include the personnel to operate them. The last thing you want to do is to have to order an ambulance and then separately order the people to operate it. You can easily imagine how quickly such a process can get out of control.

Resource Status

From the time a resource (personnel or major tactical equipment) checks into an incident until it's demobilized, it will always be in one of three statuses: **assigned, available,** or **out-of-service**.

Assigned Resource: A resource that has checked into the incident and is currently tasked and working to support incident operations.

Available Resource: A resource that has checked into the incident and is available for tactical assignment.

Out-of-Service Resource: A resource that has checked into an incident, but is not ready for assignment. A resource can be out-of-service for several reasons:

- Mechanical (vehicle or equipment needs repair)
- Rest (personnel need to sleep and relax)
- Staffing (insufficient personnel to operate the equipment)
- Environmental (darkness or weather)

The Resources Unit keeps track of the status of each resource.

Kind and Type

Resources can be described by both kind and type.

Kind is easy to understand. It's simply what the resource is – for example, a fire engine, a bulldozer, a helicopter, a rescue boat.

Type on the other hand is a little less obvious, but it's an excellent concept that is seldom used outside of the wildland firefighting community. Type classifies resources that are of the same kind (e.g., fire engines) into different categories based on performance capabilities, staffing, and other factors.

The best way to understand resource typing is to look at an example. We'll use a helicopter in this example. First off, we know that all helicopters are not engineered to be equal. Some are small and can carry only two people, while others are large and can carry lots of people and cargo. If, during an incident, you tell Logistics that you want a helicopter, you would have to provide several details regarding the capability you need from the helicopter. If you do not, you can be sure that the helicopter that shows up will not be what you needed. Typing resources by their capability does several things:

- Minimizes confusion on exactly what capability you want
- Streamlines the ordering process
- Greatly assists the tracking of resource cost estimates

Helicopters have already been typed by the wildland firefighting community for their operations, so we have decided to use them as an example of how it's done. *Figure 1.5* provides an example of helicopter resource typing. If you want a helicopter that has the cargo-carrying capacity of at least 2,000 pounds, then you would order a Type 2 helicopter. As you can probably tell from the example in *Figure 1.5*, a Type 1 resource will always be used to describe the resource with the most capability going down in capability as you go from Type 1 to Type 2, and so on.

Helicopter Typing Matrix

TYPE	NUMBER OF SEATS (including pilot)	CARGO WEIGHT CAPACITY (in pounds)
Type 1	16	5000
Type 2	10	2500
Type 3	5	1200
Type 4	3	600

Figure 1.5. Helicopters, like other common resources used in firefighting, have been typed to facilitate ordering, reduce confusion on what is actually needed, and assist in incident cost estimates.

Organizing Tactical Resources on an Incident

There are three ways that resources are organized on an incident:

- As a Single Resource
- As a Task Force
- As a Strike Team

Single Resource: As the name implies, a single resource is an individual piece of equipment with an identified work supervisor that can be used in a tactical assignment; for example, fire engine, ambulance, crane, dump truck. In *Figure 1.6*, the single equipment resource is a dump truck that should come with the personnel to operate it.

Figure 1.6. Single equipment resources should have an identified operator in charge.

Task Force: A Task Force is any combination of different kinds of single resources that are assembled to accomplish a certain tactical assignment (keeping in mind the span-of-control, which we will cover next). The Operations Section Chief (OSC) assembles and disassembles Task Forces as needed. For example, the OSC may assemble a Task Force to remove debris from destroyed buildings. The Task Force may be comprise of dump trucks, an excavator, and a crane (see *Figure 1.7*). The makeup of a Task Force is very flexible but every Task Force must:

- Have a leader
- Have communications between the resources and the leader
- Have transportation
- Be within the span-of-control limits

Task Force

Figure 1.7. Task Forces are made up of different kinds of resources and have an identified leader.

Strike Team: A Strike Team is similar to a Task Force except that all of the single resources on a Strike Team are of the same kind and type. For example, a Strike Team to fill in a failed levee may be made up of three dump trucks that all have same carrying capacity (see *Figure 1.8*).

Strike Team

Figure 1.8. Strike Teams are made up of the same kind and type of resource and have an identified leader.

Resource Check-in

One of the most important things to remember at an incident, which is often forgotten in the heat of the battle, is to check in. It's extremely important that all resources are checked- in so that the management team knows what resources are on-scene to be properly assigned. There are five locations where check-in can occur on an incident:

- Incident Command Post
- Incident Base
- Staging Area(s)
- Helibase
- Camps

Facilities

Common terminology is used to identify the various facilities used in supporting incident operations. By standardizing the name of the various facilities everyone involved in the incident response instantly understands where certain response functions are located.

ICS facilities include: Incident Command Post, Helibases, Helispots, Staging Areas, Bases, Camps, and the Joint Information Center.

We are going to take the next few pages to go into greater detail on incident facilities. We will discuss what they are, how they are named, and other information that will give you a better understanding of the types of facilities you're likely to see on an incident. The duration, complexity, and size of the incident will determine which facilities are activated.

Incident Command Post (ICP)

The Incident Command Post is the location where the primary command functions are performed and where the Incident Commander resides. Figure 1.9 provides a quick guide to the ICP.

Incident Command Post (ICP)

- All incidents must have a designated ICP
- The Incident Commander determines the location of the ICP
- The location of the ICP can range from a vehicle hood to a large hotel. Factors that influence the size of the ICP include:
  - Complexity of the incident
  - Multi-agency response
  - Duration of the incident
- There is only one ICP for each incident
- The ICP is where the primary command and most of the General Staff functions are performed, including:
  - Planning function
  - Finance/Administration
  - Interagency coordination/liaison
  - Communications center
- The Incident Command Post will be designated by the name of the incident, e.g., *Bayfield ICP*

Figure 1.9. The Incident Command Post is where the Incident Commander is located and where the major incident functions (Planning, etc.) are performed.

Helibases

Helibases are the main location within the general incident area where helicopters are fueled, maintained, loaded, and parked (see Figure 1.10). Helibases are designated by the name of the incident, for example, Bayfield Helibase. An incident can have more than one Helibase, so the additional Helibase would be designated Bayfield Helibase #2. Helibases are managed by a Helibase Manager (HEB).

Helibase

Figure 1.10. Helibases are incident facilities where helicopters are fueled, maintained, loaded, and parked.

Helispots

Helispots are temporary locations within an incident where helicopters can safely land and take off to load and unload personnel, equipment, supplies, water, and so on. Helispots are designated by number such as H-1 (see *Figure 1.11*).

HelispotFigure 1.11. Helispots are temporary locations used to load and unload equipment, personnel, etc. There can be more than one Helispot based on the needs of the incident.

Staging Areas

Staging Areas are established by the Operations Section Chief (OSC). These areas provide the OSC with the ability to have tactical resources immediately available for deployment in the event that more resources are needed to manage the situation. Some things to remember about Staging Areas are listed in *Figure 1.12*.

> **Staging Areas**
>
> - ☐ Staging Areas are temporary locations where personnel and equipment are kept while awaiting tactical assignment
> - ☐ An incident may have more than one Staging Area
> - ☐ The Operations Section Chief establishes and disestablishes Staging Areas as he or she deems necessary
> - ☐ Staging Areas will be managed by a Staging Area Manager (STAM) who reports directly to the Operations Section Chief
> - ☐ Resources in staging must be ready to respond within minutes
> - ☐ Staging Areas are designated by the name that describes their general location, e.g., *4th Street Staging*

Figure 1.12. Staging Areas are established by the Operations Section Chief and have tactical resources that are immediately available for assignment.

Incident Bases

The Incident Base is where the primary logistics functions are coordinated and administered. There's only one Incident Base and it will be designated by the name of the incident such as Bayfield Base. The Base is where all of the out-of-service resources are located. The Incident Base falls under the Logistics Section Chief (LSC) and is managed by a Base Manager (BCMG). The BCMG oversees all of the primary services and support activities that take place at the Base. Some of these activities are listed in *Figure 1.13*.

> **Incident Base Services and Support Activities**
>
> - Equipment maintenance and repair
> - Supply and ground support
> - Personnel comforts such as:
>   - Food
>   - Water
>   - Eating areas
>   - Sanitation facilities
>   - Showers
> - Waste management and recycling pickup
> - Fuel
> - Personnel and equipment rehabilitation
> - Billeting (sleeping arrangements)

Figure 1.13. The Incident Base is where the primary logistical functions are performed. Responders can get equipment repaired, food to eat, and take showers at the Base.

Camps

Camps are temporary locations within the general incident area that are equipped and staffed to provide sleeping, food, water, and sanitary services to incident personnel. Camps are designated by geographic name or by number, such as Texas Creek Camp or Camp #3. Camps fall under the Logistics Section Chief (LSC) and are the responsibility of the Camp Manager.

Joint Information Centers

On larger, more complex incidents a Joint Information Center (JIC) may be established to coordinate all incident-related public information activities at the incident. The public information representatives from participating agencies collocate at the JIC and work together to support the media demands of the incident.

Common Facility Symbols

Each of the ICS facilities has a unique symbol that is used to denote its location on the incident map (see *Figure 1.14*).

Facility Symbols

- ◤ Incident Command Post
- Ⓢ Staging Area
- Ⓗ Helibase
- Ⓑ Base
- Ⓒ Camp
- ● Helispot

Figure 1.14. Common terminology extends to the map symbols that the Situation Unit Leader will use to display the location of all incident facilities on the incident map.

Figure 1.15 is an example of an incident map showing some of the facilities that might be used in the response to an incident. Regardless of the incident size or complexity all incidents will have an Incident Command Post even if it's only a vehicle. All other ICS facilities are used based on the needs of the incident.

Incident Map

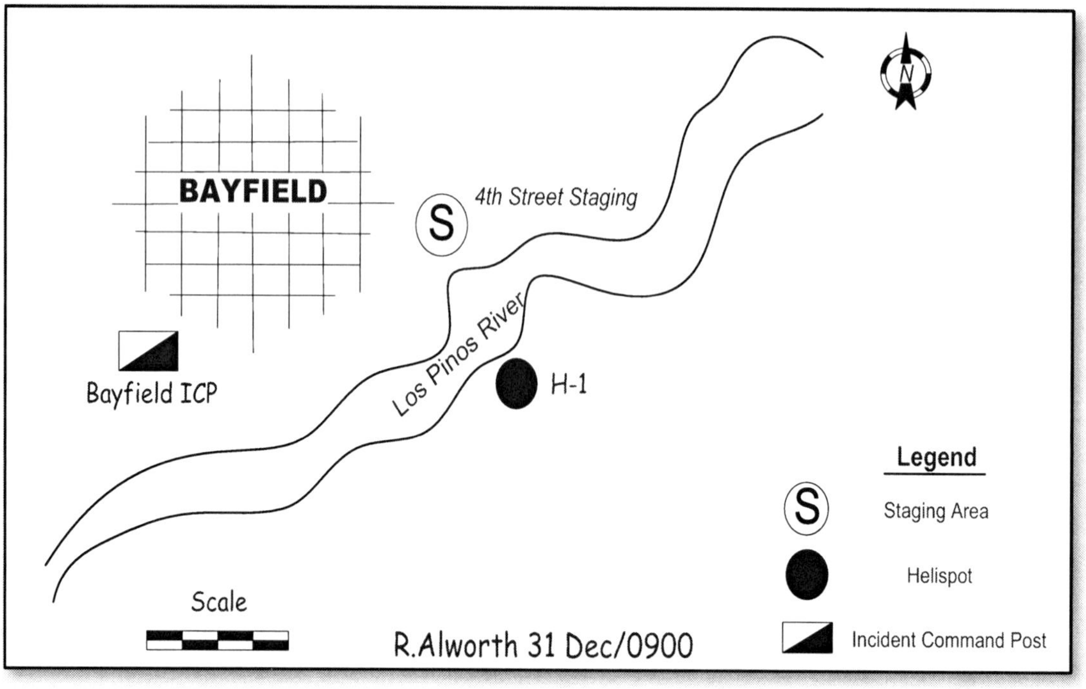

Figure 1.15. Common facility symbols are used on incident maps.

Span-of-Control

We want to take a few minutes here to talk about the concept of span-of-control before launching into our discussion of the ICS modular organization. Span-of-control simply refers to the number of individuals who are managed by another person. The developers of the Incident Command System recognized that, during an emergency, the number of people reporting to a supervisor had to be limited. The optimal ratio is five individuals reporting to one supervisor as shown in *Figure 1.16*. However, span-of-control may range from three to seven individuals reporting to one person. In emergency situations, information and decisions are occurring rapidly, and it's easy to get overwhelmed. Add in fatigue, and you have a recipe for disaster if you do not pay attention to your span-of-control.

The ICS organization has built-in mechanisms that enable you to maintain span-of-control while simultaneously expanding your organization to meet the needs of an incident. The span-of-control mechanisms are: Branches, Divisions, Groups, Task Forces, and Strike Teams.

Span-of-Control

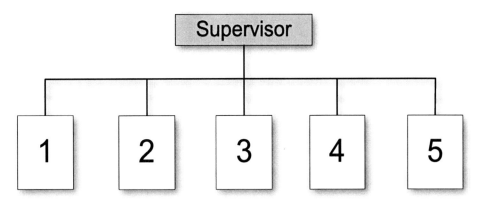

Figure 1.16. Span-of-control is an important principle of the Incident Command System.

Modular Organization

Every incident or event has certain activities or actions that must be performed to manage it. If an incident or event is small, one person might perform all of those managing activities or actions. If the incident is large, many people may be brought in to manage the response activities and each person needs to understand his or her role within the organization. The ICS organization expands or contracts to meet the needs of the incident. In other words, the organization needs to wrap around the incident, not the incident wrap around the organization. It seems like a straightforward idea, but it's often overlooked by those who do not properly implement the principles of ICS.

Rules to Organize By:

- The incident shapes the organization
- Do not "over-organize" or "under-organize"
- Leverage the inherent flexibility that the ICS offers

Five Management Functions

The ICS organization is built around five major management activities: Command, Operations, Planning, Logistics, and Finance/Administration. These five major management activities shown in *Figure 1.17* are the building blocks upon which the ICS organization develops and applies regardless of incident size or complexity.

Management Activities in the Incident Command System

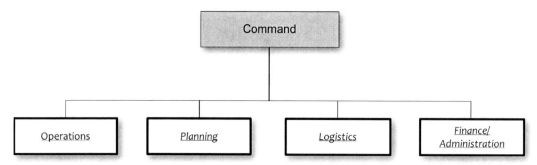

Figure 1.17. There are five major management activities under the Incident Command System.

Command (Incident Commander)

The Incident Commander (IC) is in charge of the overall management of all activities at the incident, and until the IC delegates a management function (Operations, Planning, Logistics, or Finance/Administration) to another person, the IC must perform the required function for each position. On a small scale, short duration incident, the IC may choose not to delegate any positions or only a few. However, as an incident grows in size or complexity the IC will delegate responsibilities. For example, when first on-scene, the IC will most likely be directly involved in managing tactical operations, but as the incident escalates, the IC will likely delegate this responsibility to a newly designated Operations Section Chief or the IC will quickly become overwhelmed.

Some of the primary responsibilities of the Incident Commander are:

- Establish incident priorities
- Determine incident objectives
- Establish an Incident Command Post
- Establish an appropriate response organization
- Authorize release of information to the news media
- Approve and authorize implementation of the Incident Action Plan
- Ensure adequate safety measures are in place
- Coordinate with key people and officials

In addition to the Operations, Planning, Logistics, and Finance/Administration general staff positions, the IC has Command Staff Officers to help manage incident safety, communicate with the public and incident personnel, conduct outreach to other agencies, and advise on intelligence and investigation issues. Collectively these positions in an Incident Command System organization are called the Command Staff. The Command Staff comprises the Safety Officer, Public Information Officer, Liaison Officer, and Intelligence/Investigations Officer (see *Figure 1.18*).

Command Staff

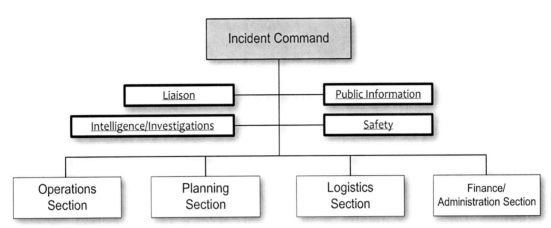

Figure 1.18. *The Safety, Public Information, and Liaison Officers are collectively referred to as the Command Staff. * The Intelligence/Investigations function on an incident management team can be done by an Intelligence/Investigations Officer or, if necessary, in other parts of the organization such as Planning.*

The Command Staff

The Command Staff reports directly to the IC and provides critical support to both the IC and others on the command team. If the IC does not bring in someone to handle the Command Staff responsibilities, then the IC must perform those duties. There's only one Safety Officer, Public Information Officer, Liaison Officer, and Intelligence/Investigations Officer designated on an incident. This is true even for multijurisdictional responses. However, each Command Staff position may have as many assistants as necessary to carry out their responsibilities. A detailed explanation for each Command Staff position can be found in Chapter 5 of this book.

Safety Officer

The Safety Officer (SOF) ensures that the safety of responders and the public is not compromised while carrying out response operations. The SOF evaluates proposed strategies and tactics and works closely with the Operations Section Chief to implement safeguards if necessary. Chapter 5 has more information that is dedicated to the Safety Officer and should be read if you have been assigned as an IC, an Operations Section Chief, a Planning Section Chief or, of course, as the Safety Officer.

Public Information Officer

The Public Information Officer (PIO) has two primary customers: the public and incident personnel. The Public Information Officer works closely with the news media to ensure that accurate incident information is conveyed and the PIO disseminates incident information to responders to keep them informed.

Liaison Officer

The Liaison Officer represents the Incident Commander in communicating with agencies that are providing support to the incident. These agencies can support incident operations in one of two ways. First, as an assisting agency that provides tactical resources to assist at the incident, or, second, as cooperating agencies, that provide nontactical support like the Red Cross or Salvation Army. In addition, the Liaison Officer (LOFR) interfaces with stakeholders to ensure that their concerns are brought to the attention of the Incident Commander. For example, a stakeholder may be a private landowner whose property is threatened by a wildland fire.

Intelligence/Investigations Officer

The Intelligence/Investigations Officer provides the Incident Commander with a conduit to intelligence and investigation information that can have a direct impact on the safety of response personnel and influence tactical decisions. The Intelligence/Investigations Officer also ensures that sensitive information (e.g., classified information) is handled in accordance with the prescribed safeguards.

The General Staff

The General Staff is comprised of the Operations, Planning, Logistics, and Finance/Administration Section Chiefs (see *Figure 1.19*). In addition to the Command function, the General Staff represents the other four management functions within the Incident Command System. The General Staff reports directly to the Incident Commander.

General Staff

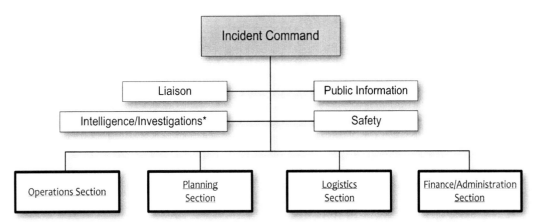

Figure 1.19. The General Staff is made up of the Operations, Planning, Logistics, and Finance/Administration Section Chiefs.

Operations Section Chief

The Operations Section Chief (OSC) is responsible for directing and coordinating all tactical operations on the incident. The OSC grows the operations organization from the bottom up always ensuring that operations remain within the span-of-control limits. Some of the major responsibilities of the Operations Section Chief include:

- Managing tactical operations
- Ensuring tactical operations are conducted safely
- Maintaining close communications with the Incident Commander
- Assisting in the development of the Incident Action Plan
- Identifying required tactical resources to accomplish incident objectives
- Identifying staging areas
- Assembling and disassembling strike teams and task forces

The Operations Section Organization

The Operations Section organization in *Figure 1.20* is designed to be highly flexible so that it can be used during any type of emergency or planned event.

Operations Section Organization

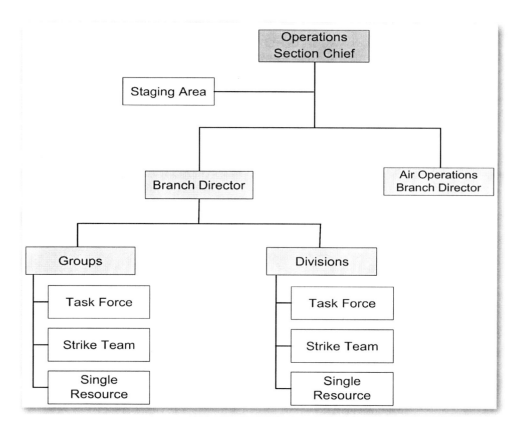

Figure 1.20. The Operations Section organization is tailored to the needs of the incident.

We have already discussed some of the organizational elements within operations such as staging areas, strike teams, task forces, and single resources. Now we will cover Divisions, Groups, and Branches.

Divisions

Divisions are one way to break up an incident into manageable pieces. Divisions are always established using geographic boundaries. Examples of divisions are: the area between points on the ground such as that shown in *Figure 1.21*, or the east side of a stadium. Each division is managed by a supervisor.

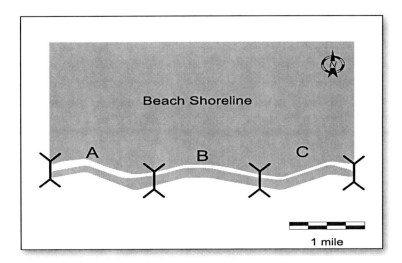

Figure 1.21. Divisions are a way for the Operations Section Chief to divide an incident geographically into manageable pieces.

Things to remember about Divisions:

- They are used to divide an incident geographically
- They are designated by letters of the alphabet (A, B, C, etc.)
- They are established by the Operations Section Chief or Incident Commander (IC) if there is no OSC
- Divisions are managed by supervisors
- Division supervisers do not have deputies

Groups

Another way to divide a response organization is functionally by groups. Functional Groups are not limited by geographic boundaries. A group performs a certain activity or function such as: search and rescue, decontamination, security, marine salvage.

Things to remember about Groups:

- They are used to divide work at an incident by function
- They are designated by their functional assignment, for example, Disposal Group
- They are established by the Operations Section Chief (or IC if there is no OSC)
- Groups are managed by supervisors
- Group supervisors do not have deputies

Both divisions and groups may be established during an incident.

Branches

Branches are primarily established to ensure that span-of-control is not exceeded. For example, if the Operations Section Chief has established eight divisions, he or she should create a branch to reduce his or her span-of-control. Branches can be either geographic and/or functional and are managed by a Branch Director who, as we learned earlier, can have deputies.

Things to remember about Branches:

- They are primarily used for span-of-control
- They are designated by Roman numerals or by function (e.g., Branch I or Mitigation Branch)
- They are established by the Operations Section Chief
- Branches are managed by Branch Directors
- Branch Directors can have deputies

Air Operations Branch

Aviation assets are highly complex machines, limited in quantity, and very expensive to operate. Because of these factors, an Air Operations Branch Director (AOBD) may be activated to manage these critical resources. There is no specific number of aviation assets that will trigger when the OSC establishes an Air Branch, but the sooner they are brought into the response, the better. This is especially true when the number of air assets requires additional management support or when the incident requires both tactical and logistical aircraft to support operations.

A review of the Operations Organization naming conventions:

- Branches (Roman numeral or functional name): I, II, III, etc. or Air Operations, Mitigation, etc.
- Divisions (geographic with alphabetical identifiers): A, B, C, etc.
- Groups (functional): "Triage Group," etc.
- Strike Teams: ST1, ST2, ST3, etc.
- Task Forces: TF1, TF2, TF3, etc.
- Single Resources (original identifier): Ambulance 146

Operations Section (building the tactical organization)

The rest of the ICS organization builds from the top down, but the Operations Section builds from the bottom up. We are not trying to add confusion here, but we are highlighting the simple fact that when an incident occurs, only a few tactical resources may respond. If the needs of the incident dictate, the Operations organization will expand accordingly, and the cascade of additional tactical resources must be organized and managed to ensure a safe and efficient response. For example, a response to a serious car accident may involve a police car, an ambulance, a fire truck, and a helicopter (see *Figure 1.22*). In most incidents, this is all of the ICS organizational structure that is required to deal with the situation. The Incident Commander for this incident may be the initial responding police officer. This type of incident requires minimal management.

Initial Response Organization

Figure 1.22. The Operations Section organization builds from the bottom up. Initially, tactical resources report directly to the Incident Commander.

In the event that an incident grows in complexity, which may include such considerations as hazardous materials or crowd control, a larger operations organization may be required so that the IC can manage his or her responsibilities. The IC may use one of the span-of-control mechanisms to better manage the incident and allow more time to address concerns such as media and safety. In *Figure 1.23*, the IC has created a Task Force to reduce the number of individuals reporting directly to him or her. Remember that the IC maintains all of the ICS responsibilities until he or she delegates that responsibility to someone else.

Expanding Response Organization

Figure 1.23. A task force has been established to reduce the number of individuals directly managed by the Incident Commander.

As an incident becomes more complex, the IC may delegate the tactical responsibility for the incident. The person designated assumes the title of the Operations Section Chief (OSC). Directing and managing tactical operations on an incident that is escalating can quickly spin out of control if the OSC is not managing the growth of the tactical organization. To help manage the influx of tactical resources, the OSC will continue to grow the operations organization. In *Figure 1.24*, the OSC has added a group and two divisions to better manage the tactical response. In addition, the OSC has created a staging area that places additional resources at his or her fingertips for immediate deployment.

Operations Organization Continued Growth

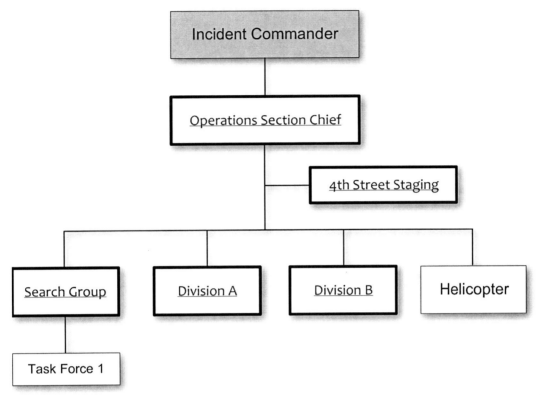

Figure 1.24. The Operations Section Chief has established two Divisions and a Group to manage tactical operations. The OSC's span-of-control is 5, keeping it within the recommended range.

Figure 1.25 is an example of a large Operations Section. The incident is of such magnitude and complexity that the OSC used branches to ensure that the span-of-control was kept within manageable limits. The needs of the incident dictate the size and structure of the Operations Section's organization.

A Large Operations Organization

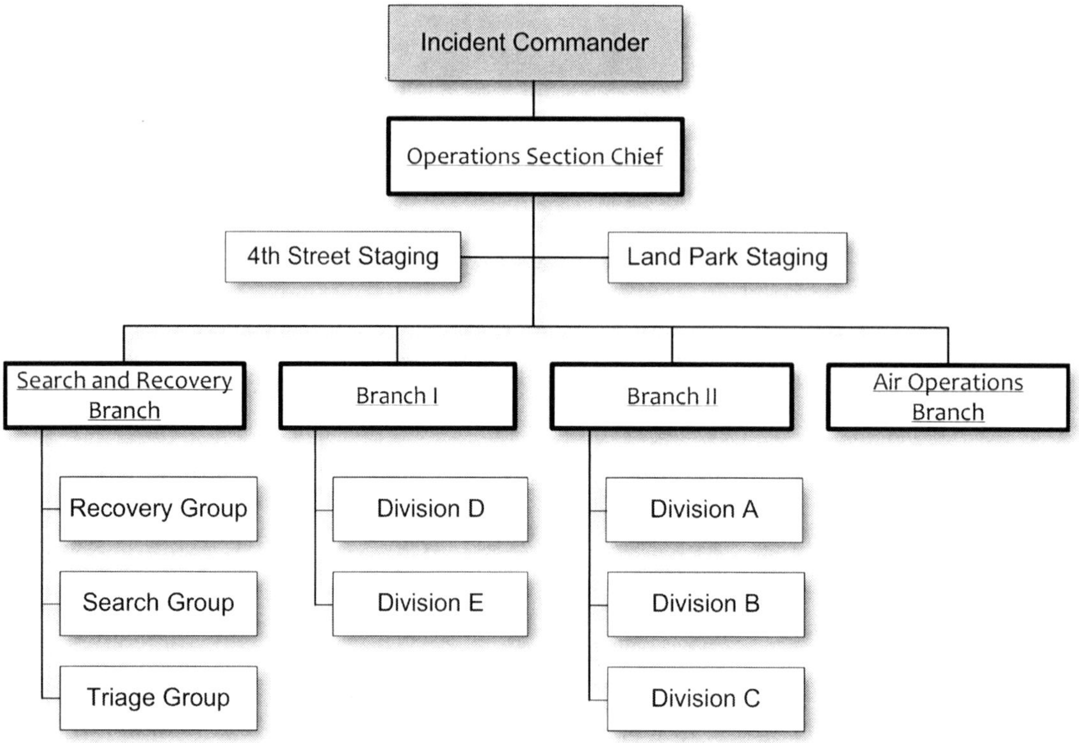

Figure 1.25. The Operations Section can become immense when the size and complexity of an incident requires a large number of tactical resources to bring the situation under control.

You should be able to look at *Figure 1.25* and apply many of the ICS principles that you have learned. Let's review some of those principles. Organizationally, you know that the Operations Section Chief is a member of the General Staff and that the OSC can assign deputies to help perform his or her duties. We have mentioned span-of-control and you can see that the organization in *Figure 1.25* is within acceptable range with respect to this concept. When you see the terms Division and Group you know that Divisions represent geographic areas of the incident and the Groups are functional ways to divide a response up into manageable work units. Finally, you will notice that the Divisions are labeled by letters of the alphabet, Groups are labeled by the function they perform, and Branches are identified by Roman numerals or by function.

If you look at *Figure 1.25* and are comfortable with the terminology and understand the organizational structure that you see, then you are on your way to speaking the ICS language.

The Planning, Logistics, and Finance/Administration Section Chiefs exist to support Operations. These sections are built from the top down, meaning that Section Chiefs are brought in to manage all of the particular responsibilities of that section and will order additional staff as needed. We are going to take a look at these sections and get a glimpse of what they provide to support response operations. Chapter 6 has more detailed information on the roles and responsibilities of the OSC

Planning Section Chief

Depending on the incident complexity and duration, the Incident Commander/Unified Command (more than one Incident Commander working together as a team) may assign a Planning Section Chief (PSC) to provide the critical planning support necessary to move incident operations from a reactive response to a proactive response. The Planning Section Chief keeps all of the ICS organizational elements moving together as a cohesive team and oversees the ICS Planning Process and development of an Incident Action Plan. Major responsibilities include:

- Establishing and maintaining situational status
- Managing the development of Incident Action Plans
- Developing functional short-range plans (e.g., transition plans, cargo offload plans, evacuation plans, mass triage plans)
- Developing long-range plans
- Developing plans for demobilization of resources
- Maintaining resource status information on all incident resources
- Maintaining incident documentation
- Providing the primary location and oversight for Technical Specialists

To accomplish all of the Planning Section responsibilities, the Planning Section Chief can activate units designed to support planning activities. The four units in *Figure 1.26* form the Planning Section. Based on the size and complexity of the incident the Planning Section Chief (PSC) will determine the need to activate all, some, or none of the units. In addition to these units, an incident may require that specialized skills are brought in to assist the command team. In the ICS, personnel providing the specialized skills are collectively called Technical Specialists.

Planning Section Organization

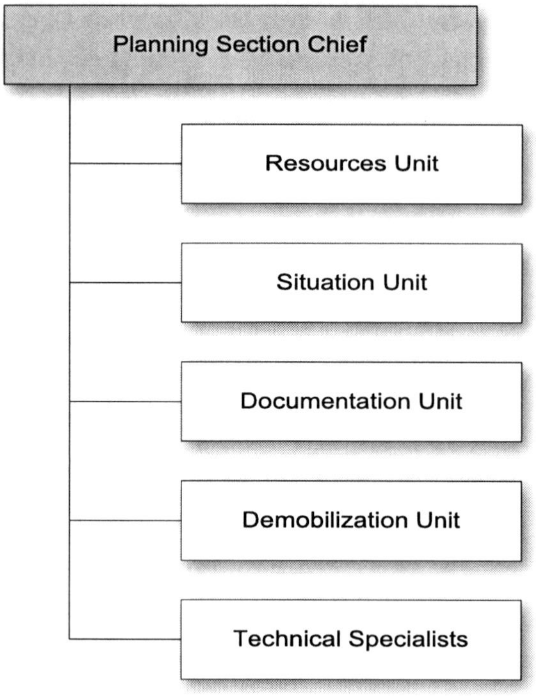

Figure 1.26. The Planning Section organization.

Resources Unit

The Resources Unit is responsible for maintaining the status and location of all resources (personnel and major tactical equipment) at the incident. The unit is also responsible for establishing check-in locations so responding personnel and tactical resources can properly check into the incident. The Resources Unit will have a master list of all resources on the incident, including all primary and support resources.

Situation Unit

The Situation Unit collects, processes, and organizes incident information and provides the command team with both current situational information and future projections of incident potential. The Situation Unit may have two positions established within the unit to support the Situation Status function. These positions are Display Processors and Field Observers.

Documentation Unit

The Documentation Unit maintains the critical incident documentation that will be kept for legal, historical, and case-history purposes. The Documentation Unit will also provide any duplication services needed to support the response. All Command and General Staff should be aware of the Documentation Unit so that they can ensure that they submit their important documents for the incident archive.

Demobilization Unit

The Demobilization Unit ensures the safe and orderly release of personnel and equipment from the incident when they are no longer required. On a large incident, this may involve an entirely separate planning activity. Planning for demobilization should start as early as possible for an incident to ensure a most efficient and safe demobilization process. The Demobilization Unit will generate the Demobilization Plan, and once approved, will help oversee its implementation.

Technical Specialists

Technical Specialists provide specialized knowledge and experience that may be necessary to support incident operations. Technical Specialists can be assigned anywhere on the incident where their skills are needed. Some examples of Technical Specialists may include:

- Structural Engineer
- Chemical Engineer
- Fuels and Flammability Specialist
- Legal Specialist

Chapter 7 has a detailed discussion of the role and responsibilities of the Planning Section Chief and each of the Unit Leaders.

Logistics Section Chief

The Logistics Section Chief (LSC) is responsible for providing facilities, transportation, communications, supplies, equipment, food services, billeting, and first-aid medical services in support of the incident and incident personnel. The LSC major responsibilities are:

- Managing all incident logistics
- Providing input into the development of the Incident Action Plan
- Developing the communications and medical plans
- Managing and maintaining incident facilities except for staging areas

To help manage the logistics responsibilities, the LSC can establish six units and on large incidents may elect to designate Support and Service Branch Directors to help manage the extraordinary logistical demands (see *Figure 1.27*). If you want to learn more about the Logistics Section Chief, read Chapter 8.

Logistics Section Organization

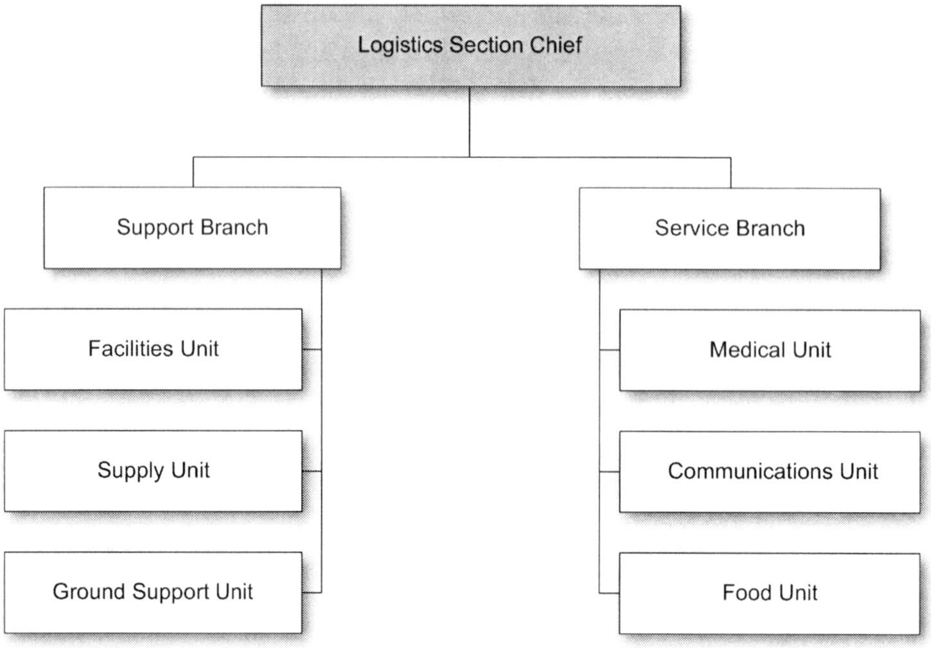

```
                    Logistics Section Chief
                    /                    \
          Support Branch            Service Branch
           |                          |
         Facilities Unit            Medical Unit
           |                          |
         Supply Unit                Communications Unit
           |                          |
         Ground Support Unit        Food Unit
```

Figure 1.27. The Logistics Section comprises six units, and, if necessary, the LSC can designate Support and Service Branch Directors.

Facilities Unit

The Facilities Unit is responsible for set up, maintenance, and demobilization of all incident support facilities with the exception of Staging Areas, which are under the operational control of the Operations Section Chief, but are supported by the Facilities Unit.

Supply Unit

The Supply Unit is responsible for ordering, receiving, processing, and storing all incident-related resources. All off-incident resources are ordered through the Supply Unit, including tactical and support resources and all expendable (e.g., batteries) and nonexpendable support supplies (e.g., fire hose).

Ground Support Unit

The Ground Support Unit is primarily responsible for the maintenance, service, and fueling of all mobile equipment and vehicles with the exception of aviation assets.

Medical Unit

The Medical Unit provides medical services for personnel assigned to the incident. This unit develops the Medical Plan for inclusion in the Incident Action Plan.

Communications Unit

The Communications Unit is responsible for developing plans for the use of incident communications equipment and facilities; installing and testing of communications equipment; distribution and maintenance of communications equipment; and supervision of a communications center, if one is established.

Food Unit

The Food Unit is responsible for supplying the food needs for the entire incident.

Finance/Administration Section Chief

The Finance/Administration Section Chief (FSC) oversees the financial aspects of the response. Some of the specific responsibilities of the FSC are to:

- Provide financial and cost analyses
- Ensure an accurate accounting of all financial documents
- Keep the Incident Commander/Unified Command apprised of any financial or administrative concerns
- Establish a system to document compensation for workers who are injured and claims for property damage that is associated with the incident
- Ensure that all personnel and equipment time records are accurate
- Provide financial input for demobilization

The Finance/Administration Section is made up of four units as seen in *Figure 1.28*. Chapter 9 discusses the role of the Finance/Administration Section Chief in greater detail.

Finance/Administration Section Organization

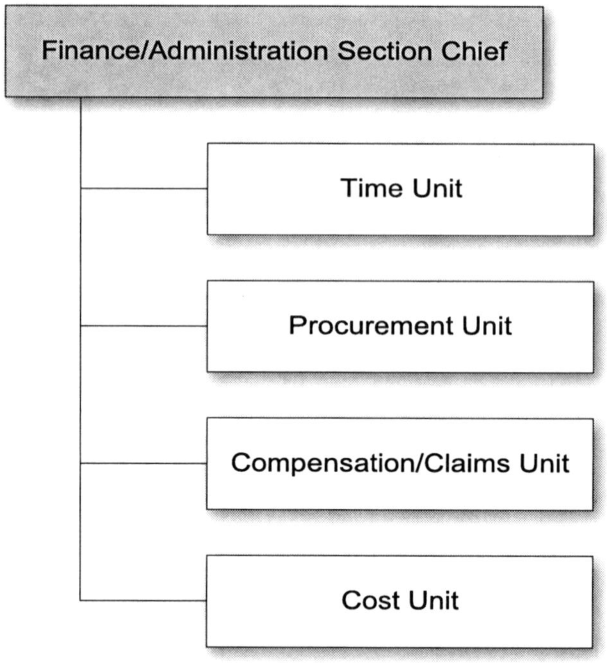

Figure 1.28. The Finance/Administration Section Chief may establish four units to support finance and administration activities on the incident.

Time Unit

The Time Unit is responsible for ensuring accurate recording of daily personnel and equipment hours and that agency-specific time-recording policies are met.

Procurement Unit

The Procurement Unit manages all financial matters that deal with vendors, including contracts, leases, and fiscal agreements. In addition, the Procurement Unit also is responsible for maintaining all equipment time records.

Compensation/Claims Unit

The Compensation/Claims Unit oversees the completion of all forms required by worker's compensation. This unit is also responsible for investigating all claims involving property associated with or involved in the incident.

Cost Unit

The Cost Unit maintains accurate information on the actual cost of all assigned resources. This unit also estimates current and projected incident costs and conducts incident cost analyses.

Unified Command Structure

Although a single Incident Commander normally handles the command function, an ICS organization may be expanded into a Unified Command. The Unified Command is a management structure that brings together the "Incident Commanders" of all major agencies and organizations involved in the incident to coordinate an effective response while at the same time carrying out their own jurisdictional or functional responsibilities.

The advantages of Unified Command are:

- A single integrated Incident Organization
- One set of objectives
- Coordinated information flow
- Collocated (shared) facilities
- A single joint-planning process and Incident Action Plan
- A coordinated process for ordering resources
- Duplicated efforts are reduced or eliminated

The difference between an ICS organization under a single Incident Commander and a Unified Command is at the very top of the organization (see *Figure 1.29*). You will continue to have only single Operations, Planning, Logistics, and Finance/Administration Section Chiefs.

Actual make up of the Unified Command is based on the type of incident and the jurisdiction(s) where the incident occurred. Participation in a Unified Command is done without giving up authority, responsibility, or accountability. For more information, see Chapter 4, Unified Command.

Unified Command

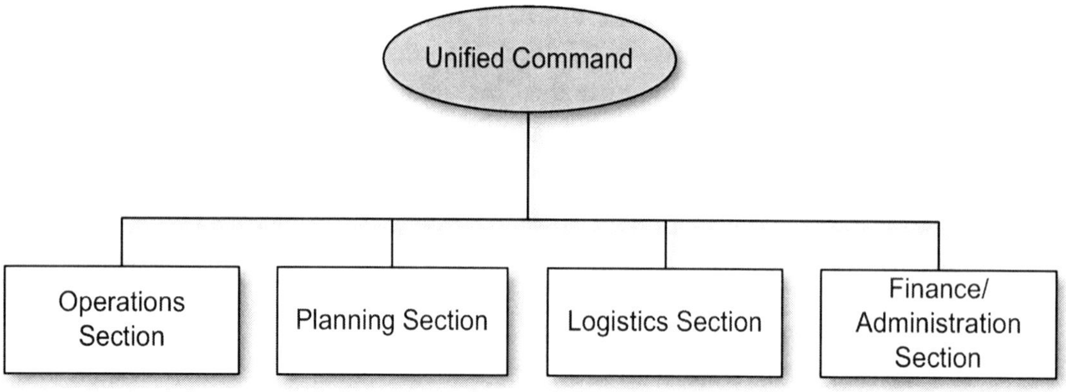

```
                    ┌─────────────────┐
                    │ Unified Command │
                    └────────┬────────┘
         ┌──────────────┬────┴─────┬──────────────┐
    Operations      Planning    Logistics       Finance/
     Section        Section      Section     Administration
                                                Section
```

Figure 1.29. ICS Unified Command.

Incident Action Plan

Every incident needs a verbal or written plan. Our focus is going to be on the Incident Action Plan (IAP). The IAP is designed to move response operations from a reactive mode to a proactive mode. It provides the responders with direction on what to accomplish in a certain period of time (operational period) and the resources necessary to support the operations. However, before we can get into our discussion there are two concepts that we need to explain up front. The first is the concept of an operational period, and the second is the ICS Planning Process.

Operational Period

Simply put, the operational period is the length of time (e.g., 12 hours, 24 hours) that the Incident Action Plan (IAP) can accommodate. Once that time has expired another IAP must be ready to go to cover the next operational period.

ICS Planning Process

The ICS Planning Process is a disciplined way to help responders go through various *steps* or activities and build a workable Incident Action Plan. To help illustrate the ICS Planning Process we are going to use the visual representation that is in *Figure 1.30*. It's known as the Operational Planning "P" and shows the different *steps* that make up the ICS Planning Process. You can read Chapter 2 for a full discussion of the ICS Planning Process.

The Operational Planning "P"

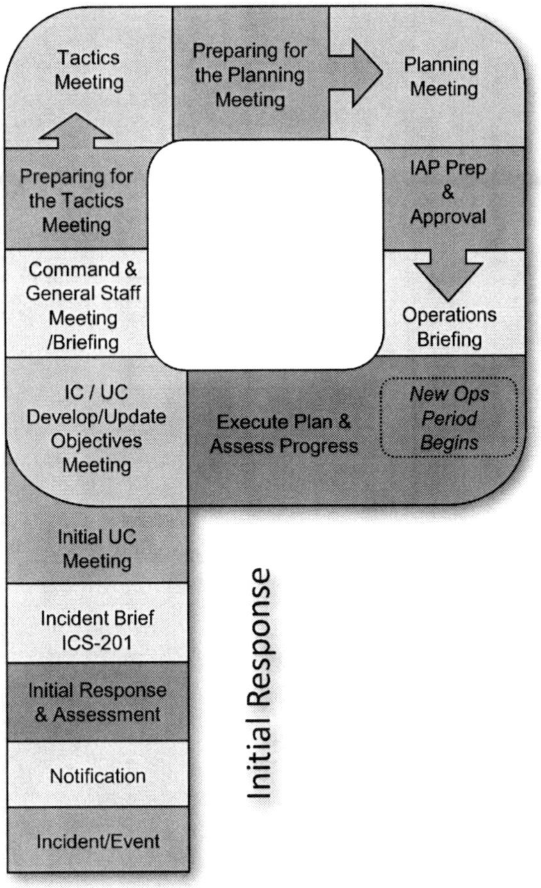

Figure 1.30. The Operational Planning "P."

By working through the ICS Planning Process, the command team will have methodically developed an Incident Action Plan (IAP). The IAP is comprised of various components (see *Figure 1.31*) that together provide the responders with a clear direction of what has to be done, how it's going to be done, the resources that will be used to accomplish the work, the communications protocols, and other important information to keep the entire response team moving in the same direction.

Components of the Incident Action Plan

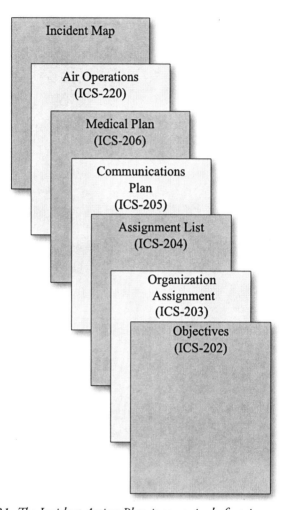

Figure 1-31. The Incident Action Plan is comprised of various components.

Hopefully, the information in this overview has refreshed your memory and whetted your appetite for more! The remainder of the book is focused on specific Incident Command System processes and positions and will provide you with many tools to help you do your ICS job more effectively.

CHAPTER 2
INCIDENT COMMAND SYSTEM PLANNING PROCESS

When most people picture the Incident Command System (ICS) they usually have the ICS organizational chart squarely situated in their mind. Common terminology, including standard terminology for organizational elements, is one of the cornerstones of ICS and it's drummed into our heads early and often. However, another equally important cornerstone of the Incident Command System is the ICS Planning Process. Like common terminology, the ICS Planning Process establishes a common method for developing and implementing tactical plans to efficiently and effectively manage an emergency response or a planned event such as a city parade.

The ICS Planning Process transcends the different processes that agencies use in their day-to-day operations and provides a common process for all responders to use to work toward the successful resolution of an incident. Most of us have probably experienced the uncomfortable feeling you get when you're engaged in an activity where you do not know the "rules of the game." Not only are you uncomfortable, but you're not nearly as effective, standing on fringes of the activity instead of being a part of the work effort. The goal of this chapter is to provide you with an overview of the ICS Planning Process that will give you an overall understanding of how it works. Read some of the ICS position-specific chapters such as Chapter 6 on the Operations Section Chief and Chapter 7 on the Planning Section Chief to learn about their specific responsibilities in the planning process.

Reactive Response

Every incident begins with the dispatch of resources to the scene. Emergency responders use their training, experience, and time-tested standard operating procedures to bring the incident under control, and for the majority of incidents these fundamentals are successful. The initial response to any emergency can accurately be called a reactive response. *Figure 2.1* is a graphic depiction of how an initial response to an emergency is conducted—where resources are dispatched to the incident and upon arriving, decisions are made as to how the resources are employed. A reactive response is reality and there is no avoiding it. The fact is that first responders often do not know what they are dealing with until they arrive on-scene and they adapt their initial tactics based on available resources and situational awareness.

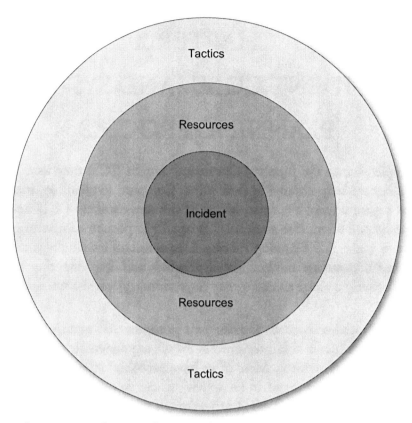

Figure 2.1. In the reactive mode, responders "encircle" the problem (incident) with resources, which in turn drives tactical decisions. Although a reactive response cannot be avoided, shifting to a proactive mode should be accomplished as soon as possible.

Proactive Response

As you see later in this chapter, the Incident Command System Planning Process is designed to move the incident response team out of a reactive response to a well-planned, proactive response that maximizes responder safety, tailors incident resources to meet response objectives, and allows for constant assessment of progress toward resolving the incident. The challenge for emergency responders is to have the ability to transition from a reactive response to a more methodical response when the complexity of an incident exceeds their standard operating procedures. *Figure 2.2* is an illustration that shows you how a proactive response is conducted. Unlike the reactive response, in the proactive mode, resources are employed to meet specific response objectives. This employment enables the deliberate use of resources that are appropriate for the task; and by establishing objectives, the Incident Commander/Unified Command can measure the progress achieved toward bringing an incident under control.

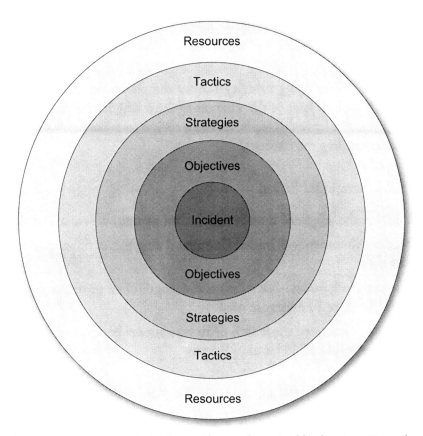

Figure 2.2. A proactive response encircles the problem with attainable objectives. From these objectives, strategies are developed as well as tactics, which are used to bring an incident to a safe and successful conclusion. The tactics also drive the demand for the appropriate number and kinds of resources needed to accomplish the tasking.

The Operational Period

Before we actually begin our discussion on the ICS Planning Process there's an important concept that you'll need to understand: the operational period.

Think of an operational period as the length of time in which you want to achieve a certain amount of work. For example, on Friday evening you sit down at the kitchen table and list the chores you want accomplish on Saturday such as mow the lawn, power wash the house, clean the inside of the house, and wash the car. You may give yourself eight hours to do everything on your list. Those eight hours are your operational period. After the eight hours are up, you'll be able to determine quickly whether you achieved all of the planned work items. If not, Sunday is looking less like a day of leisure. Operational periods can vary in length from a few hours to 24 hours and longer. Responders often use either 12 or 24 hours to define the length of time of their operational period.

The Incident Commander/Unified Command (IC/UC) determines the length of the operational period. If the incident can be brought under control quickly by the first responders, defining an operational period is not necessary; however, if the incident is going to require some time to resolve, the IC/UC will define an operational period and start moving toward a proactive response. You cannot enter the ICS Planning Process without defining the operational period.

Are you with us so far? We're going to delay further discussion on the operational period for now, but we'll come back to it later in this chapter to show why it's critical to the ICS Planning Process. However, before we leave this topic, let's do a review to make sure we're on the same page with respect to understanding the concept.

Review of the Operational Period

- It defines the length of time in which a set amount of work is to be accomplished
- Incident Commander/Unified Command determines the operational period and start time
- There are no hard-and-fast rules on the length of an operational period but generally 12 or 24 hours are used
- You must determine the length of an operational period and start time or you cannot enter the ICS Planning Process

The Operational Planning "P"

There have been many attempts to visually depict the ICS Planning Process to both facilitate teaching the process and to guide responders in its use during an emergency or in event planning. We have chosen to use what is referred to as the Operational Planning "P." The Operational Planning "P" in *Figure 2.3* was developed for all-hazard responses.

To get you through the ICS Planning Process we're going to briefly discuss each *step* using the Operational Planning "P" as a guide. Although we use the term *step* for ease in our discussion of the ICS Planning Process, it's not an entirely accurate way to view the process. The reality is that many actions are being conducted simultaneously. For example, if you look at *Figure 2.3* you find the words "Execute Plan & Assess Progress." The assessment of the effectiveness of your plan in bringing the incident under control never stops—so plan on always assessing your progress. Assessment continues through all *steps* in the process. Another example is the Tactics Meeting. The Operations Section Chief (OSC) does not wait for that meeting to begin thinking about what resources he or she requires and how best to employ those resources. The OSC is thinking about those things well before the established meeting time.

The Operational Planning "P"

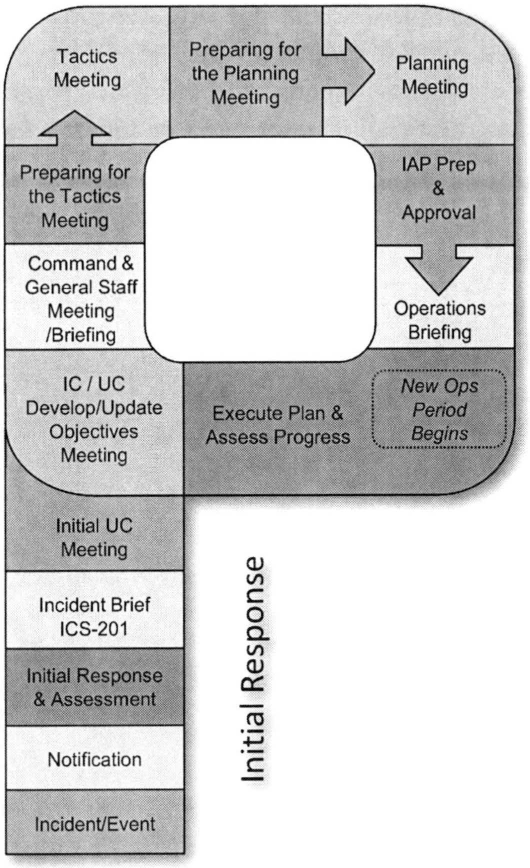

Figure 2.3. The Operational Planning "P" is a visual illustration of the ICS Planning Process. The planning process is used by ICS command and control teams to proactively respond to an emergency. It's also used to guide event planning such as a July 4th parade.

The ICS Planning Process

Before we go any further, we want to set some expectations. We have often seen, both in actual responses and in exercises, that Incident Commanders or Unified Commands place unrealistic time requirements on their management teams to develop and implement an Incident Action Plan (IAP). Development of an Incident Action Plan is both time- and resource-intensive. You need to have adequate staff on the incident and you must give the staff sufficient time to produce a usable plan. Incident Commanders who arrive on-scene at an incident at 0800 should not be mandating that an IAP be in their hands by 1100. That demonstrates a complete lack of understanding and appreciation for what it takes to plan for and develop an Incident Action Plan. The ICS Planning Process is an awesome tool, but it does not come free. It costs both time and personnel. Enough lecturing, let's move on.

The first stop on the Operational Planning "P" is on the stem of the "P." An incident has occurred and emergency response resources are responding as they do every day. The initial response is totally driven by operations and no formal planning is conducted except for what has been done in developing responding agencies' standard operating procedures. Planning, Logistics, and the other ICS positions you're familiar with do not typically come into play early on in an incident. All functions are done by the Incident Commander until he or she delegates those responsibilities. Fortunately, the majority of emergency response operations never reach the point where the ICS Planning Process is necessary. These responses start and end in the stem of the "P" in *Figure 2.4*.

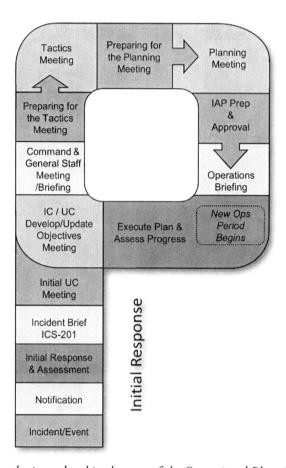

Figure 2.4. Most incidents begin and end in the stem of the Operational Planning "P." Operations totally drive the response and no formal planning is conducted.

The ability to transition from the reactive mode (the stem of the "P") to the proactive mode (that is the part of the Operational Planning "P" that makes it a P) with as minimal amount of chaos as possible is dependent on how well the initial responders document their objectives, response actions taken, organizational structure, resources on-scene, and incident situation. Taking the time to record this type of information places the initial Incident Commander in the best position to rapidly and accurately brief incoming responders and in the event that the initial IC is to be relieved by someone more senior, they have at their fingertips an excellent tool from which to smoothly transition command. The developers of the Incident Command System learned the hard way just how important it is to document incident information. In response to this need, they created a user-friendly form known as the Incident Briefing Form, ICS-201.

Incident Brief (ICS-201)

The Incident Brief *step* in the ICS Planning Process is focused on transfer-of-command, and the tool used to conduct the briefing is the Incident Briefing Form, ICS-201 (*Figure 2.5*). The ICS-201 functions as the Incident Action Plan and documents the incident situation, actions taken, and decisions made during the early stages of a response. It also provides the initial Incident Commander with a means to capture decisions. In addition to facilitating transfer-of-command briefings, Incident Commanders use the ICS-201 to provide briefings to incoming responders. See Chapter 3, Incident Commander, for more information on the ICS-201, Incident Briefing Form.

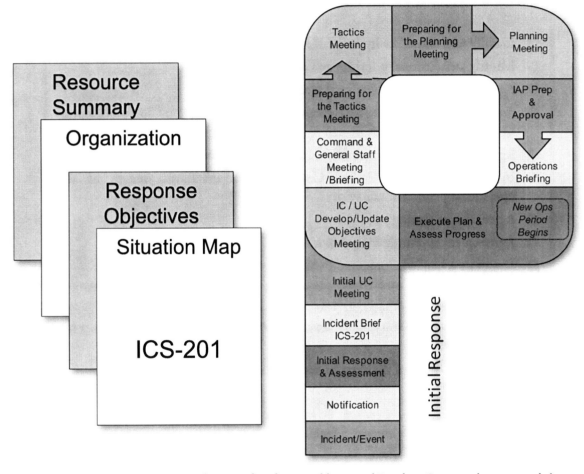

Figure 2.5. The ICS-201, Incident Briefing form enables initial Incident Commanders to record the situation, response objectives, actions taken, response organization, and resources on the incident.

Initial Unified Command Meeting

If the response to an incident will be managed by a single Incident Commander then this *step* in the Operational Planning "P" is not necessary. However, responses using a Unified Command are common and the initial Unified Command meeting is important because it will set the tone for the remainder of the response. This meeting is where the Unified Commanders meet to discuss and agree on important response issues prior to entering an integrated planning process. The last thing you want to add to an emergency operation is more confusion. By increasing coordination you're increasing complexity, so clarification of critical response issues is essential. Ideally, contingency planning and relationships developed before an incident occurs will make the Initial Unified Command Meeting go more smoothly. However, many times Incident Commanders may be working with someone they do not know, and this meeting provides the opportunity to engage on critical incident issues such as those shown in *Figure 2.6*.

Initial Unified Command Meeting

At a minimum the Initial UC Meeting Agenda should:

- ☐ Identify who should be in the Unified Command
- ☐ Identify jurisdictional priorities and objectives
- ☐ Agree on the basic organizational structure
- ☐ Agree on the best-qualified and acceptable Operations Section Chief and Deputy Operations Section Chief
- ☐ Agree on who fills the remaining General Staff positions
- ☐ Incident Commander/Unified Command determines the length of the operational period and start time
- ☐ Agree on cost-sharing
- ☐ Agree on resource-ordering procedures
- ☐ Designate a Unified Command Public Information Officer

Figure 2.6. The Initial Unified Command Meeting will set the tone for the remainder of the response. There must be agreement on many crucial issues in order to create a successful multi-agency team.

We're now going to move out of the stem of the Operational Planning "P" and focus on the ICS Planning Process.

IC/UC Develop/Update Objectives Meeting

To effectively manage an incident, the IC/UC must set objectives and establish priorities that guide the efforts of all response personnel. At this *step* in the planning process the IC/UC is getting together to evaluate the current incident status, what needs to occur next, and how they're going to achieve those things. This is where they develop the objectives that will drive the work activity for the next operational period. These objectives may be the same as those currently in effect or completely new. It all depends on the incident situation. One thing for sure is that if the IC/UC does not develop realistic and focused objectives the command team will struggle unnecessarily, and the incident could take longer to bring under control.

Command and General Staff Meeting

This meeting is very important. During this meeting, the members of the IC/UC get together with the Command and General Staff and brief them on their decisions, objectives for the next operational period, priorities, limitations/constraints, and expectations. This meeting takes place early in the planning cycle to ensure that everyone is heading in the same direction. This is a prime opportunity for the Command and General Staff to get clarification on issues. For example, if this is the first meeting, the Public Information Officer may ask for clarification on his or her authority to release press information or the Logistics Section Chief might verify where the IC/UC would like the incident support facilities to be located.

Two parting thoughts before we move to the next *step* in the planning process. First, although the Command and General Staff Meeting is a structured time for the IC/UC members to meet with their primary command personnel, Command and General Staff personnel always have direct access to the IC/UC and you should not hesitate to get clarification or brief on important issues anytime during the operational period. Second, come to this meeting prepared to receive direction and, if necessary, present any issues requiring IC/UC clarification.

Preparing for the Tactics Meeting

During the time between the Command and General Staff Meeting and the Tactics Meeting, the Operations Section Chief (OSC) supported by the Planning Section Chief (PSC) will take time to develop draft strategies (how they will accomplish an objective) and tactics (the equipment and personnel required to implement the strategy) for each operational objective that the IC/UC has set for the next operational period. The OSC, working with the Planning Section Chief, will outline work assignments and develop an operations organization. It's easy for the OSC to become completely focused on the immediate operational needs of the incident, but taking the time to "rough out" the future tactical plan before the meeting will significantly reduce the length of the Tactics Meeting.

Tactics Meeting

The Tactics Meeting is facilitated by the Planning Section Chief and is attended by Operations Section Chief, Resources Unit Leader, Logistics Section Chief, Situation Unit Leader, Safety Officer, and the Communications Unit Leader. In addition, you may see the Finance Section Chief in attendance as well, due to cost considerations. This is the time the OSC sets aside to discuss how he or she will organize and conduct operations during the next operational period. The OSC presents the planned strategies and tactics to ensure that those present can support the tactical plan. *Figure 2.7* is a checklist that the OSC uses to guide the discussion and to present a thorough tactical plan for the next operational period.

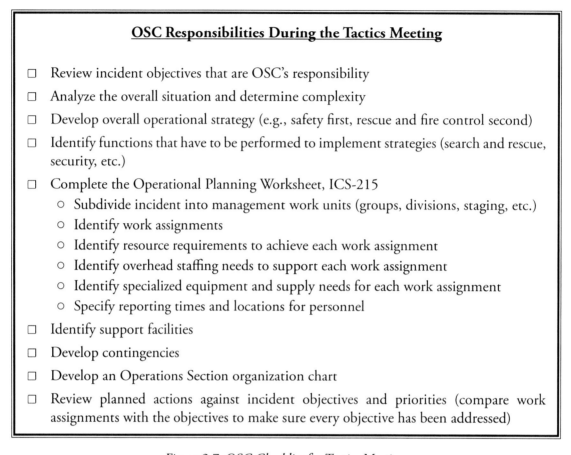

Figure 2.7. *OSC Checklist for Tactics Meeting.*

Before going forward let's review where we are on the planning cycle:

- The IC/UC met to develop and update objectives and discuss other issues such as contingency plans
- The IC/UC then met with the Command and General Staff to communicate the direction they want to head with the response and to clarify any issues or concerns. The IC/UC's goal in that meeting was to keep everyone working toward a common goal and communicate clear expectations
- Using the objectives set down by the IC/UC, the OSC developed strategies and tactics to accomplish the IC/UC objectives and developed the operations organization. This was not done in a vacuum, but with the support of the Planning Section Chief, Logistics Section Chief (LSC), Safety Officer (SOF), and others.

Preparing for the Planning Meeting

During this *step* in the planning cycle, the incident management team is spending time gathering current incident information, confirming availability of resources, and verifying that the information they will present at the Planning Meeting is as accurate as possible. All of this is occurring simultaneously while managing current operations. The reason for all the preparation is that the IC/UC should have the latest incident information and to ensure that there are no surprises at the Planning Meeting.

Planning Meeting

This meeting is the culmination of all the meetings that have taken place prior to this one. The result of this meeting is to be ready to produce a coordinated and sustainable Incident Action Plan (IAP) that everyone agrees they can support. The tactical plan presented in the Planning Meeting must be of such quality that it guides and supports the activity of those who actually implement it on the ground. The IC/UC will give tentative approval of the proposed plan of action at this meeting. Final approval will come later.

The Planning Meeting is attended by members of the IC/UC, Command and General Staff, the Resources Unit Leader (RESL), Documentation Unit Leader (DOCL), Air Operations Branch Director (AOBD), and technical specialists as required. You may be tempted to think, after reading these last few pages on the planning cycle, that there are many meetings and that they seem redundant. This would be an incorrect assumption. You have to remember that many hours have passed since the IC/UC developed their objectives for the next operational period. The dynamic nature of emergencies means that changes in the incident situation may result in changes in the response objectives, and as a result, changes in strategies and tactics. Everything is linked. Also, not everyone attends every meeting. It's not until the Planning Meeting that all the primary players get together and see the proposed plan in its entirety.

Incident Action Plan Preparation and Approval

Once the Planning Meeting has ended and the tactical plan (plan of action) has received tentative approval, the next *step* in the Incident Command System (ICS) Planning Process is to prepare the various components of the Incident Action Plan for final approval by the Incident Commander/Unified Command. The Planning Section Chief will oversee the assembly of the IAP.

Incident Action Plan

The Incident Action Plan (IAP) is comprised of various forms that when put together provide the user with a wealth of information that will enable them to carry out their tactical assignments. We will look briefly at what information is normally included in an IAP and which ICS positions are responsible for preparing the various forms (see *Figure 2.8*).

A few important notes here on preparing the specific parts of the IAP:

- ICS-202, Incident Objectives: The Planning Section Chief prepares the ICS-202, but does not establish the objectives, which are the responsibility of the IC/UC.
- ICS-203, Organization Assignment List: The Resources Unit Leader prepares the ICS-203, which lists the names and positions of the management team.
- ICS-204, Assignment List: The ICS-204 contains information on the operations organization and the work to be accomplished. That information comes directly from the Operations Section Chief. The Resources Unit ensures that the information is accurately transferred to the form. The ICS-204 also shows the tactical resources that are assigned to each part of the operations organization. The Resources Unit works closely with the OSC to ensure that tactical resources are assigned correctly, with the Communications Unit Leader (COML) to ensure that the communications information on the ICS-204 is correct, and with the Safety Officer for any safety information to be added to the form. The number of ICS-204s included in an IAP depends on how the Operations Section Chief organized the section. For example, if the OSC established two Divisions, two Groups, and a Staging Area, there will be five ICS-204s in the IAP.

Principal Components of the Incident Action Plan

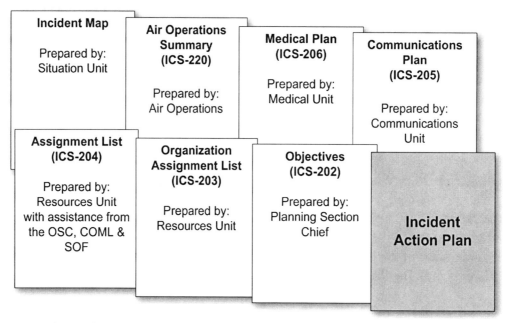

Figure 2.8 The Incident Action Plan is comprised of several components that together provide responders with the necessary information to carry out their responsibilities in a coordinated and efficient manner.

If necessary, other documents may be included in the IAP such as a Demobilization Plan, disposal instructions, and decontamination procedures. However, the goal is to not have an IAP that resembles the telephone book for New York City. See Appendix A for a sample Incident Action Plan.

Evolution of an Incident Action Plan

The Incident Action Plan (IAP) has its beginnings back when the initial Incident Commander (IC) put pen to paper and completed the ICS-201, Incident Briefing Form. When the initial IC or his or her relief recognized that the incident was going to take many operational periods to resolve, he or she shifted the incident management team to the ICS Planning Process. Using the ICS-201 as a reference for what had occurred on the incident, the IC/UC developed incident objectives that would further frame the development of an IAP. The Operations Section Chief and Planning Section Chief looked at different strategies and tactics necessary to accomplish the objectives and documented them so that all response options were laid out and could be evaluated. Choosing the best options, the OSC developed the Operational Planning Worksheet, ICS-215, defining work assignments and breaking the work up into manageable units. These work assignments then became the core of the IAP. *Figure 2.9* illustrates the evolution of an Incident Action Plan.

Beyond Initial Response   *Incident Command System Planning Process*

Evolution of an Incident Action Plan

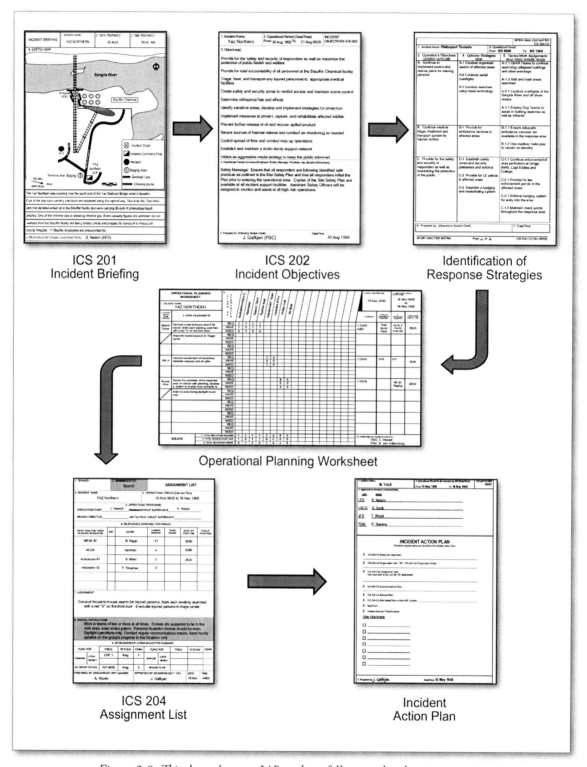

Figure 2.9. *This shows how an IAP evolves, following the planning process.*

Operations Briefing

Operations Briefings are absolutely critical. This is the first time that those who have to carry out the plan are going to be told of their assignment. You have to remember that they have not been privy to all of the meetings and discussions that have taken place during the planning process. They most likely have been in the field conducting tactical operations or perhaps they have just arrived on the incident. The Operations Briefing provides the opportunity for the OSC to brief the organization and to provide clarification on any of the tactical assignments. It's attended by the IC/UC, Command Staff, General Staff, Branch Directors, Division/Group Supervisors, Staging Area Managers, Task Force/Strike Team Leaders, and Unit Leaders.

Execute the Plan and Assess Progress

The IAP should be treated as a living document that is subject to change based on the incident situation. Constant assessment of how well the plan is designed to meet the current situation must be conducted. This will ensure that the objectives are still viable and that the tactical direction and resources are supporting attainment of the IC/UC objectives.

The Operational Period and Operational Planning "P"

At this stage in the discussion of the ICS Planning Process, we are hopeful that you're at least somewhat comfortable with the concept of an operational period, and that by using the Operational Planning "P" you understand the planning process and appreciate the power it brings to emergency response and event planning. Now we proceed with the tricky part of applying the concept of the operational period to the ICS Planning Process.

When you're working through the planning process, you're developing an Incident Action Plan for the next operational period, not the operational period you are currently working in. Does that make sense? Look at it this way: you cannot stop response operations while you're developing a plan so both planning and response have to occur simultaneously. While Operations staff are predominately focused on the here-and-now, conducting tactical operations (*responding*), Planning is overseeing the development of a response plan that will be implemented at the beginning of the next operational period. Let's look at *Figure 2.10* to help clarify any possible confusion. During the first operational period, an Incident Action Plan is being developed that will guide response operations during the second operational period and, you guessed it, during the second operational period an IAP is being developed for the third operational period and so forth.

Operational Periods

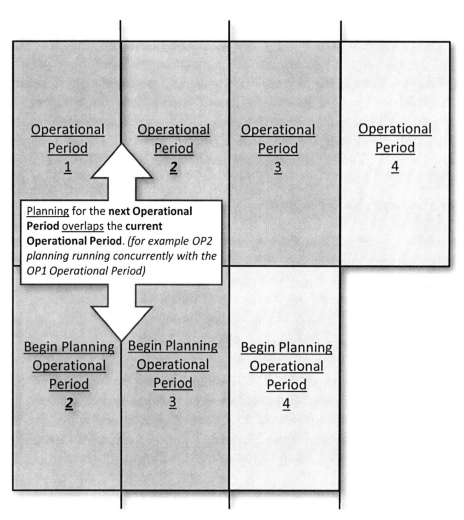

Figure 2.10. While Operations is conducting tactical operations during the current operational period, Planning is overseeing the development of the Incident Action Plan that will guide response operations during the following operational period.

Determining the Meeting Schedule

Once the Incident Commander (IC) determines that the nature and extent of the incident will require a shift to a formal planning process and development of an Incident Action Plan (IAP), he or she will set the start time for the first operational period using an IAP. Experienced ICs know that it takes both time and energy to develop an IAP and this is especially true when transitioning from a reactive mode to a proactive mode. Once the time for the next operational period is set, the Planning Section Chief will determine the times that certain *steps* in the ICS Planning Process should occur. This is necessary to ensure that all *steps* are accomplished and that an IAP is ready for the IC's approval in time to begin the operational period. Let's work through an example to illustrate how the Planning Section Chief does their magic.

Example: At approximately 0700 the Incident Commander tells her Command and General Staff that the start of the next **operational period** will begin at 1800 (eleven hours from now or 6 PM).

The Planning Section Chief (PSC) will work backward from 1800 (see *Figure 2.11*) to determine when each *step* in the planning process needs to start. The PSC does this to make sure that all *steps* in the planning process are conducted to ensure the timely delivery of the IAP.

Starting at 1800, the PSC allows for a 30–40 minute **Operations Brief,** which means that the briefing must start no later than (NLT) 1700. Moving further back in the planning cycle, the PSC allows two hours for **IAP Preparation and Approval** setting the time at NLT 1500 to accomplish the various tasks and still be ready for the **Operations Brief**.

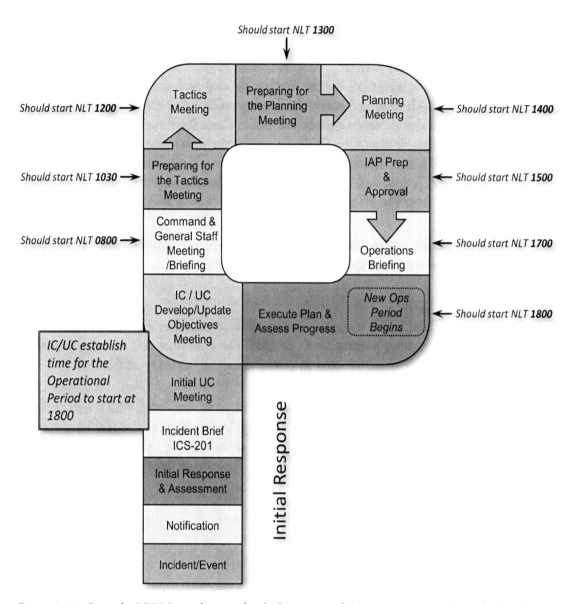

Figure 2.11. Once the IC/UC sets the time for the beginning of the next operational period, the Planning Section Chief works backward to determine when each step in the planning process needs to start to ensure the IAP is ready in time for the oncoming shift.

The **Planning Meeting** must take place NLT 1400 because it can take up to 45 minutes to complete. This is the first Planning Meeting, so it most likely will take more time than subsequent Planning Meetings. To ensure there's adequate time to prepare for the planning meeting the PSC sets the time of the **Tactics Meeting** at NLT 1200. To hold an effective Tactics Meeting the OSC and PSC will prepare for the tactics meeting around 1030. The kick-off meeting that gets this whole ball rolling is the **Command and General Staff Briefing,** which needs to happen at NLT 0800 or the team will be hard-pressed to deliver an IAP on time.

Once the PSC has determined the times that various *steps* in the planning process will occur, the information is disseminated to the command team. The Situation Unit Leader will post meeting times.

Use the example times shown in *Figure 2.11* as a guide. If you're operating with a 24-hour operational period, you can allot more time for each *step*. Conversely, if the operational period is less than 12 hours, you'll have to move faster to accomplish all of the *steps*.

A Word on Work Shifts

You may be on a response where the IC/UC has established a 24-hour operational period. This, as you remember from your reading, means that the Incident Action Plan (IAP) is viable for 24 hours. It would be unsafe to work responders for 24-hours straight, so shifts should be established to conduct around-the-clock operations. In this case, the IAP is briefed to each oncoming shift, resulting in the need for two Operations Briefings.

Closing Thoughts

We have covered a lot in these pages and it's possible your head is spinning from all of the new information. So let's take stock of where we are:

- The ICS Planning Process is used to move us from a reactive response to a proactive response
- Shifting to an ICS Planning Process requires personnel and time to do it effectively
- The operational period is the length of time in which the IC/UC wants to accomplish a set amount of work
- The ICS Operational Planning "P" is a visual representation of the ICS Planning Process, with *steps*. It guides the response team toward development of an Incident Action Plan (IAP)
- The IAP outlines the tactical assignments to be achieved for the next operational period and is constantly evaluated to ensure that it addresses the current incident situation
- Once the Planning Section Chief knows the time that the next operational period will begin the PSC will set the meeting schedule to ensure that all of the *steps* are accomplished in time for the Operations Briefing.

CHAPTER 3
INCIDENT COMMANDER (IC)

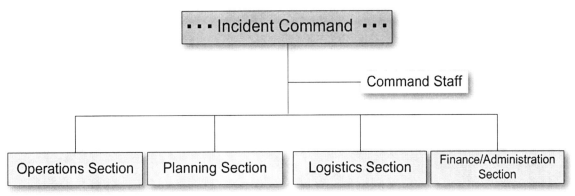

Figure 3.1. Incident Commander in relation to the Command and General Staff.

The one position in the Incident Command System that is on every response regardless of the size, type, complexity, or duration of the incident is the Incident Commander (IC), as in *Figure 3.1*. Although not always acknowledged, the first responder who arrives on-scene is acting in the capacity of an Incident Commander regardless of whether he or she usually carries that title. The minute you arrive at the scene of an incident, such as an auto accident, you'll be using all of your senses to assess the situation. This is almost an automatic reaction. Are there physical hazards that are preventing me from approaching the injured? Is there a need to secure the area to prevent further injuries? Is the fuel that I smell going to pose a fire or explosion hazard? Based on your assessment you'll be making some immediate decisions and will continue to do so until you successfully resolve the incident or are relieved by a higher official in your agency or by another agency with more jurisdictional authority. However, until relieved, you need to have that Incident Commander mindset that responds to immediate issues, ensures safety of responders and bystanders, orders help to assist, and asks "What if?" with respect to different contingencies that may impact the response.

Depending on the type of incident and its complexity, the number of issues that must be managed by the IC can range from few to many. The problem is that you never know whether that garden-variety incident response will resolve quickly or grow into something much larger and more complex. Even incidents such as a minor traffic accident can rapidly increase in complexity when a speeding vehicle inadvertently collides into the incident scattering debris and bodies in its wake. A house fire can quickly spiral out of control when chemicals in the basement explode injuring responders. As an Incident Commander, you must be ready to assume control over the unexpected until the incident is resolved or you are relieved by a more experienced IC.

The role of Incident Commander can be quite extensive and requires a variety of skills to perform. A job advertisement announcing the vacancy for an Incident Commander position would probably be written like the one below:

> **Incident Commander Position Available**
>
> Looking for a leader, decision maker, risk manager, multitasker, and communicator who is as equally adept at managing immediate challenges as well as predicting future events. Individual must thrive on adrenaline, and be prepared on occasions for sleepless nights, inadequate nourishment, and over-stimulus. Absolutely must be an exceptional team player. If interested apply within.

It's fair to say that the majority of incidents responded to every day do not require all of the skills and traits in the advertisement, but at some point you may come face-to-face with an incident that requires every bit of your experience, knowledge, and leadership to successfully manage.

If you answered yes to the advertisement or otherwise find yourself in the role of Incident Commander, this chapter will be a big help to you whether you're responding to those day-to-day incidents that take only an hour or two to resolve or when facing a more complex and dynamic situation.

If you'll be filling the role of IC during an incident, we highly encourage you to read all of the Command and General Staff positions that are included in this book. The reason is twofold. First, you should always keep in mind that as the IC, all responsibility for the incident management lies with you unless you assign someone else to take care of a particular aspect; for example, safety. If you do not assign a Safety Officer, then you retain that responsibility and are accountable for developing a Site Safety Plan and monitoring all activities for potential safety issues, along with the myriad responsibilities that a Safety Officer must perform. This is similar for any Command and General Staff position. The second reason to read up on those positions is that if you do assign someone to perform one or more of those jobs, you need to understand what work they will be managing.

There's a lot to cover in this chapter so we decided to break it into four sections. Section One is focused on your role as Incident Commander during the initial response to an incident. Section Two discusses documenting your initial response actions. Section Three is about transferring the role of IC (either you're transferring IC to a more qualified person or you are the oncoming IC). Section Four covers the IC's responsibilities when management of the incident requires the use of the ICS Planning Process. So let's get started!

Section One: Initial Response

There are many things you have to do when responding to an incident as the Incident Commander. In addition to the technical skills and experience that you bring to the incident, you must also be proficient at managing all aspects of the incident. The Incident Command System was designed to help you succeed. The information we discuss in this section is focused on helping you accomplish your management responsibilities by leveraging the strengths of the ICS in the early minutes and hours of the response.

As the initial IC you have many responsibilities, such as the need to:

- Determine incident priorities
- Establish incident objectives
- Manage tactical operations
- Assure the safety of responders and the public
- Determine the need to expand your organization
- Ensure that appropriate facilities are established to support the response
- Identify and order the necessary tactical resources to accomplish response objectives
- Keep your agency or organization management briefed
- Identify staging areas
- Ensure scene integrity and evidence preservation
- Communicate with stakeholders
- Evaluate "What if?" contingencies
- Assemble and disassemble strike teams and task forces
- Maintain an ICS-201 Incident Briefing Form

To help you remember the critical management actions that you should take after arriving on-scene we have included a checklist (see *Figure 3.2*). Most likely, when you get a chance to review the checklist you'll find that you've already addressed many of the items; however, like pilots on an aircraft, use the checklist to make sure that nothing is inadvertently forgotten.

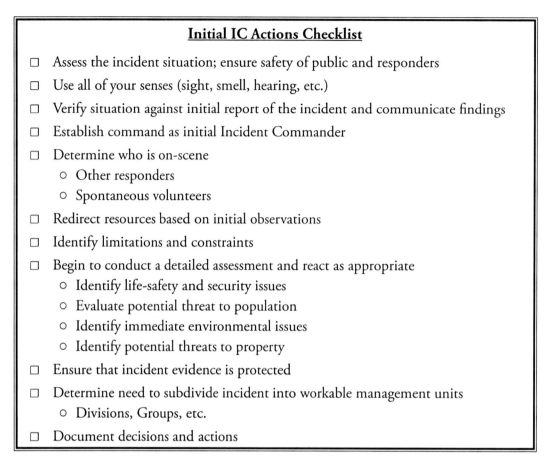

Figure 3.2. Checklist of actions that the Initial Incident Commander should consider taking.

Establishing Incident Objectives

One of the core principles of the Incident Command System is management by *objectives*. Simply put, this is determining what it is you want to accomplish, and working toward that goal.

During your initial response to an incident, you often use your agency's internal guidance and training to determine your initial actions, but no two incidents are alike and an incident that at first appears under control can get ugly real quick so be ready to adapt.

Objectives are not always easy to develop, but there's a speedy way to evaluate whether the ones you created are on the right track. To evaluate your objectives ask yourself the following questions:

- Is your objective **Specific**? (Is the objective focused enough that responders know what you want done?)
- Is your objective **Measurable**? (Can you measure progress toward completion of the objective)

- Is your objective **Attainable**? (Do you understand the situation clearly enough that the objective isn't impossible to accomplish given the constraints of resources, weather, geography, etc.)
- Is your objective **Realistic**? (Can your objective be accomplished?)
- Is your objective **Time Sensitive**? (Did you place a time frame for when the objective should be accomplished? (e.g., 11:00 AM 31 December))

Here are a couple of examples to help make the point:

Objective: *Lower the water in Vallecito Reservoir by 24 inches by 0800, 16 October.* This is focused on an operations task and is an appropriate objective that meets the criteria above.

Objective: *Provide portable sanitary facilities for the work crews in Division A by 0800, 16 October.* Although this objective meets the criteria, it is not an operational objective. Remember that the objectives you have developed are driving the direction of your operations. If you brought in a Logistics Section Chief (LSC), the LSC will identify support issues such as portable sanitary facilities and will ensure that they're addressed. You have more pressing concerns.

Objective: *Lift the sunken railway tank car from the Buffalo Bayou trestle bridge by using a floating 200-ton heavy lift crane so that the navigation channel can be opened by 1000 on October 17.* This objective may meet the criteria, but it's too specific. The real intent is to clear the navigation channel. By specifying a heavy lift crane as the tactic, the IC has limited the options available to the Operations Section Chief. Perhaps the tank car can be floated by using divers and compressed salvage airlift bags. Or, perhaps it can be pulled out of the channel by using tracked bulldozers on the beach. The point is, don't make the objective extremely specific unless that is exactly what you want your responder to do.

Objective: *Be safe.* This does not meet the criteria for an objective, but we often see it anyway because the IC wants to emphasize safety. There may be better wording; perhaps the objective could be: *Conduct a hazard risk analysis and develop safeguards to protect responders and the public.* Optimally, safety is stated in the messages from the IC, and safety is integrated into all response tactics through operational briefings, including those done on-scene by the responders (known as tailgate briefings). Another way to ensure safety is by assigning an Assistant Safety Officer at the site of high-risk activities, and through prepared Health and Safety Plans and Job Safety Analyses. When safety is a proper part of a response organization, it shows up in both the incident priorities and objectives, as well as in all of the incident work products.

Remember that the objectives you set for the response will be used to guide the actions of every responder who comes to assist in the response. Objectives are critical to the response effort. If objectives are unclear, your team will have to guess at what it is you want and may not accomplish what you want. If objectives are too narrow, your team will not have the flexibility necessary to adapt to a changing situation.

Now that we've discussed your initial responsibilities and reviewed objectives with you, let's move onto documenting your initial response.

Section Two: Document Your Initial Response

The ICS-201 Incident Briefing Form is an excellent tool for the initial Incident Commander to help organize and manage the initial response phase of an incident. It's a simple four-page document that has a place to draw a quick sketch of the incident, record the initial objectives and actions taken, jot down a rough sketch of the organization and list resources that have responded to the incident. Your agency may have another form that captures similar information, but we advocate that you use the ICS-201 because it's a part of the ICS and should be a part of your response kit.

Page 1 of the ICS-201 (Sketch Map)

You are not required to have artistic talent to draw a simple layout of the incident, such as that in *Figure 3.3*, and the benefits of a picture will enable others to quickly grasp the scope of the incident and important details. The map should show the current situation, any incident facilities that have been established such as staging areas; the wind direction, and other critical aspects of the incident. In addition, at the bottom of Page 1, take a few minutes to summarize what you and the other responders face.

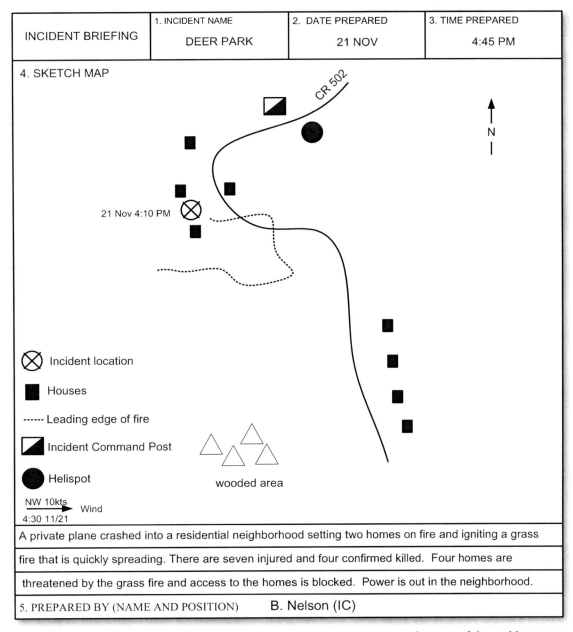

Figure 3.3. A sketch map of the incident is an excellent way to convey the scope of the problem.

Page 2 of ICS-201 (Summary)

An important part of incident documentation is recording the decisions you make, the objectives you have set for the response organization, and the actions you have already taken or plan to take. Page 2 of the ICS-201 is designed to enable you to quickly capture this type of information. *Figure 3.4* provides an illustration of what you might record on the Summary of Current Actions part of the ICS-201.

6. SUMMARY OF CURRENT ACTIONS
Objectives:
(1) Ensure response operations are conducted in accordance with safe work practices
(2) Remove, triage, and transport the injured
(3) Evacuate nearby residents in the path of the grass fire
(4) Protect the remaining homes from fire damage
(5) Establish perimeter control and secure the incident area
(6) Contain the fire west of CR 502
Current Actions:
Medical Group - Stabilize injured and transport to Durango Hospital
Fire Group - Conduct fire suppression operations and protect remaining structures.
Hold the fire west of CR 502 and north of the wooded area.
Law Enforcement Group - Evacuate residents in the path of the fire. Establish
traffic control point at the entrance of CR 502.

ICS-201   Page 2

Figure 3.4. The decisions you make as an Incident Commander during the initial response to an incident can be recorded on Page 2 of the ICS-201.

Page 3 of the ICS-201 (Organization Chart)

The Incident Briefing Form provides a place to jot down your organization. List any Command and General Staff positions that you have assigned to someone else and include his or her name and agency. Also, list any Divisions, Groups, Staging Areas, Task Forces, or Strike Teams that you have established and include the name of the person in charge of each and the resources that are assigned to that organizational element. *Figure 3.5* is an example of how Page 3 of the ICS-201 might look.

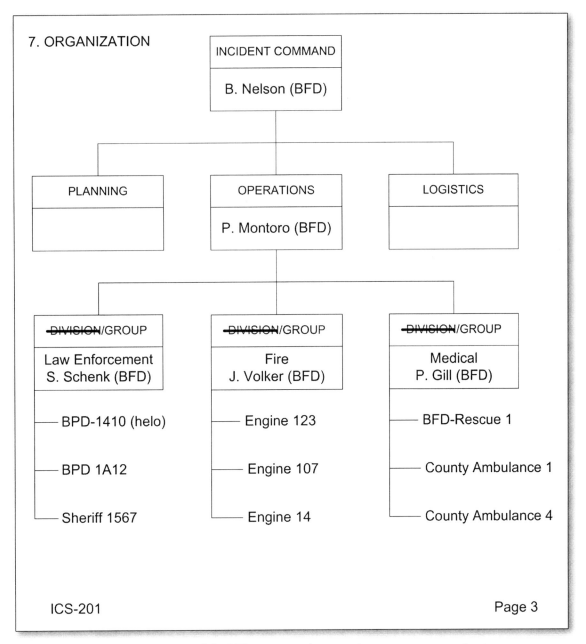

Figure 3.5. Page 3 of the ICS-201 is used to record your response organization.

Page 4 of the ICS-201 (Resources Summary)

In several places throughout this book we discuss the importance of resource management. Effective resource management starts with you, the initial response Incident Commander. If you do not have a good handle on what resources you have on-scene and where you have assigned them, your response has a very good chance of not operating as efficiently as it should and the safety of that resource could be in jeopardy. As an IC you must have situational awareness and part of that awareness is to know where all of your resources are working.

In addition, what if you wanted to hand off the responsibility for tracking resources to someone else? Having an accurate resource summary such as that shown in *Figure 3.6* would make that handoff much smoother and the person who assumes the resource tracking responsibilities would have a good start for supporting your operations.

8. RESOURCES SUMMARY				
RESOURCES ORDERED	RESOURCES IDENTIFICATION	ETA	ON-SCENE (X)	LOCATION ASSIGNMENT
	B. Nelson		X	IC/ICP
	P. Montoro		X	OSC/ICP
Helicopter	BPD-1410		X	Law Enforcement Gp.
Fire Engine	Engine 123		X	Fire Group
Fire Engine	Engine 107		X	Fire Group
	S. Schenk		X	LE Group Supervisor
LE Unit	Sheriff 1567		X	Law Enforcement Gp.
LE Unit	BPD 1A12		X	Law Enforcement Gp.
	P. Gill		X	Medical Gp. Sup.
Ambulance	Ambulance #1		X	Medical Group
Ambulance	Ambulance #4		X	Medical Group
	J. Volker		X	Fire Gp. Supervisor
Fire Engine	Engine 14		X	Fire Group
Rescue Unit	BFD Rescue 1		X	Medical Group
ICS-201				Page 4

Figure 3.6. The final page of the ICS-201 is for listing the resources that are assigned to the incident and their location.

By now, you should understand everything you should be doing to make sure that your response starts right. Typically, a response that starts right goes right. Using the tools of the ICS will be a big help. Now let's discuss how a transfer of command would occur.

Section Three: Transfer of Command

Being prepared to conduct a transfer of command is an important responsibility of an Incident Commander. Changing the leadership at the top during an emergency response has to be done as seamlessly as possible with minimum disruption to the response effort. If you carry around some type of emergency response field guide, we strongly encourage you to take this section of the chapter and insert it into that guide. In the middle of an incident it's easy to forget to brief key information so use checklists where possible to assure that no important details are lost.

Whether you're the one handing off the responsibilities of Incident Commander to someone else more qualified or you're the one assuming the duties of IC, the information that follows will help you as command is shifted from one to another.

Responsibilities of the Off-going Incident Commander

The quality of the command transfer rests on your shoulders as the initial response Incident Commander, so be prepared. The briefing should be methodical and well thought out. The information that we're providing below is to help you give structure to the briefing. The more structure you can bring to an emergency response, the less reactive you will be. When preparing for a transfer of command briefing try to address the following:

- Negotiate a time and location for the transfer of command briefing with the oncoming IC
- Designate someone to manage the on-scene operations while you're briefing the oncoming IC
- Determine who should attend the briefing in addition to the oncoming IC
- Ensure that the ICS-201 is current and up-to-date with latest incident information
- If possible, provide a large chart or map of the incident area
- Organize your thoughts so that you know what you want to say
- If possible, make a copy of the ICS-201 Incident Briefing Form for your relief

Remember that ICS-201 that you, as the initial Incident Commander, spent time filling out? You're going to be very glad that you completed that form. It contains much of the information you'll need to pass on, and it will serve as a memory jogger as you start to review it with the oncoming IC. *Figure 3.7* is a checklist of what you want to cover as you give your brief. The list does not cover everything, but it provides a foundation that you can build on.

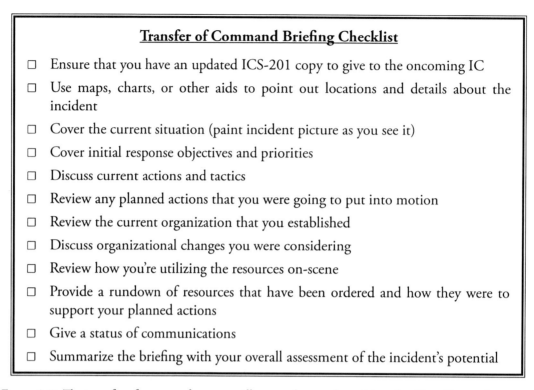

Figure 3.7. The transfer of command process will go much smoother with a checklist of items that you as the off-going IC should cover.

It is our hope that as you read through the briefing items in *Figure 3.7* that you see the benefit of having a briefing document. If you conduct a briefing from memory, it likely will not be as thorough as the oncoming IC deserves.

Responsibilities of the Oncoming Incident Commander

As you park your vehicle, you notice the smoke obscuring the sun, the pieces of plane wreckage dotting the landscape, and the gesturing of responders as they work to control the situation. You're there to relieve the initial Incident Commander due to the jurisdictional nature of the incident and its complexity. From experience you know that this response is going to grow in difficulty and size before you and other responders can get a handle on it. Luckily, you know the initial Incident Commander you're going to relieve and she is both thorough and highly professional. Before assuming command you'll be receiving a transfer-of-command brief from the initial Incident Commander.

The briefing you receive should provide you with several key things: first, it will give you a feel for the size and complexity of incident operations and everything going on pertaining to the tactical response; second, it provides you with essential information on resources currently involved in the response and the incident potential (how bad is bad?). Finally, it provides you with the direction and priorities of the initial or current IC, and you can determine if those should continue or change, using your experience to guide you. Use the checklist in *Figure 3.8* to remind you of some of the more important issues that you should try to resolve, as well as some things that you should do after assuming the responsibilities as the Incident Commander.

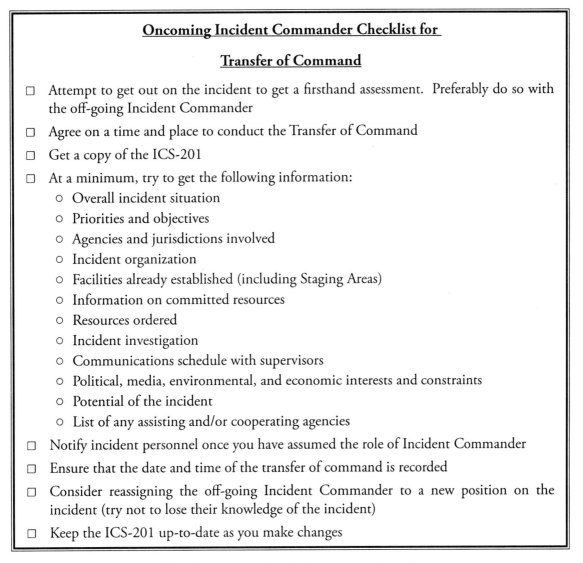

Oncoming Incident Commander Checklist for Transfer of Command

- ☐ Attempt to get out on the incident to get a firsthand assessment. Preferably do so with the off-going Incident Commander
- ☐ Agree on a time and place to conduct the Transfer of Command
- ☐ Get a copy of the ICS-201
- ☐ At a minimum, try to get the following information:
  - Overall incident situation
  - Priorities and objectives
  - Agencies and jurisdictions involved
  - Incident organization
  - Facilities already established (including Staging Areas)
  - Information on committed resources
  - Resources ordered
  - Incident investigation
  - Communications schedule with supervisors
  - Political, media, environmental, and economic interests and constraints
  - Potential of the incident
  - List of any assisting and/or cooperating agencies
- ☐ Notify incident personnel once you have assumed the role of Incident Commander
- ☐ Ensure that the date and time of the transfer of command is recorded
- ☐ Consider reassigning the off-going Incident Commander to a new position on the incident (try not to lose their knowledge of the incident)
- ☐ Keep the ICS-201 up-to-date as you make changes

Figure 3.8. Checklist for oncoming IC to help you remember what you need to ask.

With the briefing complete and any nagging questions hopefully answered, you're now in overall charge of the incident and responsible for every responder under you. As you survey the scene in front of you, a lot is going through your mind. Is the current organization configured to meet the needs of the incident? What if the grass fire makes it to the wooded area or we cannot stop it west of County Road 502? Are there any safety concerns with how the response is being conducted? Can we get the response under control within a reasonably short period of time? These few examples are just the tip of the iceberg when it comes to the issues and concerns that you have to think about and, if necessary, take action on.

If you determine that the incident will not be resolved quickly you'll have to transition your response team to a proactive planning process that allows you to plan and then execute tactical operations. This is almost the opposite of how we conduct initial response operations where we react and immediately direct resources to the scene of the incident and then figure out the best way to employ them. This transition from reacting to an incident to a methodical planning of future tactical operations can be challenging, and it requires solid leadership skills and a firm grasp of the ICS Planning Process. You have to manage the very dynamic situation that you're currently facing while simultaneously shifting into the planning process. In order to make the shift without losing focus on current operations, you need to build an organization that can support both the current demands as well as plan for future operations.

Building Your Organization

We are going to limit our discussion on building your organization to the Command and General Staff positions (see *Figure 3.9*). If you want to know how the Section Chiefs staff their respective sections, go to chapters 6, 7, 8, and 9.

Command and General Staff

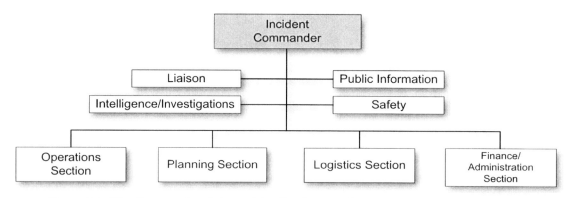

Figure 3.9. The Command and General Staff is the core of your primary management team.

Don't underestimate the incident potential. Even if you're unsure of the exact situation you face, but you know it will be difficult to bring the incident to closure, order key management staff positions in to support you. The demands on your time will be extreme and you need their help to accomplish all of your responsibilities. The longer you delay ordering staff, the longer it will take them to arrive. You can always turn them around and send them home if you guessed wrong. Either that or you may find yourself racing to get ahead of the incident. Although this approach might seem risky, there are many examples of incidents that were underestimated initially. The result was that it took twice as long to get the incident under control than it should have taken had the incident management team been properly resourced and managed. Everyone remembers the incident response that was slow or inefficient, but no one usually remembers a response that went well and had a few too many responders than perhaps were necessary.

Remember that every Command and General Staff position that you do not assign to someone you are responsible for doing that job. For example, if you know that the media is going to be demanding your seriously limited time you may want to get a Public Information Officer in to help you deal with that aspect of the response. What about logistics? You know your response effort cannot last long or expand if you do not have adequate personnel, equipment, and supplies. If you do not bring in a Logistics Section Chief, then you are going to be the one dealing with logistics when a report of a responder injury is brought to your attention. Because you are also the Safety Officer, logistics is going to have to wait until you deal with the injury. Building an effective ICS organization starts with recognizing what you can manage on your own and what you need to delegate. There are no hard and fast rules; it comes down to your experience.

We do not encourage or suggest to you that every incident you respond to requires you to staff all Command and General Staff positions. In fact, for the majority of incidents you face the opposite is true. What we're trying to illustrate is that you have to be ready to expand, and to do so in an efficient and effective manner.

As your response team mobilizes and checks in, it's in your best interest to properly assign, brief, and support them. This is easier said than done, because team members will be arriving while the response efforts are very intense and are demanding all of your attention, but the sooner you get them on-line, the better you will fare. One of the common threads that we have written into every Command and General Staff position chapter in this book is that when they arrive at the incident to relieve you of a specific responsibility, they will need a thorough briefing from you before getting started with their work. We recommend that you invest the time to get them quickly operational. Be organized and methodical. Have the Incident Briefing Form, ICS-201, properly filled out and ready to use. Remember how it helped you in your transfer of command briefing.

Section Four: Incident Commander and the ICS Planning Process

We're going to walk you step-by-step through the ICS Planning Process so that you understand how it's designed to get you and your management team out ahead of the incident. Our discussion of the planning process in this chapter touches only on the highlights. We have a whole chapter that covers this process, so make sure you take a look at Chapter 2 if you haven't already. You are, after all, the Incident Commander and you need to understand how the planning process works.

The ICS-201 Incident Briefing Form that we looked at earlier is really a reactive plan. You took action and then recorded what you did. By contrast the ICS Planning Process that we're going to look at next is a way to methodically develop a proactive plan called an Incident Action Plan (IAP). You assess the situation and build a tactical plan tailored to the needs of the incident. A sample IAP can be found in Appendix A.

Before you can shift your team to the ICS Planning Process you must have two things: first, you have to have enough management personnel within the command team to do the work that is necessary to plan and build an Incident Action Plan; and second, you have to define the operational period.

We have a great visual aid that we use throughout the book to illustrate the various *steps* in the ICS Planning Process. It's called the Operational Planning "P" (*Figure 3.10*) and it is our guide. We go through each meeting and briefing with you, pointing out your responsibilities and things to consider as you work to resolve the increasingly complex incident you're managing.

The Operational Planning "P"

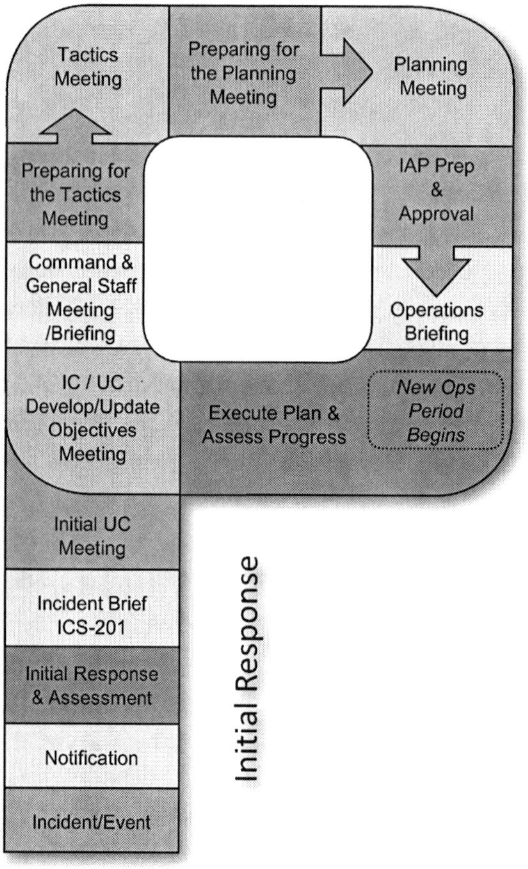

Figure 3.10. The Operational Planning "P" is a visual illustration of the ICS Planning Process.

If we start at the bottom of the Operational Planning "P" and walk up the stem, we make our first stop at the IC/UC (Incident Commander/Unified Command) Develop/Update Objectives Meeting. This is the first stop because you have already completed the Incident Brief ICS-201 *step* prior to assuming the responsibilities of the IC. The *step* above that, titled Initial UC Meeting, applies only if there's going to be more than one Incident Commander collectively known as Unified Command. For our discussion in this chapter we're going to assume that you're the only Incident Commander. Chapter 4 covers the concept of Unified Command.

IC/UC Develop/Update Objectives Meeting

First, because there's only a single Incident Commander in our scenario this *step* is not a meeting, but it's a time for you to gather your thoughts. Based on the best information that you have, you must make several decisions here that will impact the future actions of you and your team. Consider these decisions:

- Determine priorities for the incident (e.g., evacuate injured, secure area, and suppress the fire)
- Determine the incident response objectives. (The objectives form the foundation of the IAP. A poor foundation means a poor plan.) Are the objectives:
    - Specific
    - Measurable
    - Attainable
    - Realistic
    - Time Sensitive
- Determine the operational period. The operational period is the period of time that an Incident Action Plan (IAP) is designed to cover. For example, if you want 12-hour-long operational periods, an IAP will have to be developed twice a day. Take into account:
    - Pace of the operations
    - Rate of change in incident situation
    - Weather or other criteria and trends that impact ability to work
    - Safety and well-being of responders
- Establish an incident organization that is capable of meeting initial and long-term challenges required to mitigate the incident
    - Is your span-of-control within limits
    - Did you consider the need for a Deputy
- Identify and select incident support facilities
    - Incident Command Post
    - Base
    - Staging Areas
    - Camps
- Ensure scene integrity and evidence preservation
- Identify constraints and limitations (e.g., will respond during daylight hours only and without aircraft support)
- Identify Incident Management Team operating procedures for the incident (e.g., organization's protocols for dealing with next-of-kin notifications)
- Make any other key decisions that will impact the overall response

Command and General Staff Meeting

This meeting takes place early in the planning cycle and is designed to give you time with your Command and General Staff. It's a great opportunity for you to set the framework on how the incident will be managed, communicate your expectations, assign tasks, brief the staff on how you see the response operations going and what you anticipate. It's also a good time for the staff to ask for clarification on any issues or concerns that they may have. For the example, the Safety Officer may want to get clarification on their authority. You want open dialogue so that everyone is working toward the same goals.

You always manage two concurrent areas of focus when you're in the ICS Planning Process: current operations and the myriad of activities that are necessary to develop an Incident Action Plan for future operations. The focus of this meeting is to set the tone for the next operational period. Don't lose sight of that need; it's easy to have the focus of the meeting shift to current operational issues. Although current operational issues may be a necessary discussion for the assembled team, work to keep the focus on the future.

Before the meeting, review the list of items below to see if there are specific things you need to communicate at this meeting. This list might include setting a media briefing time, discussing facility or Incident Command Post space needs with the Logistics Section Chief, or discussing spending ceilings with the Finance/Administration Section Chief. Give your Planning Section Chief guidance on your expectations for meeting facilitation, but realize that this meeting may need to allow for discussion as the response effort is growing and changing.

Some items you want to cover during the meeting:

- Your priorities (e.g., the highest priority is the protection of structures)
- Present any limitations and/or constraints
- Review the incident objectives for the next operational period so your management team can begin work on the IAP
- Define the hours of work and operational period
- Identify staffing requirements
    - Who will be the Operations Section Chief
- Any directions (e.g., ensure that tactical operations are planned to minimize impact to wildlife)
- Any work tasks that you want done and who is to do it (e.g., you may task the Public Information Officer to establish a Joint Information Center)
- Clarify any staff roles and responsibilities
- Expectations of the team for staff communications

Preparing for the Tactics Meeting

During this time in the planning process, you may want to meet one-on-one with Command and General Staff members to follow up on any work assignments that are still incomplete, receive an update on response operations, and prepare any further guidance and clarification that you think are necessary.

Tactics Meeting

Normally, the Incident Commander does not attend the Tactics Meeting. During this meeting, the Operations Section Chief (OSC), Planning Section Chief (PSC), and other members of the incident management team get together to help the OSC develop a tactical plan for the next operational period. This plan is built to meet the objectives you laid out during the Command and General Staff Meeting. Specifically, the OSC:

- Reviews the priorities and objectives with the PSC and considers the incident's limitations and constraints
- Identifies functions that have to be performed
- Divides the Operations Section's work into manageable units (e.g., Divisions, Groups)
- Assigns work tasks for each identified organizational element
- Lists the resources required to accomplish the work assignment

Preparing for the Planning Meeting

It has been several hours since you sat down with your staff and they are busy carrying out current operations and planning for the next operational period. This is a time when you:

- Check in with the Operations Section Chief (OSC) to find out how the current operations are going and whether there are any concerns about the ongoing or future operations
- Talk with the Planning Section Chief (PSC) to see how the team is functioning and determine if there are any problems that the PSC thinks may be on the horizon
- Meet with the Logistics Section Chief for a few minutes to discuss any issues about getting resources and supplies to support the incident
- Spend time with the Safety Officer and find out if there are any concerns

Planning Meeting

This meeting is central to the entire planning process. A good six hours have probably gone by since you last met with your entire Command and General Staff and they have been hard at work developing a plan that will meet your established objectives, priorities, limitations, and constraints. During this meeting, the plan for the next operational period will be briefed to you for tentative approval. Your role is to make sure that you're in agreement with the strategies, tactics, and other activities. Here are some items that you need to do to support the meeting:

- Provide opening remarks (remember that the focus is on the future, so keep your comments forward leaning)
    - Discuss any changes, deletions, or additions to the original incident objectives that you gave your team earlier
    - Discuss where you see progress and where improvements can be made
- Make sure that the response plan the OSC has briefed follows your directions and the incident objectives have been properly addressed
    - Are you comfortable that the responder and public safety risks are properly balanced with the priorities you established?
    - Is the OSC-proposed organization adequate to meet the needs of the incident?
- Provide any necessary further guidance and resolve any issues that come up
- Give tentative approval of the proposed Plan
- Agree on a time when the Planning Section Chief will be ready to give you a written Plan for your review and final approval

If your command team has done its job, you should have a positive sense of momentum that you're getting ahead of the incident. The first one or two Planning Meetings may not go as smoothly as you would like, but that is to be expected. Remember that you're shifting from a reactive response to a proactive response and that can be difficult. Everyone on your team may not know each other, and it can take a little time to gel as a unit. Information on the incident situation may be very dynamic, making it difficult to get accurate information in time to make key decisions. That said, you should see and expect a rapid improvement in subsequent meetings.

Incident Action Plan Preparation and Approval

At this *step* in the Operational Planning "P" the Planning Section Chief will be overseeing the development of an Incident Action Plan (IAP). When completed, the PSC will meet with you to review the IAP. Several hours have passed since the Planning Meeting so take time to go over the plan and ensure that it still meets the needs of the incident. Review each aspect of the IAP carefully as this document will reflect all decisions, strategies, safety concerns, and facility information. The IAP is the document that will be referred to when post-incident questions arise about cost and legal issues, the efficiency and effectiveness of you and your team's performance, and other incident-related issues. It's in your interest to make sure the IAP reflects what you want it to reflect. Have the PSC make any changes and then approve the plan with your signature. When reviewing the IAP, look for:

- Accurate recording of the objectives
- Reflection of organization and key personnel in the command team
- Clear and concise assignments for each Division, Group, Staging Area, etc.
- Logical use and assignment of resources
- Integrated safety and communication procedures for each Division, Group, Staging Area, etc.
- A logical and useable Medical Plan
- A Communications Plan that addresses command, tactical, and support communications needs
- Situation maps that are useful and accurately reflect status, organization assignments, facilities, weather, and other trend information
- Support plans specific to the incident, including hazardous waste operations safety, waste disposal plans, transportation plans, decontamination and demobilization plans

The original copy is placed in the incident documentation files.

Operations Briefing

You and your team have been immersed in the development of an IAP that within an hour or so will be implemented. The IAP has been built to meet the objectives that you gave the team for the upcoming operational period. The majority of the people who have to conduct the work assignments outlined in the plan have not been involved in the plan's development, nor have they necessarily been involved in the response. The Operations Briefing is the first time they hear what they are supposed to do. The Planning Section Chief makes sure that all Branch Directors, Division and Group Supervisors, Staging Area Managers, Air Operations Branch Director, and others responsible for specific work assignments have a copy of the IAP. The Operations Section Chief briefs each individual on their assignment and encourages questions to clarify tactical assignments. Other members of the command team conduct a brief on safety and logistical issues. Your role in this briefing is to:

- Provide a leadership presence
- Provide overall guidance
- Provide motivational remarks
- Emphasize your response philosophy
    - Teamwork
    - Safety

Let's stop and take a breath here. We have covered a ton of stuff. You have just participated in the Operations Briefing and when it was completed the responders headed into the field armed with a plan that will guide their activities for the next 12 hours (we assume your plan is good for a 12-hour operational period).

The reason the responders have a plan to execute is because after you assumed the role of Incident Commander, you recognized that the incident was not going to be resolved quickly and that many operational periods would be needed to bring it under control. You brought in more staff to help manage the cascading resources that were arriving on-scene and to get the ICS Planning Process up and running. While the Operations Section Chief that you designated is organizing and directing current tactical operations, the Planning Section Chief and others are moving forward on the development of an IAP (with critical input from the OSC).

The IAP that your staff put together was built on the objectives that you set. Those objectives were based on where you thought the response would be 12 hours in the future. By using the planning process, your team came together and methodically designed a tactical plan that could be logistically supported and met any constraints or limitations that you placed on the responders.

The moment that the new operational period begins, the planning process starts all over again, and in a short time you have to give your team the objectives for the next operational period so that they can begin work on the next IAP.

Execute Plan and Assess Progress

No matter how good your Incident Action Plan (IAP) is, incidents always have a way of throwing you curve balls, so vigilance in assessing the effectiveness of the plan, given the realities taking place in the field, is a must. The plan is just a plan and therefore nothing is set in stone, so be flexible and make adjustments at any time to ensure that the plan helps you get where you want to go. Make sure you document changes and anyone involved in the response who needs to know is briefed on the changes. Some things to consider as you assess incident operations:

- Review progress of assigned tasks with your OSC
- Receive periodic situation briefings
- Review work progress
- Identify changes that need to be made during current and future operational periods
- Ensure that the organization you have in place is adequate for what you're facing

Look in Appendix B to see a one-page summary of the Incident Commander responsibilities during the various *steps* of the ICS Planning Process. The page contains the Operational Planning "P" and calls out specific command activities that we presented above.

Closing Thoughts

As an Incident Commander, the responsibility placed on you is tremendous. You most likely are managing incidents where lives are on the line: the lives of your responders, and the lives of the public. Every capability and skill that you have is called upon. You must lead, you must make intelligent decisions under pressure, you must anticipate the incident's potential, and you must have a competent team of responders carrying out your intentions. All this synergy is not assembled at the spur of a moment. It takes time and dedication. The best Incident Commanders in the world got to be that way through training, experience on incidents, and selection by the knowledgeable ICs that came before them.

Throughout this chapter we have concentrated on you, the Incident Commander, not acknowledging the fact that in today's response climate multi-agency responses are becoming more the norm than the exception. We did this intentionally. Everything you learn here is applicable when you're involved in a response where there's a Unified Command and you're just one of the Incident Commanders who make up the Unified Command. Objectives still have to be developed, but now you have to work in a consensus environment and agree on what the objectives are. Chapter 4 is dedicated to Unified Command, for an IC this should be required reading because you never know when you might be a member of a Unified Command.

CHAPTER 4
UNIFIED COMMAND (UC)

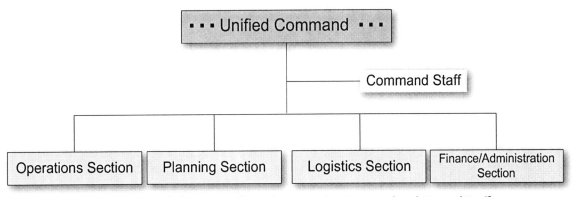

Figure 4.1. Unified Command in relation to the Command and General Staff.

Commuters are on their way home for a long holiday weekend when, without notice, a section of the bridge crowded with vehicles collapses into a river, blocking a major shipping channel that lies beneath the bridge. The cause of the collapse is not known, but in today's world, terrorism cannot be ruled out. The list of agencies that have a role in responding to this type of incident is long. Some of the issues that would have to be addressed are: search and rescue, rerouting transportation, incident security, loss of critical infrastructure, the blocked channel, body recovery, bridge integrity, and evacuation of stranded motorists. Add terrorism to the mix and the response complexity rises dramatically. Even with no more information about the incident, we can all make one safe assumption: in this type of incident there are going to be several Incident Commanders from various organizations all with a legitimate claim to a piece of the response.

Fortunately, the developers of the Incident Command System (ICS) recognized the need to integrate multiple agencies with jurisdiction over the same incident into a single team and included as part of the ICS the management concept of Unified Command (UC) as seen above in *Figure 4.1*.

In Chapter 3 we covered the roles and responsibilities of an Incident Commander and everything that you learned in that chapter holds true for our discussion on Unified Command. However, instead of making decisions solo as the IC, you join forces with other Incident Commanders to direct the response activities. You need all of the talent, skill, and experience that it takes to be an effective Incident Commander when you become part of a Unified Command. What you do not need is an overabundance of ego. The Unified Command's success rests heavily on the ability to work together and resolve all of the complex and competing demands the UC faces.

Your participation in a Unified Command occurs without your agency or organization giving up its authority, responsibility, or accountability.

What Is Unified Command?

Unified Command is about teamwork at the top of the ICS organization (see *Figure 4.1*). It's about the shared responsibility of command among several Incident Commanders. Unified Command links the organizations responding to an incident and provides a forum for those agencies to make consensus decisions; however, Unified Command is not "decision by committee." Those who are sitting in the UC position are there to command the incident and have to come together and agree on how to best respond to the incident they face. The UC can disagree all they want behind closed doors, but when they emerge and face their management team, they must be united.

When Is a Unified Command Necessary?

The type of incident, its complexity, and location influences whether a Unified Command is established to manage response operations. As the Incident Command System continues to be adopted for use in all-risk, all-hazard incidents, the use of Unified Command is becoming more commonplace. Some indicators that the response should be managed by a Unified Command include when an incident:

- Crosses geographic boundaries (e.g., two states, Indian Tribal Land)
- Involves various governmental levels (e.g., federal, state, local)
- Involves private industry, or public facilities (e.g., oil company, school)
- Impacts different functional responsibilities (e.g., fire, police, hazardous materials response, emergency medical service)
- Includes different statutory responsibilities (e.g., Federal Land Managers)
- Has some combination of the above

Makeup of the Unified Command

Although there are no limits on how many Incident Commanders can make up the Unified Command, smaller is always better. Much is expected of those in the Unified Command and they operate in a time-critical and mentally demanding environment. So make every attempt to keep participation on the UC to a manageable level that enables it to operate in a dynamic environment and moves the response operations forward. The actual makeup of the Unified Command for a specific incident must be determined on a case-by-case basis taking into account:

- The specific needs of the incident (e.g., type of incident)
- Determinations outlined in existing contingency plans
- Decisions reached during the initial meeting of the Unified Command

To help sort out whether your organization (government or industry) should be included in the Unified Command, use the following as guidance. If you answer "yes" to all four questions for the particular type of incident that you are responding to, then your organization belongs in the Unified Command:

- My organization has jurisdictional authority or functional responsibility under a law or ordinance for the type of incident
- My organization is specifically charged with commanding, coordinating, or managing a major aspect of the response
- My organization has the resources to support participation in the response organization
- The incident or response operation impacts my organization's area of responsibility

Let's use our guidance and see if the fire chief for the town of Bayfield would be in the Unified Command when responding to a fire incident in the nearby City of Durango. The chief answers "yes" to the first question because he carries functional authority to put out fires; question two is a "yes" because fire departments are responsible for command and control of fire incidents; he answers "yes" to the third question because the Bayfield fire chief has tactical resources to support operations; however, for question four, the chief answers "no" because the response is outside of his area of responsibility. However, the Bayfield fire chief may become an Agency Representative or support activities in the Operations Section.

The makeup of the Unified Command can change over time to meet the needs of the incident. Several agencies that are in the Unified Command initially may not be involved after their jurisdictional or functional responsibilities are no longer involved or needed. The bridge-collapse scenario that we used to introduce this chapter is a good example of the type of incident where representation in the Unified Command changes as the response shifts from rescue to recovery.

Making Unified Command a Success

You have made it through the gauntlet of questions and determined that you indeed belong in the Unified Command and you want the UC to be a success. You have placed any excess ego on the hook by the door; you're clearly a team player and can compromise when necessary; and you're comfortable with your organization's authorities and capabilities to support the UC efforts.

As an individual Incident Commander you have done your part, but what about the team? A successful UC must be a team. A successful Unified Command skillfully uses the strengths of each Incident Commander and acknowledges each representative's unique capabilities and authorities. A successful Unified Command has a shared understanding of the situation and agrees on common objectives to bring the incident to closure. A successful Unified Command is open to different perspectives. A successful Unified Command knows that contentious issues may arise, but trusts the UC framework to provide the forum to resolve problems and find solutions.

If you find yourself in a Unified Command that is operating with the spirit of cooperation outlined above and there's respect among the Incident Commanders, then that Unified Command is going to make a positive difference for both the responders and the community impacted by the incident.

Unified Command Responsibilities and Expectations

The responsibilities of the Unified Command are no different than those for an incident where there is only a single Incident Commander. The Unified Command is responsible for the overall management of the incident, providing direction, establishing objectives, and approving the order and release of incident resources just like the single IC. The difference is that decisions that are reached on how to manage the response just became a unified effort. If your agency has identified you as someone who may be a member of a Unified Command, then the agency needs to provide you with the authority to do the job. That authority should include the ability to:

- Agree on common incident objectives and priorities
- Have the capability to sustain a 24-hour, 7-day-a-week commitment to the incident
- Commit agency or organization resources to the incident
- Spend agency or organization funds
- Agree on an incident-response organization
- Agree on the appropriate Command and General Staff position assignments to ensure clear direction for on-scene tactical resources
- Commit to speak with "one voice" through the Public Information Officer, Liaison Officer, and through all off-site reporting
- Agree on logistical support procedures (e.g., resource ordering procedures)
- Agree on cost-sharing procedures, as appropriate

You cannot be an effective Incident Commander, representing your agency in a Unified Command if you have to use your cell phone to call your boss every time a decision is made that impacts the response or your agency's participation. We all have bosses and have to operate within the confines of our organization's rules and procedures, but you're going to need some latitude to do your job. If you're not given that latitude, then perhaps you're not at the right level in your organization to be in the Unified Command. Try to work out ahead of time with your boss the level of authority that you'll be given, and make sure you understand what he or she expects from you. Their reputation is on the line as well as yours so appreciate your boss's hesitation to give you the keys to the kingdom. He or she will, however, need to provide you with enough tools to do the job.

As a member of a Unified Command, you can assign a deputy to assist you in carrying out your Incident Commander responsibilities. UC members may also assign individual legal technical specialists and administrative support from their own organizations.

Unified Command Decisions

Decisions that can be reached by consensus are generally the best, but when consensus cannot be reached on a particular issue, the Unified Command does not come to a screeching halt. Members of the Unified Command will voice their concerns, but the Incident Commander on the Unified Command that represents the agency with primary jurisdiction and expertise over the specific issue is deferred to for the final decision.

Not a Member of the Unified Command

We all know that not everyone (and depending on the incident that could be you) who wants to can actually belong in the Unified Command. It does not take an active imagination to appreciate the rapid demise of the Unified Command that does not control its size. The goal of establishing a Unified Command is not to be exclusive, but to increase the effectiveness of multi-organizational responses. If you find yourself outside of the Unified Command, the Incident Command System was designed to ensure that your organization's concerns or issues are addressed. What are your options? Some include:

- Serve as an Agency Representative or company representative with direct access to the Liaison Officer (LOFR)
- Provide stakeholder input to the Liaison Officer (e.g., environmental, economic, political issues)
- Serve as a Technical Specialist in the Planning Section or in other sections as needed

The Incident Command System is inclusive by nature, but how that inclusion is structured requires knowledge on the part of the responders as well as agency and organization representatives.

Unified Command and the ICS Planning Process

The ICS Planning Process does not change because the incident response is being managed by a Unified Command instead of a single Incident Commander. So, as we encouraged you earlier, please read Chapter 3 because we're going to cover just a single *step* in the ICS Planning Process that is unique to Unified Command. In *Figure 4.2* that *step* is called the Initial Unified Command Meeting. If you were the sole Incident Commander, you would jump right over this *step* because this is a meeting of the minds where some very important joint decisions have to be made.

The Operational Planning "P"

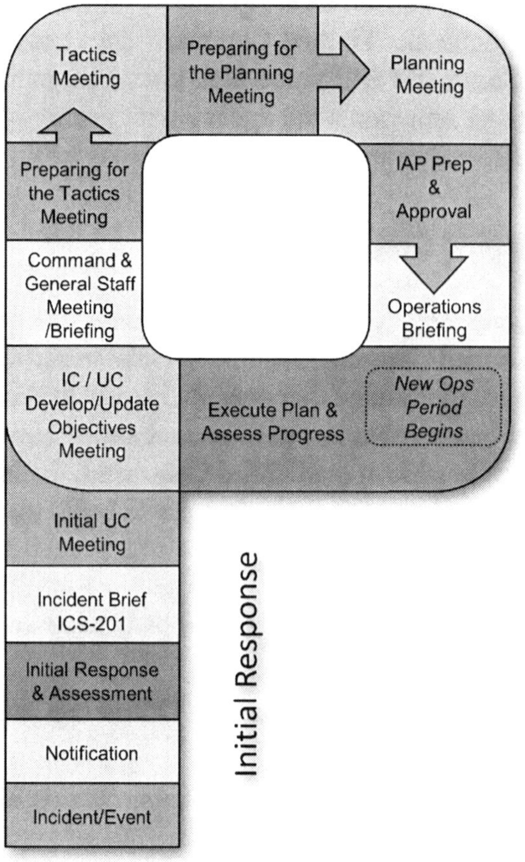

Figure 4.2. The Operational Planning "P" helps guide the response team in the ICS Planning Process. The one step in the process that is unique to the Unified Command is the Initial Unified Command Meeting.

Initial Unified Command Meeting

Once the Unified Command (UC) has assumed responsibility for the incident from the initial response Incident Commander, UC members will need to meet quickly to discuss some important issues and make some key decisions. The Unified Command will be under substantial pressure as members will be working to manage current operations, setting the command team's direction for future activities while simultaneously trying to come together as an effective and efficient team.

The checklist in Figure 4.3 sets the agenda for the Initial Unified Command Meeting. Use the checklist to guide your UC discussions and ensure that critical decisions are recorded. You have to work quickly to cover the items in the checklist. At this point in the incident response effort, the responders are probably still reacting to events and working on the momentum of the initial response Incident Commander's objectives.

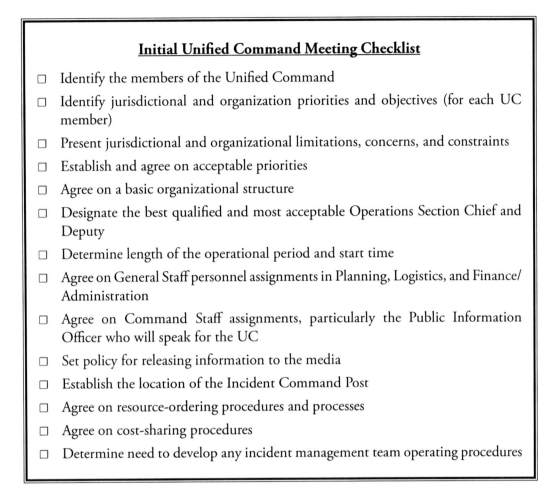

Figure 4.3. Checklist for the Initial Unified Command Meeting.

Some of the decisions made in this Initial Unified Command Meeting are changed as the UC members gain a better handle on the situation and as the situation itself changes. For example, the Operations Section Chief (OSC) who is selected for the initial days of the response may not be the person who remains for the entire incident. UC members need to constantly fine-tune their organization to best meet the situation they confront.

As you probably noted, the Initial Unified Command Meeting occurs only once, early on in the incident, and is intended to enable the Unified Command members to discuss and agree on important issues that will impact their command team.

Pre-Incident Planning and Exercises

If there's any way you can get to know and exercise with those you believe you may work with in a Unified Command, make every attempt to do so. The most effective Unified Commands are those where people already know each other and have practiced how they intend to respond together.

Incidents often have the bad habit of occurring when least expected. Do you want to have to step into an incident-response operation as a member of the Unified Command and see nothing but new faces? At two o'clock in the morning, all of your energy and focus needs to be on managing the incident and not on trying to develop new relationships.

Meeting and forging critical relationships during the heat of an emergency response is a necessary skill for any member of the Unified Command; but the better scenario is for the Unified Command to have experience working with each other. For critical response scenarios that involve significant consequences and require highly complex agency and organization interactions, a practiced and experienced Unified Command is optimal.

Closing Thoughts

The management concept of Unified Command has a proven track record of success. The ICS organizational structure that sits below the Unified Command, along with the ICS Planning Process that the command team uses to move an incident from a reactive response to a proactive response, is identical whether there's a single Incident Commander making all of the command decisions or a Unified Command.

Through joint exercises, training, and using ICS for event planning and contingency planning you can pre-identify who will be in the Unified Command if a particular incident occurs. By successfully doing this, instead of strangers at the response table, you are surrounded by faces you know. We have covered information on Unified Command that will help you and other Incident Commanders to manage jointly an incident. However, the true effectiveness of a Unified Command rests with its members. With a keystroke we cannot suddenly make you a team player, you have to do that. We cannot force you to seek consensus if you're bent on the thought that it has to be only your way. The skills and experience you bring to the table and your willingness to work as a member of a team will ultimately be the deciding factors as to how well the Unified Command performs its responsibilities.

CHAPTER 5
COMMAND STAFF

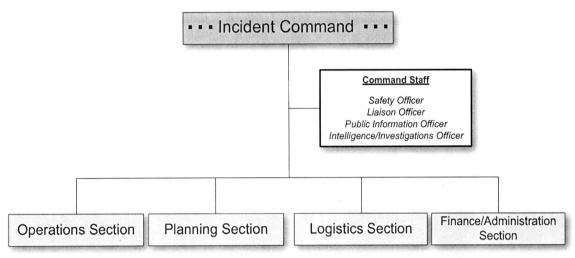

Figure 5.1. The Command Staff in relation to Incident Command.

The specialized staff positions that comprise the Command Staff include the Safety Officer, Liaison Officer, and Public Information Officer. The Intelligence/Investigations function also can be assigned as a Command Staff position on the incident management team and would be called the Intelligence/Investigations Officer.

Safety Officer

Emergency response operations are an inherently risky business. As a responder, you're going into an area that everyone else is trying to leave. In addition to the hazards faced by the victims of the incident, the hazards to the responders working to save lives, property and the environment can be significant. Managing that risk falls on the shoulders of the Incident Commander/Unified Command (IC/UC). On a small incident with few responders and few issues that require the Incident Commander's attention, the IC/UC can manage the safety of the response operations without a designated Safety Officer. However, on complex or larger incidents the IC/UC will be unable to effectively provide the level of oversight and direction necessary to ensure the safety of responders. The responsibility for safety will shift to a dedicated Safety Officer (SOF) who reports directly to the IC/UC as seen in *Figure 5.1*.

The Incident Commander/Unified Command assigns a Safety Officer not only because they're unable to give safety the attention it requires, but because they recognize that complex incidents require a high degree of technical expertise that a skilled safety professional brings to the table.

The Safety Officer position is a difficult job even under the best of times. Somehow you have to maintain a 360-degree view of response operations. You cannot afford to get focused on one concern to the extent that all of the others disappear from your vision. It's easy to get sucked into looking at how best to safely conduct a dangerous and difficult aspect of the response and forget about all of the more mundane ways that responders can get injured.

There are many skills that are going to be important to how well you succeed in ensuring that response operations are conducted safely. Primarily, your technical expertise (e.g., understanding regulations, assessing hazards, risk management), your experience in the response discipline, and your knowledge of how to participate in the ICS organization and processes in support of incident operations will assure the success of a safe, injury-free response.

As a Safety Officer, there are three important skills you need: (1) the ability to recognize the hazards that responders are facing; (2) understanding how bad those hazards are (this is known as severity); and, (3) being able to determine the likelihood of how often the hazard will occur. The combination of hazard recognition, hazard severity, and hazard occurrence is called *risk management*. See Appendix C, Managing Risks for more information on managing the risks using the processes and tools of the Incident Command System.

Some of the primary responsibilities of the Safety Officer are to:

- Work with the Operations Section Chief to identify and mitigate safety hazards associated with planned strategies and tactics
- Ensure that responders have adequate and appropriate safety training under local, state, and federal regulations
- Participate in the planning process
- Identify hazardous situations associated with the incident, and develop strategies to mitigate or manage those hazards
- Participate in the development of the Incident Action Plan
    - Review every Assignment List, ICS-204
    - Review the Communications Plan, ICS-205
    - Review the Medical Plan, ICS-206
- Exercise STOP WORK authority to stop or prevent unsafe tactics
- Ensure that incident support facilities are safe
- Investigate accidents and injuries that occur in the incident area or at incident facilities
- Assign Assistant Safety Officers as needed
- Develop the appropriate safety plans for the response, addressing job hazards and risk to responders (e.g., Site Safety Plan and use of safety equipment and gear)
- Monitor the compliance of response agency or organization safety requirements, as well as Occupational Safety and Health Administration (OSHA) regulations

Actions to Take upon Arriving at the Incident as the Oncoming SOF

Once you arrive at the incident, check in and then meet with the IC/UC to get a briefing. If you're relieving another Safety Officer, you can get the briefing from them. You need to understand the incident situation thoroughly. The checklist in *Figure 5.2* can help remind you of questions you would like answered before assuming responsibility for the safety function for the incident. Once you are briefed, you can make a determination on the number of Assistant Safety Officers you'll need to oversee safety of the incident.

Safety Officer Briefing Checklist

At a minimum the briefing should include:

- ☐ Incident situation: magnitude and potential of the incident
- ☐ Information on any current safety activities
- ☐ Results of any risk assessments
- ☐ Incident priorities, limitations, and constraints
- ☐ Incident objectives
- ☐ Current incident organizational structure
- ☐ Expected incident duration
- ☐ Estimate on the potential size of the response organization
- ☐ Operational Period
- ☐ Agencies and jurisdictions involved
- ☐ Review of any accidents, injuries, or near misses
- ☐ Validate stop work authority

Figure 5.2. Safety Officer Checklist for initial briefing upon arriving at the incident.

Managing the Safety Officer Responsibilities

The nature of the Safety Officer position requires that you spend time in the field ensuring that operations are being conducted safely, visiting incident facilities to make sure they're safe for personnel, participating in the ICS Planning Process, and developing required safety plans. If anyone gets injured on the incident, you're back in the field investigating the accident. If you're wondering how you're going to do all of this and keep responders safe, the answer is through the use of Assistant Safety Officers.

Although in the Incident Command System (ICS), the title of assistant means that the individual does not have to be as qualified as the primary person they're supporting, on a complex response you'll want to be surrounded by excellent safety professionals who are as capable as you are. Some of the reasons you would want this type of talent to support you as the Safety Officer are: the incident scope, incident duration, incident complexity, or an Assistant Safety Officers' technical expertise that is in an area where you are not as strong (e.g., hazardous materials knowledge).

Incident scope

On a large response with complex challenges, real-time decisions that affect operations will have to be made. You cannot be at all places at all times, so you need to have highly capable Assistant Safety Officers with delegated authority. Remember, the goal of the SOF is not to impede operations; it is to facilitate operations and ensure the safety of the responders. This means the Assistant Safety Officers must have the essential knowledge and ability necessary to work with operations personnel; recognize and anticipate emerging and evolving hazards; identify and understand proper mitigation of identified hazards; and know when operations must be terminated or modified to prevent unacceptable risk to response personnel or the public.

Incident duration

Any large incident has the potential to run for an extended period. As a Safety Officer, you must be prepared to provide safety support to the Incident Commander/Unified Command 24 hours a day, 7 days a week.

Incident complexity

Some larger incidents may require multiple response disciplines to manage the safety aspects associated with various response operations. The expertise to address safety for a diverse array of operations may easily exceed your breadth of knowledge and experience as a safety professional. The Assistant Safety Officers that you bring in to support your efforts should be able to provide the skills and expertise that you require but do not possess yourself.

You want to take a hard look at what safety issues you're facing to determine the level of support you'll require. Below are some rules you can use to help determine the number of Assistant Safety Officers required:

- One Assistant Safety Officer for each high-risk activity
- One Assistant Safety Officer for every 100 responders
- One Assistant Safety Officer for completing the Site Safety Plan and providing input into the Incident Action Plan
- One Assistant Safety Officer to coordinate air monitoring or other specialized assessments
- One Assistant Safety Officer available to assist the Operations Section Chief with real-time tactical decisions
- One Assistant Safety Officer to support multiple incident support facilities

As you read the scenario below, think about the scope of the incident, its complexity, and duration. What issues do you face as the Safety Officer? Along with the operational safety issues you consider, decide how many Assistant Safety Officers you need for support.

After you have checked in to the incident, you meet with the Unified Command and receive your briefing on the incident situation.

Floodwaters have caused the Florida River to overrun its banks, flooding low-lying neighborhoods and uncovering buried petroleum pipelines. The pipelines have broken and spilled oil into the racing water, spreading oil throughout neighborhoods and into environmentally sensitive ecosystems. Some of the oil caught fire and burned. The command team is facing the following issues: evacuation of residents in flooded areas; firefighting operations; oil-spill recovery operations; pipeline repairs; reopening of the shipping channel; and supporting an expanding response organization estimated to reach 1,100 responders.

Some of the safety concerns that you may immediately recognize are: displaced animal life (snakes); oil recovery efforts in fast-moving water; hot and humid weather; oil removal in neighborhoods; firefighting hazards; and difficult terrain.

Here are some of the issues that we came up with and the minimum support that the Safety Officer will need to do his or her job effectively:

- Rescuer safety (going house-to-house by small boat and helicopter to rescue stranded residents)
- Underwater hazards in flooded areas
- Fires associated with spilled product
- Poisonous snakes driven from their normal environment due to flood, and found everywhere in the response areas
- Dogs protecting their flooded homes
- Oil recovery and cleanup crews working on water and on shore
- Pipeline repair crews working in free-standing product (flammable hazards)
- Confined space entry (during decontamination of response vessels)
- Fall protection activities associated with removing oil from structures
- Chemical exposure from both light product oil and crude oil
- Personnel protective equipment challenges in a hot and humid environment (heat exhaustion)

There's no "classroom" answer to the number of Assistant Safety Officers you decide to bring in to support you during a response. To begin to determine the number of additional safety personnel that you will need to support incident operations, you really have to have a sense of which of the hazards you have identified present the greatest risk. We determined that you would need about 15 Assistant Safety Officers for the hazards that we identified in the Florida flood scenario above.

Safety Officer's Role in the ICS Planning Process

One of the main reasons that the Incident Command System was developed was because of the unacceptable loss of responders' lives that were occurring on wildland firefighting operations. As we walk through the ICS Planning Process, you will see how your involvement as Safety Officer is absolutely essential to the response team in developing and implementing a safe and effective tactical plan. The Safety Officer must be involved in every *step* of the ICS Planning Process to ensure that the Incident Action Plan (IAP) being developed is done so with your input. Almost without fail, the Incident Commander/Unified Command will make their number one incident objective solely focused on the safety of the responders and the public. The IC/UC has assigned you as their Safety Officer and is relying on you to ensure that safety is integrated into every aspect of the response.

Next, we will take you through the various *steps* that make up the ICS Planning Process and discuss your responsibilities at each *step*. *Figure 5.3* is a visual representation of the ICS Planning Process called the Operational Planning "P." We start our discussion at the *step* called IC/UC Develop/Update Objectives Meeting.

In Appendix B you will see a one-page summary of the Safety Officer responsibilities during the various steps of the ICS Planning Process.

The Operational Planning "P"

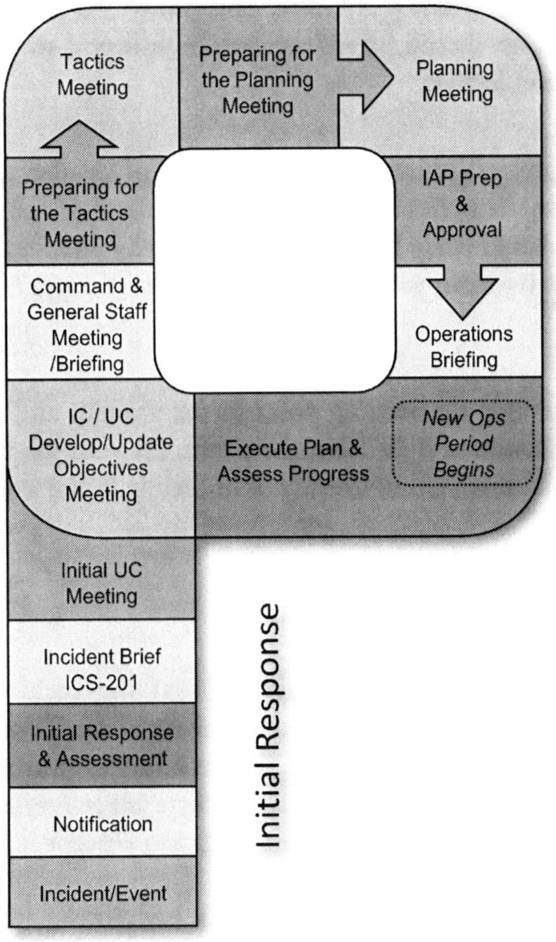

Figure 5.3. The ICS Planning Process is represented in the Operational Planning "P" as a series of activities or steps.

IC/UC Develop/Update Objectives Meeting

The ICS Planning Process begins with the Incident Commander/Unified Command (IC/UC) establishing objectives for the response team to focus on for the upcoming operational period. These objectives are critical to ensuring that everyone's efforts are focused and heading in the same direction.

As the Safety Officer (SOF), you want to be familiar with the Command's priorities, limitations and constraints, and objectives. Here's an example of why it's important for you to understand each operational objective.

Objective: *Secure the release of chlorine from the derailed chlorine tank car by 1800 hours.* As the SOF, you must have a firm grasp of the incident situation so you can evaluate the objective from a safety standpoint. In this example, the IC/UC may be worried about upcoming weather that will shift the wind direction and present a new exposure problem for the local community. If this potential weather problem is the case, as the SOF you understand the importance of the objective in terms of potential human exposure. This provides a framework for the amount of risk the response organization should be willing to take in regard to the types of strategies and tactics used to achieve the objective.

On the other hand the same objective without the threat of weather and subsequent exposure of the public may not have the same need for urgency. The level of risk for this scenario that the IC/UC is willing to take would likely be much lower. Only by understanding both the situation and the intent of the objectives can you determine how best to support the IC/UC in achieving the goals.

Setting the objectives is simply the starting point in the process, and that starting point must address safety and health concerns right from the outset. It's your job to make sure those safety and health concerns are addressed either directly or indirectly in the wording, tone, and purpose of the objectives.

Command and General Staff Meeting

It is at the Command and General Staff Meeting that the Incident Commander/Unified Command will sit down with you and the other members of the Command and General Staff and provide the direction that they want the response effort to pursue. The IC/UC members provide the incident objectives (ensure that one of these objectives is a safety objective), set priorities, assign tasks to individual members, and discuss their expectations.

This meeting is an excellent forum for you to ask for clarification on any issues with which you are uncomfortable and to discuss any safety concerns. Once you leave this meeting you're going to be in "the thick of it," dealing with ongoing operations and being involved in the planning of future operations. So make sure you have the information you need to support the effort.

Preparing for the Tactics Meeting

Sometime after the Command and General Staff Meeting, the Operations Section Chief (OSC) and the Planning Section Chief (PSC) will get together to discuss how Operations is going to accomplish the objectives that the IC/UC has set for the next operational period. The goal is to have most of the tactical plan (plan of action) roughed in before attending the Tactics Meeting. As the Safety Officer, you must help the OSC and PSC as they begin to think through the strategies and tactics that are needed to meet the objectives.

To help examine the various options available and ensure that they stay in alignment with the objectives, the OSC and PSC may develop a quick table of alternative strategies and tactics such as that shown in *Figure 5.4*.

The scenario for the below identification of response strategies is a flood with potential survivors who must be located and supported.

Identification of Response Strategies and Tactics

Objectives	Strategies (How)	Tactics (Who, Where, What, When)
Conduct thorough search for survivors	• Use aerial assets • Use on water assets	On water teams are to go house-to-house in search of survivors. Transport the injured to triage center.

Figure 5.4. Recording the various strategies and tactics available to meet the IC/UC objectives.

As the Safety Officer, by participating at this early stage in the development of the tactical plan for the next operational period, you can conduct a risk assessment on each proposed tactic and help the OSC make changes now instead of at the upcoming Tactics Meeting.

Tactics Meeting

Your responsibility as the Safety Officer in this meeting is very clear: help the Operations Section Chief create a safe plan to meet the IC/UC objectives. This meeting is totally focused on the development of the Operations Section Chief's tactical plan for the next operational period.

The Tactics Meeting is facilitated by the Planning Section Chief (PSC) and attended by the Operations Section Chief (OSC), Logistics Section Chief (LSC), Resources Unit Leader (RESL), Communications Unit Leader (COML), and you, the Safety Officer. Also, depending on the operations that must be planned, you may find Technical Specialists (THSP) in attendance at this meeting.

As the Safety Officer, you're evaluating the risks and benefits of the various strategies and tactics that the OSC would like to use to accomplish the objectives for which he or she is responsible. If you determine that a specific strategy or tactic presents an unacceptable risk to the responders, then you have the opportunity to take up the concern with the OSC and the PSC. This should be a discussion that communicates the perceived risks and mitigations that may be incorporated to protect responders. Some specific responsibilities that you have at the Tactics Meeting are to:

- Conduct a hazard/risk analysis
- Provide mitigation recommendations to control hazards (e.g., personal protective equipment, air monitoring)
- Evaluate the need to embed Assistant Safety Officers in Operations to monitor hazardous operations

Using the ICS-215, Operational Planning Worksheet

The Tactics Meeting works by the Operations Section Chief using an ICS-215 Operational Planning Worksheet to develop a tactical plan. The worksheet is designed to walk the OSC through the development of the plan. As the SOF, you provide input as the plan is laid out. *Figure 5.5* is an ICS-215.

Operational Planning Worksheet (ICS-215)

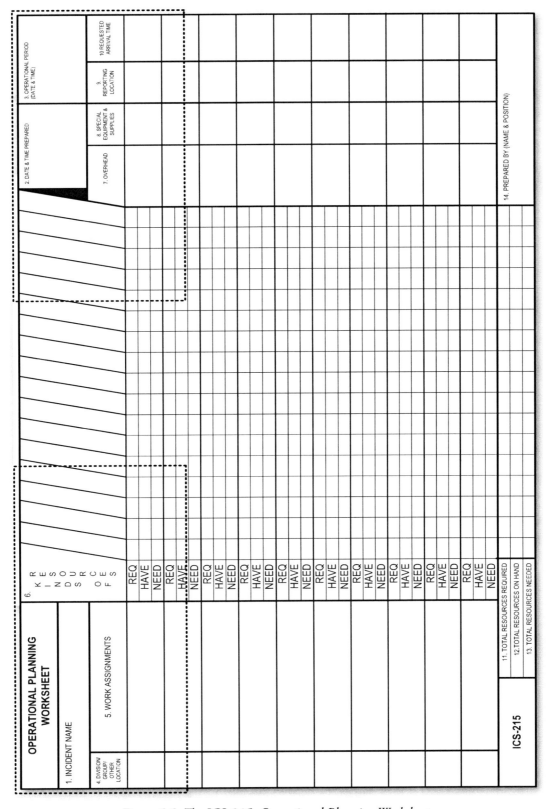

Figure 5.5. The ICS-215, Operational Planning Worksheet.

We have taken two areas of the ICS-215 and expanded them to show what information is on the form. Recognizing this form will help you to better understand how it's completed and how you can provide the support that the Operations Section Chief (OSC) needs. *Figure 5.6* is a cutaway of the upper left corner of the Worksheet. Let's discuss the information captured there.

On the far left you see where the OSC has created a *Search Group* as part of his or her organization. Next to the Search Group, the OSC has developed a work assignment that the Search Group is to accomplish. Pay particular attention to this block. Evaluate what the OSC needs to have done from a safety perspective and let the OSC know when and where you see a potential problem and have any recommendations for mitigating actions. The rest of the information on this part of the form deals with tactical resource requirements that the OSC has determined the Search Group Supervisor must have to do the job.

Upper Left Side of the Operational Planning Worksheet

OPERATIONAL PLANNING WORKSHEET		6. KINDS OF RESOURCES	Ambulance (type 2)	Helicopter	Search Team (six person)	Rescue Boat
1. INCIDENT NAME	MERIDIAN FLOOD					
4. DIVISION/ GROUP/ OTHER LOCATION	5. WORK ASSIGNMENTS					
Search Group	Conduct house-to-house search for injured persons. Mark each dwelling searched with a red "X" on the front door. Evacuate injured persons to triage center.	REQ	1	1	1	1
		HAVE				
		NEED				

Figure 5.6. A cutaway of the ICS-215 shows organization element, work assignment, and required resources.

The wall-size version of the Operational Planning Worksheet measures 3 feet by 5 feet. *Figure 5.7* is the upper right corner of the form.

Upper Right Corner of the Operational Planning Worksheet

Ambulance (type 2)	Helicopter	Search Team (six person)	Rescue Boat	2. DATE & TIME PREPARED 15 Nov 2100	3. OPERATIONAL PERIOD (DATE & TIME) 16 Nov 0600 to 16 Nov 1800		
				7. OVERHEAD	8. SPECIAL EQUIPMENT & SUPPLIES	9. REPORTING LOCATION	10. REQUESTED ARRIVAL TIME
1	1	1	1	Supervisor / Assistant Safety Officer	Red Spray Paint	4th Street and County Road 502	0530

Figure 5.7. A cutaway of the ICS-215 shows the right side of the form.

On the right side of the ICS-215, the OSC accounts for the number of supervisors needed to manage the tactical operations. In addition, the OSC lists specialized skills that they need to support the Search Group's operations. You see that an Assistant Safety Officer has been listed specifically to support the Search Group.

The designation of an Assistant Safety Officer to serve in the Search Group came about either because the OSC is aware of the inherent risk of the work assignment they're asking the Search Group to do, or you as the Safety Officer have recommended to the OSC to consider an Assistant Safety Officer because of the safety implications. Either way, with your support, one of your safety assistants will be out there at the start of the next operational period. The ICS-215 will cover every tactical operation planned for the next operational period. We just walked you through a single activity.

Using the ICS-215A, Incident Action Plan Safety Analysis

To help you, as the Safety Officer, evaluate the hazards that are associated with each of the work assignments that the OSC identified on the ICS-215, Operational Planning Worksheet, we recommend that you use an ICS-215A, Incident Action Plan Safety Analysis (*Figure 5.8*).

A note about the ICS-215A used in this book. We looked at three versions of the ICS-215A: the wildland firefighting agencies', the US Coast Guard's, and the draft all-hazard ICS-215A from the Federal Emergency Management Agency (FEMA). We chose to go with FEMA's draft version. Our reasoning is that it is simple to use and can be used for any type of incident. The challenge that we faced as authors is that the form is only a draft with the final version due out some time in the future. We have placed an online version of the ICS-215A form we are using in this book at the following location www.emsi-ics-services.com. You can find the form under the tab labeled ICS Forms and it is titled ICS-215A all-hazards. Please download it and use it until a national all-hazard form is made available.

You probably noticed that the ICS-215A in *Figure 5.8* is similar in design to the ICS-215. The reason for the similarity is to make it easier for you to conduct a hazard analysis on each of the work assignments that the Operations Section Chief has recorded on his or her ICS-215.

In addition to identifying the hazards associated with each work assignment, the ICS-215A enables you to document various mitigation actions that you believe can be put in place to reduce or eliminate the identified hazards.

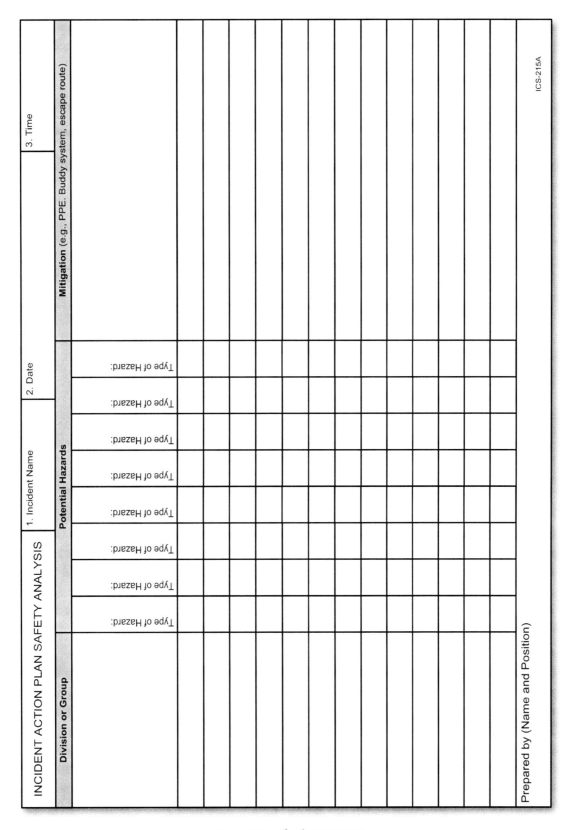

Figure 5.8. Blank ICS-215A

The best way to illustrate what we are talking about in regard to the ICS-215A is to first look it over so that you have a basic idea of the form's layout. We will go over an example of how the ICS-215A form is filled in using our simple flood scenario that we discussed earlier in the chapter.

Figure 5.9 shows the left side of the ICS-215A. You see a column labeled Division or Group and this is where you copy, directly from the ICS-215, the organizational elements that the OSC is planning to establish. In the flood scenario, the OSC established a Search Group and you note this on the ICS-215A. The next column on the form is where you record any potential hazards to which the responders may be exposed.

Upper Left Corner of the ICS-215A

INCIDENT ACTION PLAN SAFETY ANALYSIS										1. Incident Name Meridian Flood	2. Date 15 No
Division or Group	**Potential Hazards**										
	Type of Hazard: Drowning	Type of Hazard: Snakes	Type of Hazard: Blood Pathogen	Type of Hazard:	Type of Hazard:	Type of Hazard:	Type of Hazard:	Type of Hazard:	Type of Hazard:		
Search Group	X	X									

Figure 5.9. The upper left side of the ICS-215A, Incident Action Plan Safety Analysis.

In *Figure 5.9* you note three potential hazards that could threaten responders. For those responders who work in the Search Group you place an "X" below the drowning hazard and snake hazard. For the Search Group, you did not place an "X" under the identified hazard: blood pathogen since that particular hazard is not one that the Search Group will encounter as they perform their work assignment. If the OSC had established a Medical Group that was responsible for caring for the injured, a blood pathogen hazard would have been identified.

Let's move over to the right side of the ICS-215A (*Figure 5.10*) and see how you fill in the remaining information. The hazards have been identified (snakes, drowning, blood pathogens). Now you need to understand the risks associated with each of the hazards that you listed on the ICS-215A. This is where you as the Safety Officer identify and record any mitigating measures that can be used to minimize or eliminate the hazards that you have recorded and risks associated with those hazards. For the Search Group, the risks are high. You require the use of personal flotation devices, regular communications checks, snake gators, the buddy system, and daylight operations only. For the Medical Group, the risks are less because you have trained emergency medical technicians. Specifying protective equipment helps address the risks to the responders from blood pathogens.

Upper Right Corner of the ICS-215A

2. Date				3. Time	
			15 Nov		2130
Mitigation (e.g., PPE, Buddy system, escape route)					
Type of Hazard: Drowning	Type of Hazard: Snakes	Type of Hazard: Blood Pathogen			
X	X		Personal floatation devices, regular communications checks, snake gators, buddy system, daylight operations only		
		X	Use medical personal protective equipment (e.g., gloves, N-95 mask)		

Figure 5.10. The right side of the ICS-215A.

In Figure 5.11, you find a completed ICS-215A. The hazard analysis that you conduct is critical to the safety of responders. Work closely with the OSC to ensure that the measures that you have recommended to minimize injury and illness are in place prior to any operations.

INCIDENT ACTION PLAN SAFETY ANALYSIS	1. Incident Name Meridian Flood				2. Date 15 Nov				3. Time 2130
Division or Group	**Potential Hazards**								**Mitigation** (e.g., PPE. Buddy system, escape route)
	Type of Hazard: Drowning	Type of Hazard: Snakes	Type of Hazard: Blood Pathogen	Type of Hazard:	Type of Hazard:	Type of Hazard:	Type of Hazard:	Type of Hazard:	
Search Group	X	X							Personal floatation device, regular communications checks, snake gators, buddy system, daylight operations only
Medical Group			X						Use medical personal protective equipment (e.g., gloves, N-95 mask)
Prepared by (Name and Position) J. Gafkjen, PSC									ICS-215A

Figure 5.11. Completed ICS-215A.

When you leave the Tactics Meeting you want to be sure that the proposed tactical plan the Operations Section Chief has developed for the next operational period addresses any safety concerns. The tactical plan developed by the Operations Section Chief to support the objectives set out by the IC/UC must have your support or the IC/UC will not accept it.

Use your safety expertise, skills, and experience to mitigate hazards to the plan that you, or others at the meeting, identify. Depending on the incident, your responsibilities in this meeting can be challenging. You cannot compromise safety, but you have to help move the response forward.

If you cannot satisfactorily address a safety concern in this meeting, you should try to work with the Planning and Operations Section Chiefs to resolve the issue. If you still cannot resolve it, you should suggest that all of you meet with the IC/UC to obtain decision.

Planning Meeting

All of the effort that has gone into developing the tactical plan has been done by certain members of the Command and General Staff. The Incident Commander/Unified Command has not seen the proposed tactical plan up to this point in the planning process. The entire Command and General Staff, including the IC/UC will be at the Planning Meeting and the Operations Section Chief will use the ICS-215 that you helped develop to conduct the briefing, showing how the plan of action is in alignment with the IC/UC objectives.

As the Safety Officer, your input is critical. If high-risk activities are going to be conducted you need to reassure the command that you and the OSC have worked out mitigating controls to protect the responders. For example, when overhauling a large manufacturing facility following an intense fire, the firefighters are at risk from falling debris and collapsing structures. Mitigation efforts might include assessments by engineers before the firefighters enter the structure as well as the use of a "buddy system" and a rescue team with a crane. As the SOF you have to ensure that the IC/UC understand the risks and mitigations. You want to come to the meeting prepared to discuss:

- Incident hazard/risk analysis and identified mitigating factors
- The Site Safety Plan
- Your commitment to the Incident Action Plan

Because of your participation in the Tactics Meeting, you and the OSC should be in agreement on how to best conduct safe operations. There should be no surprises at the Planning Meeting. The IC/UC does not want to see conflicting views about risk and mitigation strategies. Any differences should have been resolved before the Planning Meeting. The ultimate goal of the Meeting is to have the IC/UC review and give tentative approval of the plan so that the Incident Action Plan (IAP) can be developed.

IAP Preparation and Approval

The IAP explains, among other things, the objectives for the operational period, the organization, the work activities, and safety protocols associated with response work activities. The IAP also contains support information including weather summaries, site maps, a medical plan, and a communications plan. An example of an IAP can be found in Appendix A.

Following the Planning Meeting, you're going to be intimately involved in the preparation of the Incident Action Plan. *Figure 5.12* contains the different parts of the IAP. The components shaded in gray are the ones that you, as the Safety Officer, will have input.

Components of the Incident Action Plan (IAP)

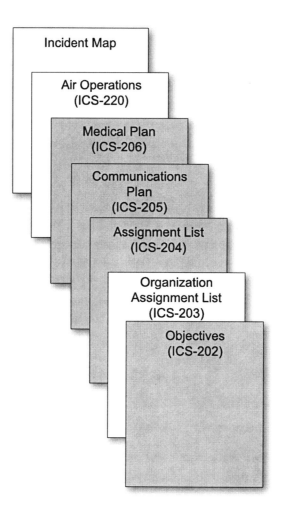

Figure 5.12. The various components of the Incident Action Plan (IAP).

Incident Objectives, ICS-202

The Incident Objectives, ICS-202 form can be used by you to put out a general safety message right below the IC/UC objectives. Consider preparing a short message for the Planning Section Chief to insert on the ICS-202. The message can refer to a more comprehensive health and safety plan, or it might simply be a reiteration of a particular safety risk appropriate for that operational period.

Assignment List, ICS-204

ICS-204 Assignment Lists provide the work assignments for Divisions, Groups, Branches, and Staging Areas. You have to review every ICS-204. There's space on the form to put down specific safety instructions that are tailored to the work that a particular Division or Group will be performing. Some examples of special instructions might include:

- An Assistant Safety Officer must be present at all times that this activity is being conducted
- Work in teams of two or more at all times
- Snakes are suspected to be in the work area, wear snake gators and carry snakebite kits
- A half-face respirator with organic vapor cartridges is required for work when workers are within 100 feet of the pipeline rupture area. Air monitoring will be conducted by the assigned Assistant Safety Officer. Only approved workers are authorized to work in areas requiring respiratory protection
- Flammable vapors are present in the work area. All equipment and communications devices must be explosion proof, spark resistant, or intrinsically safe

Figure 5.13 is an example ICS-204 with some special safety instructions. You will notice that the mitigation strategies that you identified on the ICS-215A for the Search Group are noted in Block 8. This reminds the Group Supervisor to pay particular attention to the unique safety concerns that their personnel face.

The Resources Unit Leader will prepare the ICS-204s, so you want to provide any safety instructions that you want included in the Incident Action Plan to the Resources Unit Leader early enough so that the IAP is completed on time.

In addition to placing special safety instructions on specific ICS-204s, you may want to consider inserting a one-page safety message directly into the Incident Action Plan for all to read. Consider using a different color paper so your safety message stands out.

Assignment List (ICS-204)

1. BRANCH	2. ~~DIVISION~~/GROUP Search	ASSIGNMENT LIST
3. INCIDENT NAME **MERIDIAN FLOOD**	4. OPERATIONAL PERIOD (Date and Time) 16 Nov 0600 to 16 Nov 1800	

5. OPERATIONS PERSONNEL

OPERATIONS CHIEF ___L. Hewett___ ~~DIVISION~~/GROUP SUPERVISOR ___P. Robert___

BRANCH DIRECTOR _____ AIR TACTICAL GROUP SUPERVISOR _____

6. RESOURCES ASSIGNED THIS PERIOD

STRIKE TEAM/TASK FORCE RESOURCE DESIGNATOR	EMT	LEADER	NUMBER PERSONS	TRANS NEEDED	DROP OFF POINT/TIME	PICK UP POINT/TIME
MESA #3		B. Riggs	21		0530	
44120		V. Kammer	4		0530	
Ambulance #1		S. Miller	2		0530	
Helicopter 12		T. Troutman	2			

7. ASSIGNMENT

Conduct house-to-house search for injured persons. Mark each dwelling searched with a red "X" on the front door. Evacuate injured persons to triage center.

8. SPECIAL INSTRUCTIONS

Work in teams of two or more at all times. Snakes are suspected to be in the work area, wear snake gators. Personal floatation devices should be worn. Daylight operations only. Conduct regular communications checks. Send hourly updates on the group's progress to the Situation Unit.

9. DIVISION/GROUP COMMUNICATIONS SUMMARY

FUNCTION		FREQ.	SYSTEM	CHAN.	FUNCTION		FREQ.	SYSTEM	CHAN.
COMMAND	LOCAL REPEAT	CDF 1	King	1	SUPPORT	LOCAL REPEAT			
DIV./GROUP TACTICAL		157.4505	King	3	GROUND TO AIR				

PREPARED BY (RESOURCES UNIT LEADER) A. Worth	APPROVED BY (PLANNING SECT. CH.) J. Gafkjen	DATE 16 Nov	TIME 0400

Figure 5.13. ICS-204, Assignment List

Communications and Medical Plan

You need to work with the Communications Unit Leader and Medical Unit Leader to review the Communications and the Medical Plans. You want to ensure that the plans are written to support the tactical plan that's going to be implemented at the beginning of the next operational period.

Safety Plan

As the Safety Officer you may be required to complete a Safety Plan. The Safety Plan is an important support document of the Incident Action Plan. You want to make sure that your Safety Plan addresses all hazards of the incident and reinforces proper safeguards. In addition, the Safety Plan must be in compliance with appropriate local, state, and federal regulations.

Operations Briefing

Just prior to the start of the operational period, the Branch Directors, Division and Group Supervisors, Staging Area Managers and others will be briefed on the content of the Incident Action Plan. Up to this point none of the field people who have to carry out the actions outlined in the IAP, have been involved in its development.

The Operations Section Chief will brief his or her personnel to ensure that each one understands what is expected of them. You'll have an opportunity to say a few words on safety. Your primary message should be to emphasize the hazards that the field responders are facing and the importance of following the hazard mitigation strategies that have been put in place to ensure responder safety. This might include a detailed discussion of the chemical hazards associated with styrene inhalation, a discussion of personal protective equipment, emergency procedures for medical casualties, or whatever is applicable.

Execute Plan and Assess Progress

Once the responders roll into the field and begin to implement the plan, you want to have your Assistant Safety Officers or yourself out there ensuring that work assignments are being conducted with responder safety as the priority. Those same assistants should be constantly assessing the risk. This is especially critical in a very dynamic response or one where the complete situation on the ground is unknown.

The Safety Officer should be working closely with the Division/Group Supervisors to better understand what they are up against. A plan, no matter how well built, never quite lines up with reality. You want to avoid getting into a position of having to use your stop-work authority.

In Appendix B you see a one-page summary of the SOF responsibilities during the various steps of ICS Planning Process. The page contains the Operational Planning "P" and calls out specific safety officer activities that we present through out the chapter.

Closing Thoughts

The Safety Officer must be able to address both the risks in the current operational period and anticipate risks in upcoming operational periods. You're a risk manager who must weigh various response options and advise the IC/UC and OSC on risks and the appropriate mitigations. To be able to do this, you must have detailed and extensive knowledge of the operational strategies and tactics for the response discipline. That said, do not get so wrapped up in the operational side of the response that you forget to ensure that the incident support facilities are safe for workers to do business in, eat in, and sleep in. It's the more obscure safety issues that will usually catch you off guard.

You must also interact with multiple agencies and be able to blend various safety program protocols within a response organization. This issue becomes significant when a response organization comprised of a dozen different agencies, organizations, or business entities comes together as a response team without prior exposure or practice.

As the SOF, you typically have stop-work authority. This authority is an extension of the IC/UC authority, and must be used judiciously. The consequences of stopping response activities because of an unanticipated risk or a developing problem has financial consequences that are expensive—especially when there are high-value response assets on the job. If you have developed a strong relationship with the Operations Section Chief and field personnel, you should be able to resolve safety issues before halting an operation. Prevention is always preferred over disruption. Be mindful of near-misses and potential trends. Early intervention in these cases can help avert an unwelcome safety accident.

In your role as the Safety Officer, you have to anticipate evolving risks, assess effectiveness of mitigation efforts, and make recommendations to adjust strategies or tactics to reduce risks. This requires a high degree of coordination and cooperation among you, the Operations Section Chief, the Planning Section Chief, and the IC/UC. You cannot make assumptions and decisions in a vacuum—your thoughts and concerns must be integrated into the larger incident management team's concerns and processes. Safety should be a solution-oriented function not a fault finder—an observer willing to work through a concern for the betterment of the responders and the response.

Liaison Officer

The Liaison Officer is one of the Incident Command System (ICS) positions that is the least understood and appreciated, but the reality is that even experienced Incident Commanders and Unified Commands would not be as effective without the help of a gifted Liaison Officer. The Liaison Officer (LOFR) position on an incident command team can be one of the most challenging assignments on the team especially in the hours and days where the incident is in the spotlight and the total impact of the incident on human life, the environment, and the economy are in doubt.

If you review any ICS literature on the role of the LOFR, it focuses primarily on the LOFR interaction with Agency Representatives. In the Incident Command System, the term Agency Representative (AREP) is given to those individuals who are sent to an incident to help coordinate their agency's support of incident operations. If the agency does not have jurisdiction over the incident but is providing tactical resources to the response it's called an *assisting agency*. For example, if law enforcement officers from Denver, Colorado, are providing scene security for an incident in Colorado Springs, they would be considered an assisting agency and they most likely would send an AREP to help coordinate their activities. In addition to those agencies that provide direct tactical support, agencies that provide non-tactical support to an incident are called *cooperating agencies,* and they may also send an AREP to the Incident Command Post to work with the Liaison Officer. Well-known examples of cooperating agencies would be the Red Cross and Salvation Army.

The role of the Liaison Officer is much greater than simple coordination. As the Liaison Officer, you play a critical role in helping the Incident Commander/Unified Command to effectively manage the concerns and issues raised by stakeholders during an emergency. For our discussion, stakeholders include: elected officials and their staff, government agencies, special interest groups (environmental organizations), the general public, and industry partners.

As a Liaison Officer, you need great interpersonal skills (communication), a thorough knowledge of the response discipline, must be highly organized, and have a good grasp of who the local stakeholders are and what their concerns might be. Most importantly you must have the trust of the Incident Commander or Unified Command.

As the Liaison Officer, you should be very sensitive to that fact that how an incident response is perceived by stakeholders and assisting and cooperating agencies will become the reality regardless of whether the command team is doing an outstanding job. You are on the front lines when it comes to perceptions regarding the success or appropriateness of the response effort. To minimize negative perceptions, you must establish effective communications among the Incident Commander/Unified Command (IC/UC), Agency Representatives, and the impacted stakeholders.

The role of the Liaison Officer is to "know the customer"—both internal customers (the command team) and external customers (everyone else). The LOFR is responsible to help the IC/UC manage the concerns of the elected officials, government agencies, industry and community organizations that are affected by the incident. This management of concerns is done primarily through personal communications. Communications by phone, in person, and by electronic means are a critical skill for Liaison Officer and their assistants.

The Liaison Officer's primary focus is twofold:

- To contribute to the efficiency of the response effort by assuring the best use of available assisting agency resources and cooperating agency support
- To contribute to the positive public perception of the response and to achieve stakeholder objectives by working closely with them to address their concerns

As the Liaison Officer, some of your specific responsibilities are to:

- Work closely with any assisting or cooperating agencies
    - Provide agencies and organizations with a schedule for incident updates, and determine their information needs
- Keep the IC/UC informed on issues dealing with
    - Assisting agencies
    - Cooperating agencies
    - Stakeholders
- Coordinate with the Public Information Officer
- Schedule and manage VIP visits
- Determine outreach efforts (e.g., community meetings)
- Handle external messages to stakeholders
- Be a proactive contact point for stakeholders
    - Political appointees and their staff
    - Government agencies
    - Nongovernmental organizations
    - Industry partners
- Identify public and private concerns related to the incident
- Maintain a master list of contact numbers
- Document your activities on a Unit Log, ICS-214 (see Appendix D for an example)

Liaison Officer and Pre-Incident Planning

As Liaison Officer, how effective you are in carrying out your responsibilities is directly tied to how well you establish and maintain relationships before an incident ever occurs. Involvement in pre-incident planning and exercises improves your effectiveness as a Liaison Officer. Participating in meetings and networking should be an ongoing process. In the middle of an emergency is not a good time to start meeting those with whom you will have to work closely.

Actions to Take upon Arriving at the Incident Command Post

Once you arrive on scene, you need to quickly establish yourself as a point of contact for Agency Representatives and stakeholders. Before you do that, make sure that you check in to the incident and get a briefing from the Incident Commander or a representative of the Unified Command. Use the checklist in *Figure 5.14* as a memory jogger for the questions that you would like answered.

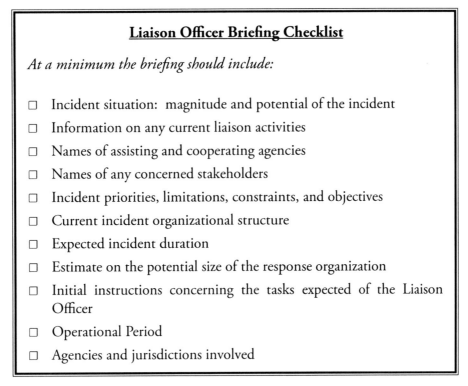

Figure 5.14. A briefing checklist will enable you to ensure that key information is covered during your in-brief.

Liaison Officer Assistants and Space Considerations

After you receive your briefing and assume the responsibilities of the Liaison Officer, you'll have a long list of items that you need to get moving on right away. The pressure to begin tackling the issues is tempting, but give yourself a few minutes to determine the number of assistants that you need to manage all of the issues that will come to you. Hopefully, the briefing gave you some good insight into what you, along with the command team, are facing. If not, try to get with the Planning Section Chief and Situation Unit Leader as they will be able to provide you with information about what is happening currently as well as projections into the future. This information helps you to prepare for emerging stakeholder concerns, gauge public and dignitary interest, and make decisions about how many assistants are necessary.

If possible, try to have representation from key responding agencies and organizations on your staff. This diversity is a big benefit if you're dealing with a wide range of stakeholders. When you select assistants, try to find people who complement your competencies and bring a wealth of information about regional politics, agency or organization concerns, or stakeholder concerns. Although the Liaison Officer function may not be needed 24 hours a day, you'll likely need a presence during the off-shift hours in the early days of the incident. Regardless of the number of assistants you finally choose, order them now. Delay ordering them now, and you'll pay the price later.

The number of personnel doing liaison work and your space requirements go hand-in-hand. You need enough space to accommodate you and your assistants, and the space must have plenty of phone lines, computers, and fax machines. You want to be accessible to those outside the response organization. Remember that you were brought to the incident to be a conduit between outside interests and the IC/UC.

Some Best Practices for Managing the Stakeholders

Prior Liaison Officers, like the rest of us, have learned some hard lessons as they carried out their duties. Below are a few tips that may help you avoid learning the same lessons the hard way.

Elected officials and their staff

- Ensure that elected officials are briefed prior to a significant press release or media event
- Be aware of who will be making the formal inquiries, who the key staff members are, and what specific concerns or "hot topics" they have
- Determine the need to develop a detailed agenda and briefing packages for VIP visitors
- Work with the affected State Incident Commander to identify the appropriate escort for the visit
- Try and group political/VIP visits together
- Government agencies
- Be cognizant of which government agencies have reported in to the response and which have not
- Initiate contact with those agencies not represented
- Offer to provide periodic situational updates, informal consultations, or requests for support as the response develops

General public

- Combine your efforts with the Public Information Officer (PIO) and their staff
- Use targeted press releases generated by the PIO
- Consider community outreach through community meetings

- Use local elected officials to help organize outreach events

Industry partners

- Use the response as an opportunity to reinforce key relationships

The Liaison Officer's Role in the ICS Planning Process

As a member of the Command Staff, you'll be participating in several meetings that occur throughout the operational period. The goal of these meetings is to work the command team through the ICS Planning Process and eventual development of an Incident Action Plan (IAP). The planning process that you'll be working through is visually represented in *Figure 5.15*. The Operational Planning "P" is designed to show the various *steps* that make up the process. We'll cover the meetings that you'll be attending and your role in those meetings.

The Operational Planning "P"

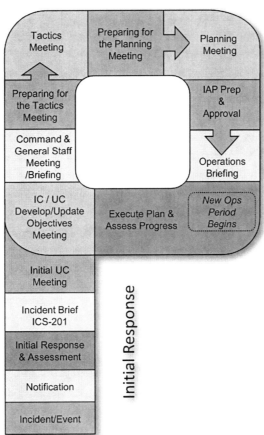

Figure 5.15. As the Liaison Officer you will be attending several key meetings.

Command and General Staff Meeting

Shortly after the beginning of a new operational period, the Incident Commander/Unified Command (IC/UC) will meet with their key staff members. In the Operational Planning "P" in *Figure 5.15*, this *step* is called the Command and General Staff Meeting. The information discussed during this meeting will set into motion the development of an Incident Action Plan. The IC/UC will lay out the objectives for the next operational period, set the priorities, discuss issues of concern, assign tasking to individual command team members, and discuss how they see the response progressing. You need to take good notes because you'll be dealing with agency representatives and stakeholders such as elected officials and you must be intimately familiar with all aspects of the response or risk losing your usefulness or credibility.

This meeting is meant to be a two-way discussion, so take a few minutes to let the IC/UC know of problems you foresee and potential solutions. As Liaison Officer (LOFR), you may become aware of concerns expressed by the elected officials, community stakeholders, or other organizations. If you cannot address any issue, you'll have to bring it to the IC/UC for resolution.

An example that highlights the importance of the Liaison Officer to the Incident Commander/Unified Command is illustrated in the following story.

In an oil spill response, the success of the cleanup is often measured by the perceptions of the media and the community about how the responders are performing. During an oil spill in San Francisco Bay in the mid-1990s, the elected officials of San Francisco were extremely worried about the impact a large spill would have on their tourist season. When the elected officials sent representatives to the Incident Command Post with a request for actions to prevent oiling of some critical shoreline in the Fisherman's Wharf area, that request came into the response team through the Liaison Officer. The UC was approached, and an objective to prevent oiling of the shoreline in a commercially important area (Fisherman's Wharf) was added to the list of objectives that had already been formulated to address environmental impacts. The Liaison Officer's role brought an important issue from the community into the Incident Command Post, and allowed for the UC to succeed in the eyes of important City Officials.

Planning Meeting

The Planning Meeting is the *step* in the ICS Planning Process where the Incident Commander/Unified Command is briefed on the proposed tactical plan for the next operational period. The entire Command and General Staff attend this meeting and will have an opportunity to provide additional information in support of the planned operations. This meeting is tightly scripted, so be prepared with your remarks. This meeting is another good opportunity to gain current information on the incident so that you have the most recent situational update and future direction. Some issues that you may want to comment on if they are applicable:

- Emerging concerns from elected officials or stakeholders
- Cooperating and Assisting Agency limitations or resource constraints

- Conflicts involving agency or organization jurisdictions, policies or procedures that may affect response management efficiency
- Requested visits by dignitaries or elected officials
- Reports on dignitary visits, expressed reactions, and anticipated follow-up actions

Demobilization of the Liaison Staff

As important as it was to arrive at the incident and go through an orderly check-in process, receive an initial incident briefing, and thoughtfully build your staff, it's as equally important to conduct an orderly demobilization of the liaison assistants and yourself. As you reduce your assistants, decisions on who stays and who demobilizes depends on such factors as performance, incident need, agency and organization interests, costs, and stakeholder relationships. Make sure you have a process in place to logically manage the downsizing of liaison support. Many stakeholder concerns emerge in the closing days of an incident so make sure you don't leave yourself shorthanded!

Closing Thoughts

If you have done your job correctly, the assisting and cooperating agencies will leave the response satisfied that their resources were used properly and were well taken care of throughout the response. The stakeholders will have felt that their interests and concerns were heard and addressed.

To be an effective Liaison Officer you should consistently take every opportunity to participate in exercises, get training from experienced responders, examine case studies, and participate in incident response. Learn other ICS functions when possible. If at all possible, try to build relationships ahead of time. The time spent getting to know the potential pool of stakeholders will be time well spent.

Public Information Officer

Emergency responders know all too well that crisis situations naturally attract the public's attention and at times the demand for information from the public can be overwhelming. Between the public and the Incident Commander/Unified Command (IC/UC) is the media (online, print, television, radio). If the incident is small and not too complex, the IC/UC team will often work directly with the media. In a large and complex incident, a Public Information Officer (PIO) will be assigned to manage communications with the public and the media. This is where you, as the Public Information Officer, come in to support the growing demand for public outreach.

The Public Information Officer is one of the few positions in the Incident Command System (ICS) organization that is primarily focused outward of the incident. Ideally, if you have been assigned the role of PIO for an incident you have accumulated experience in public affairs, risk communications, and Joint Information Center (JIC) management. If your experience is limited or you have not done the PIO job during an emergency in a while, the information below will be helpful.

As the PIO, you have the responsibility to ensure that the IC/UC get their message out to the public, keeping the public informed on the incident situation and the progress of response activities on bringing the incident under control. You help manage the public perception of how things are going. This is very important because you're always battling the phenomenon of "perception is reality." Being the PIO is particularly challenging when there's a UC with multiple members. In this case it is the responsibility of the PIO to ensure that a unified message is delivered. This means that the Unified Command must be prepared for public briefings to assure a common message to the public as well as behind-the-scenes communications where individual agency personnel are providing situational updates on Web sites, posting pictures, or using other methods to supply the public information needs.

The public and the media are not your only focus. You're also responsible for keeping the entire organization informed on the response effort. The majority of the responders see only the response from their limited position on the incident, and they do not have the larger picture unless you, as the PIO, keep them informed. Whether you establish and maintain information-display boards at key gathering locations or you use another system, do not forget about the needs of the response team.

As the Public Information Officer, some of your specific responsibilities are to:

- Support the public communication needs of the Incident Commander/Unified Command
- Oversee Joint Information Center operations
- Gather and disseminate incident information (e.g., number of responders)
- Work closely with the Liaison Officer to inform the public and stakeholders
- Assist in establishing and implementing communications requirements such as:
    - holding press conferences
    - disseminating press releases
    - answering media queries
- Attend command meetings to exchange information with the Incident Commander/Unified Command and to get approval of information to be released
- Ensure that the response organization is kept informed on the overall response efforts
- Coordinate media activities with the Command and General Staff (especially the Operations Section Chief)

- Determine need to develop an Outreach Plan
- Complete analysis of public perceptions using the Media Analysis Worksheet (see Appendix E)

Public Information Officer and Pre-Incident Planning

If you know ahead of time that you can potentially be involved in an incident as your agency's PIO, there are a number of actions you can take to prepare well before an incident occurs. Preparation is one of the keys to successfully accomplishing your responsibilities.

Below are some actions you can take before an incident occurs:

- Have existing contacts with the media outlets in your area
- Know who all of your counterparts (fellow PIOs) are before an incident occurs
- Maintain up-to-date contact lists for all forms of media
- Understand your agency's policies regarding release of incident information
- Establish a process for the release of incident information that enables rapid dissemination of that information (e.g., press releases, updates)
- Establish processes in advance to support the assessment of public perceptions during an incident

Actions to Take upon Arriving at the Incident Command Post

As the Public Information Officer, you're expected to know everything about the incident from the current situation, progress of response efforts, response costs, number of responders, and the list marches on. You most likely will not have the luxury to spend much time gathering all of the information you need, so make sure you get a good briefing after checking in to the incident, and if possible, get a copy of the Incident Briefing Form, ICS-201 (see Appendix F). The checklist in *Figure 5.16* has some of the questions that you should try to get answered. We encourage you to add to the list as you see fit.

Public Information Officer Briefing Checklist

At a minimum the briefing should include:

- ☐ Incident situation: magnitude and potential of the incident
- ☐ Political, environmental, and economic constraints
- ☐ Communities impacted
- ☐ Facilities already established (Joint Information Center)
- ☐ Stakeholders
- ☐ Level of media interest
- ☐ Any scheduled press briefings
- ☐ Gather basic facts about the incident
  - ○ **Who** is involved in incident and response
  - ○ **What** is the nature of the incident
  - ○ **When** did the incident occur
  - ○ **Where** did the incident occur
  - ○ **How** did the incident occur (if known and releasable)
- ☐ Priorities and objectives
- ☐ Command structure (single or unified)
- ☐ Operational period
- ☐ Information on committed resources
- ☐ Incident investigation
- ☐ Incident organization
- ☐ Meeting schedule, if established

Figure 5.16. Use a checklist to help guide your in-briefing.

Public Information Officer Staffing and Space Considerations

If you received a thorough briefing, you'll have a good idea whether the incident requires just yourself, or yourself and a few Assistant Public Information Officers, or the establishment of a Joint Information Center that has multi-agency representation. The demands on you can be intense and much of the information that you were given during your in-brief is going to be constantly changing, requiring you to keep it updated. If you determine that you'll need help, order the help early. The sooner your Assistant Public Information Officers arrive, the better.

There's no hard and fast guidance on how many assistants to bring in, but remember that to stay current on incident information you'll have to work with many different positions on the command team. In addition, you'll have to: attend meetings, answer phones, maintain records, and work with the media to conduct briefings and interviews, and numerous other items that are necessary for you to be successful.

The size, complexity, and number of agencies involved in the incident will help you to determine whether you should recommend to the IC/UC that a Joint Information Center (JIC) be established.

Work with the Logistics Section on your space requirements. Logistics will need to know at least the following information in order to set you up with adequate facilities:

- Anticipated number of personnel
- What media functions you'll be conducting in your spaces (e.g., video teleconferences)
- Number of computers
- Number of telephones
- Amount of wall space
- Easy access to faxes and copiers

Initial Actions of the PIO

During the early phases of a response, the actions taken by the PIO may have a significant impact not only on the success of this position, but on the positive perception of the overall response. Below are some of the initial-response actions that you as the PIO will need to consider:

- Establish a dedicated phone line(s) for inquiries from the media
- Familiarize yourself with Unified Command members
- Write initial news release and get approval from the Incident Commander/Unified Command
- Fax and/or e-mail information release to media distribution list
- Assign and staff positions in JIC for data gatherer and dissemination assistant
- Recommend to the IC/UC a location for the JIC
- Develop talking points and command messages for the incident (e.g., what message does the IC/UC want to get across to the public)
- Coordinate a press conference with Unified Command to brief the media and public about the incident
- Develop an opening statement for Incident Commander/Unified Command prior to press conferences
- Create press packages

Follow-on Actions of the PIO

Once you have taken care of your initial response actions there are a number of "maintenance" type items or processes that need to be set up to sustain the operation and ensure that you, as the PIO, are covering all of your responsibilities. These items may take a while to put in place but they will assist you in keeping up with your external outreach efforts. Items that fall into this category include:

- Set up a news-clips collection (video and paper). (Have Finance contract a clipping service for the incident, or organize some means of recording television news and print articles.)
- Establish an incident web site
- Shoot or collect photographs and video of the response for archival purposes and for releasing to the media

Actions like setting up an incident web site may significantly reduce the number of inquiries coming into the JIC by providing basic information to all who have access to the site.

Documenting the Public Information Effort

It's important to document your actions on the incident. Work closely with the Documentation Unit Leader to ensure that you capture the right information. At a minimum maintain the following records:

- Media calls (see Appendix G for an example Joint Information Center Query Record)
- Press releases
- Press packages
- Communications plans
- Questions and Answers or Frequently Asked Questions with answers developed for the incident
- Talking points
- Speaker preparations
- Fact sheets
- Video news clips
- Paper news clips with Media Analysis Worksheet (see Appendix E)
- Unit Log, ICS-214 (see Appendix D)
- Develop outreach materials for the public (e.g., claims brochures, help numbers) and the responders (e.g., phone lists, media protocols)

The Joint Information Center (JIC)

The JIC is the physical location where multiple agencies and organizations come together to respond to an emergency or manage an event with respect to information needs. Members of the JIC need to provide coordinated, timely, accurate information to the public and other stakeholders. The JIC has to be flexible and modular in order to support a wide range of incidents from smaller incidents all the way up to large, multi-agency, complex incidents. The JIC must speak with one voice as it represents the Unified Command. A JIC may also be useful in coordinating the information function for multi-agency event planning for major national or international meetings and events, such as political conventions, or gatherings like the G-8 Summit.

Risk Communications

Risk communication is maximizing public safety by presenting information to the public in a timely and professional manner during emergency situations. Maximum cooperation is needed from the public to ensure safe response efforts. In today's environment, Incident Commanders have the responsibility to communicate risks to the public concerned with terrorism, homeland security, environmental disasters, and other incidents and events of public concern. The Incident Commander/Unified Command is the trusted specialist the public is looking for to answer and address questions and concerns. A few examples of incidents that will involve risk communications:

- Natural disasters (e.g., hurricanes)
- Disease outbreaks (e.g., bird flu)
- Hazardous material releases
- Major bridge or building collapses
- Urban/wildland interface fires
- Terrorist attacks

Good Guidance When Communicating Risk

- Define all technical terms and acronyms – *Don't speak in a manner that your audience will not understand*
- Use positive or neutral terms – *Don't refer to other disasters, for example, Northridge Earthquake*
- Use visuals to emphasize key points – *Don't rely entirely on words*
- Remain calm and use questions or allegations to say something positive – *Don't let personal feelings interfere with your ability to communicate properly*
- Use examples, stories, and analogies to establish a common understanding
- Be sensitive to nonverbal messages that you're communicating – *Don't allow your message to be inconsistent with your position in the room, your dress, or your body language*

- Emphasize achievements made and ongoing efforts – *Don't guarantee anything*
- Provide information on what is being done – *Don't speculate about worst cases*
- Use personal pronouns – *Don't identify yourself as the entire organization*

Communicating Risk During the Initial Phase of an Incident (First 24-hours)

- Work with the Liaison Officer to identify stakeholders:
    - Impacted private sector entities
    - Public
    - Regulatory agencies
    - Specific population segments (old, young, certain geographic area, etc.)
    - Other agencies specifically involved in an incident/event
- Get the word out in emergency situations through widespread distribution of material to ensure effective communication (press releases, press conferences – television and radio, and public meetings)
- During an initial response, the first responders may need to brief the public on inherent safety concerns. Prepare, review, remain calm, and know your audience!

Communicating Risks During the Project Phase (24 Hours and Beyond)

- Working with the stakeholders and the Liaison Officer, develop a plan of action to organize and disseminate information to the public
- Use the following checklist in *Figure 5.17* to prepare for a speaking engagement

---

Checklist to Prepare for Speaking Engagements

☐ Time, place, and date of public appearance

☐ Incident/event name: time, place, and date of incident/event

☐ Introduction: statement of personal concern, statement of organization commitment, and the purpose and plan for the meeting

☐ Key messages: supporting data of the Incident specifically impacting the public

☐ Public involvement: names and concerns of who are helping, the organizations they represent, and their specific area of responsibility (if a volunteer group has been set up, now is a good time to mention how the community can get involved). Let the public know what they can do to help (whether that is evacuating, staying indoors, or reporting suspicious activity)

☐ Conclusion: summary statement

☐ Questions and answers: practice anticipated questions and responses

☐ Presentation material: handouts, audios, etc.

Figure 5.17. Checklist will help you prepare for your speaking engagement.

The Public Information Officer's Role in the ICS Planning Process

If the incident cannot be brought under control in a short period of time, the Incident Commander/Unified Command may begin the ICS Planning Process. The planning process enables the incident management team to systematically work toward the development of an Incident Action Plan that will be used to direct tactical operations. As the Public Information Officer, you'll be involved in several of the key *steps* that take place throughout the process. *Figure 5.18* is diagram of the planning process and shows the *steps* that the command team will work through.

The Operational Planning "P"

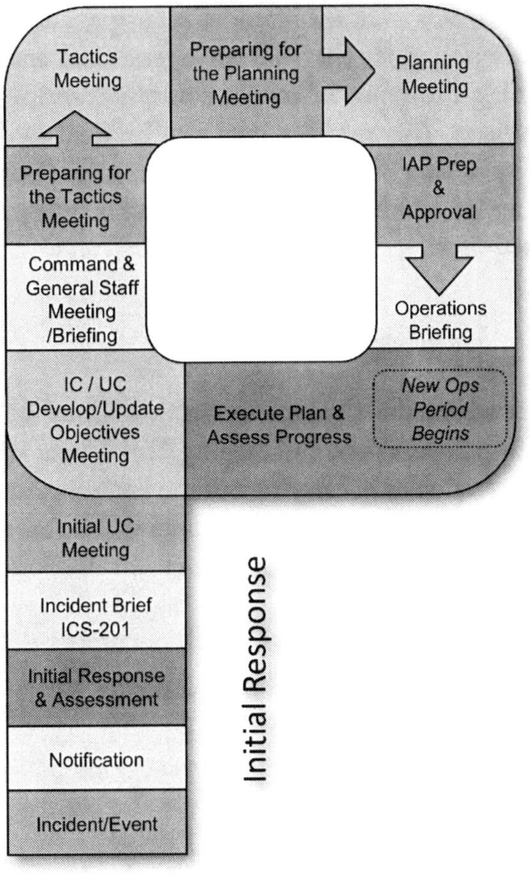

Figure 5.18. The Operational Planning "P" illustrates the various steps in the ICS Planning Process.

Command and General Staff Meeting

As the PIO, your role in the Command and General Staff Meeting is to listen to the priorities, limitations and constraints, and response objectives, and clearly understand the intent and focus of the IC/UC in order to effectively communicate that to the public and to the other responding agencies. As the response matures, you may also use this meeting to discuss how the response is being perceived by the public and any potential actions that the IC/UC may need to take to correct any misconceptions and maintain/reinforce the positive perceptions. This Meeting is designed so that the IC/UC and their staff can discuss issues. Below are a few examples of what you might want to discuss at the meeting:

- Get clarification on the media strategy and command's focus
- Agree on press interviews and timelines
- Agree on how to manage sensitive information
- Identify the location and interagency staffing of the Joint Information Center
- Report out to the IC/UC on media perception of the response

Planning Meeting

You should use this meeting to learn what the intended strategies and tactics are for the upcoming operational period. This meeting gives the PIO an operational snapshot of what is planned and can assist in determining the types of media communications that will be necessary to support the ongoing operations. This meeting may also be used to determine the appropriate time when a press availability or tour of the operational sites might be appropriate. You also have an opportunity to cover any issues with respect to responder interactions with the media and public perceptions of the response.

Operations Briefing

The Operations Briefing is where the Operations Section Chief will brief incoming Branch Directors, Division and Group Supervisors, and Staging Area Managers. The focus of the briefing is to discuss each of the attendees' roles in carrying out the Incident Action Plan and to make sure that everyone clearly understands their assignment. There will be an opportunity for you to say a few words. Some areas that you may want to cover are:

- Any media interaction opportunities with oncoming shift supervisors
- The "dos" and "don'ts" of talking with the media
- JIC contact phone numbers

Closing Thoughts

As the Public Information Officer, the Incident Commander/Unified Command is relying on you to get their message out to the public. Your message will go a long way toward defining the success of the response as perceived by the public and other stakeholders. Your skill in working with the internal response organization, and with the media will be critical to how successful you are in supporting incident operations. The media is your conduit to the public and you must make sure that you're delivering accurate, relevant, and timely incident information.

Your relationships inside the command team are as equally important as those that you establish with the media. The information you require must be obtained from the other responders and staff sprinkled through the team. By fostering these relationships and letting your fellow responders know what information you need and when you need it, you'll make your job much more manageable.

The PIO job is absolutely essential, even on small incidents. On larger incidents, a poorly functioning PIO can put enormous stress on Incident Commander/Unified Command. It's your job as the PIO to remove as many of those pressures as possible. You must get out ahead of the media and tell the responder's story. If you don't, the media will do it for you and you may not like what they have to say.

Intelligence/Investigations Function

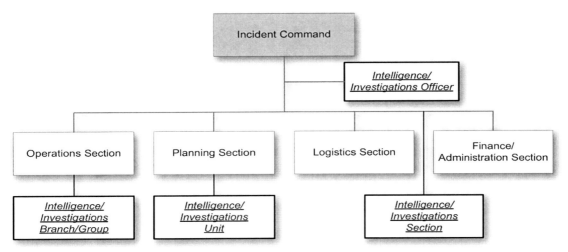

Figure 5.19. The Intelligence/Investigations Function in the ICS organization.

No other function in the Incident Command System (ICS) illustrates the benefits of the ICS's organizational flexibility better than the intelligence/investigations function. As you can see in *Figure 5.19*, the intelligence/investigations function can be located at several levels within the ICS organization: as an Intelligence/Investigations Officer, Intelligence/Investigations Section Chief, Intelligence/Investigations Branch or Group in Operations, or as an Intelligence/Investigations Unit in Planning.

The role of the intelligence/investigations function in an Incident Command System organization is to provide support to the IC/UC by establishing a system for the collection, analysis, and sharing of information developed during intelligence and investigations efforts. The two specific types of information include:

- Information that leads to the detection, prevention, apprehension, and prosecution of criminal activities or the individuals involved; and,
- Information that:
    - Leads to determining the cause of an incident (casualty investigation)

- Helps with assessing the impact of an incident (sizing up the incident)
- Assists with the selection of countermeasures for responding to the incident (need for specialized equipment or skills)

Activities undertaken by the intelligence/investigations function can have a direct impact on the safety of response personnel, and influence the response activities.

Unlike the Liaison Officer and Public Information Officer whose success relies on ensuring that incident information is shared with stakeholders and the public, the sharing of intelligence/investigations information is an entirely different matter. In general, intelligence/investigations information will be shared within the response organization only as it is deemed necessary to support effective operations. This information covers not only response operations, but the prevention, detection, and deterrence operations, which are the typical focus for information that is categorized as intelligence.

Some duties of the individual leading the Intelligence/Investigations function might include:

- Management of operational security activities
- Safeguarding intelligence/investigations sensitive information, which can include:
    - Classified information
    - Law enforcement sensitive information
    - Proprietary information
    - Witness statements
    - Case files
    - Export control information
- Collect and analyze incoming intelligence/investigations information from all sources
- Determine the applicability, significance, and reliability of incoming intelligence/investigations information
- Provide intelligence/investigations briefings to the IC/UC
- Provide intelligence/investigations briefings in support of the Incident Command System Planning Process
- Provide Situation Unit with periodic updates of intelligence/investigations issues that impact response operations
- Answer intelligence/investigations questions and advise Command and General Staff as appropriate
- Supervise, coordinate, and participate in the collection, analysis, processing, and dissemination of intelligence/investigations information
- Establish liaison with all participating law enforcement and investigative agencies
- Oversee the preparation of all required intelligence/investigations reports and plans

Guidance to Help Determine Where to Locate the Intelligence/Investigations Function

For consistency throughout this book, we have placed the intelligence/investigations function in the Command Staff, but its actual location on the command team will be determined by the Incident Commander/Unified Command (IC/UC) based on the incident or event. To help the IC/UC decide where to place the intelligence/investigations function in the ICS organization the guidance below may be helpful:

Intelligence/Investigations Officer

For incidents that require little need for tactical or classified intelligence or ongoing investigative activities.

Intelligence/Investigations Unit Within the Planning Section

For incidents that require some degree of tactical/classified intelligence and/or limited investigative activities, but where an intelligence/investigations representative is not a member of the Unified Command.

Branch Director in the Operations Section

For incidents where there's a high degree of linkage and coordination between the tactical operations and the intelligence/investigations function.

Intelligence/Investigations Section Chief

For those incidents that are heavily influenced by intelligence/investigations or when multiple intelligence and/or investigative agencies are involved, a separate section may be necessary. Especially when there is highly technical or specialized intelligence/investigative information that requires analysis. A terrorist incident would most likely require this level of intelligence/investigation organization within the incident command team.

The Intelligence/Investigations Officer's Role in the ICS Planning Process

Under certain circumstances, intelligence/investigations information could play a pivotal role in planning effective response operations. Intelligence or investigations information may help with determining and/or understanding the causal factors in an incident and as such could drive future response operations. To that end, the intelligence/investigations function within the ICS organization regardless of how it's configured (Intelligence/Investigations Officer, Branch, Unit, Section) will support the ICS Planning Process to the extent that the members engaged in the planning process have both the appropriate level of access and need-to-know for the information. The wide range of personnel that are engaged in response operations across a multitude of different federal, state and local agencies may complicate the intelligence/investigations function. There will likely be situations where not all of the individuals engaged in the planning process will have the appropriate level of access to intelligence/investigations information. In those instances it will be the responsibility of the members with access to the information to ensure that the planning process takes the information into account and establishes incident objectives necessary to address those issues.

The ICS Planning Process is illustrated in *Figure 5.20*. The various *steps* that make up the planning process are used to keep the command team working together to develop an Incident Action Plan that is used to direct tactical operations. If you're responsible for the intelligence/investigations function for the incident management team, you'll be involved in several of these *steps*. Two *steps* in particular are the Command and General Staff Meeting and the Planning Meeting.

The Operational Planning "P"

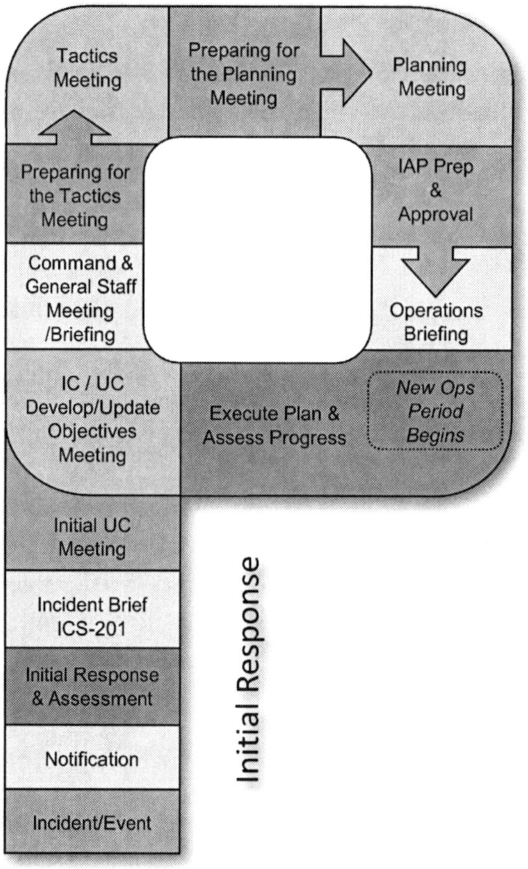

Figure 5.20. As the Intelligence/Investigations Officer, you will be attending several key meetings.

Command and General Staff Meeting

To the extent that it's feasible and allowable, all members of the Command and General Staff should be briefed on any incident-related intelligence/investigative issues at this meeting. If members of the Command and General Staff are not cleared to have access to intelligence/investigative information deemed critical to the response, then those with access should be provided a separate briefing and given specific guidelines on what can and cannot be shared with their fellow Command and General Staff members and their respective staffs.

Preparing for and Conducting the Tactics Meeting

If the intelligence/investigations functions are going to impact the tactical operations in any way, the Intelligence/Investigations Officer must provide input into the development of the proposed plan of action including update information that may impact the strategies and tactics that the OSC will consider during the development of the plan. In addition, if it is necessary to include intelligence/investigation activities within the Operations Section, intelligence/investigation technical expertise will be needed to identify: specific strategies and tactics, contingencies, work assignments, resources requirements, and technical direction for carrying out work assignments. The Intelligence/Investigations Officer will also have to identify how classified information must be dealt with along with the need for secure communications.

Planning Meeting

The broader audience and the purpose of this meeting will likely preclude the sharing of most intelligence and some of the more sensitive investigation information. The IC/UC will still need to ensure that the Incident Action Plan being briefed at this meeting addresses any issues highlighted by the additional intelligence/investigations information.

Operations Briefing

If the intelligent/investigations function is incorporated into the Incident Action Plan, the Intelligence/Investigations Officer will need to prepare and deliver a briefing so that the field supervisors are aware of any intelligence/investigation activities that may impact their ability to accomplish their assignments. Depending on the sensitivity of the information, the briefing can be done as part of the overall briefing or handled on a one-on-one need-to-know basis. This issue must be discussed with Command, Planning, and Operations prior to conducting the Briefing.

Closing Thoughts

The current guidance allows maximum flexibility for how the intelligence/investigations function is incorporated into the ICS organization. The exact nature of the incident and the intelligence/investigations needs associated with that incident will be the biggest drivers on how this function is incorporated at each individual incident/event. The responsibilities for this person are fairly broad in this discussion but are certainly more clearly defined by each individual's own agency.

CHAPTER 6

OPERATIONS SECTION CHIEF (OSC)

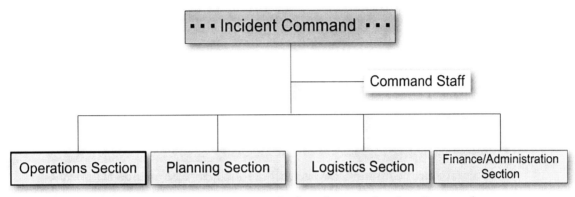

Figure 6.1. Operations Section Chief in relation to Incident Command.

The position of Operations Section Chief (OSC) is one of the most sought-after positions in the Incident Command System (ICS) organization. Depending on the type of incident, information that the OSC must manage can come at a dizzying pace. The OSC is the tactical wizard of the response and by far has the lion's share of the response resources. The Operations Section Chief (*Figure 6.1*) position is usually one of the first General Staff positions to be established. This is where the "rubber meets the road."

If the above paragraph piqued your interest, read on. The information in this chapter is designed to help you manage your enormous responsibilities.

One thing to note about the OSC position in the ICS is that it's a position that requires both a comprehensive knowledge of the Incident Command System as well as expertise and experience in the type of response you are conducting. Someone who is skilled at directing tactical operations at a wildland fire incident most likely would not have the expertise to tactically manage a radiological incident. That person would certainly possess the ICS know-how, but would lack that specialized knowledge necessary to carry out the tactical role of the OSC.

The OSC's responsibilities are complex, and the position requires an agile thinker. If you're the OSC, keep this in mind: not only must you manage the ongoing tactical operations, but you must also be thinking about the future and the next operational period because you'll be helping to develop the Incident Action Plan. You have to be an effective team builder, ensuring that your response personnel understand their jobs and the direction of the response operations. You must clearly communicate your expectations and hear out any concerns that your personnel have concerning those expectations. You must be assertive and decisive, yet flexible. Throughout this chapter, as we discuss the responsibilities of the OSC, it becomes more apparent why these skills are so vital to your success.

All incidents begin with operations. Many incidents never progress beyond the initial response as they are short term and can be quickly resolved. Short-term incidents typically require only a single-response organization and are usually resolved by using the responding organization's standard operating procedures—which should include ICS. Sometimes an incident takes days, weeks, or months to fully resolve and involves multiple agencies and numerous operational periods. A great deal of effort and operational tasking is required to resolve complex incidents. Before we go into the OSC's responsibilities for a complex incident, let's discuss the OSC's responsibilities for what we hope the majority of your incidents will be—simple and quickly resolved scenarios.

Some of the primary responsibilities of the OSC are to:

- Manage tactical operations
- Ensure tactical operations are conducted safely
- Maintain close communications with the Incident Commander/Unified Command
- Identify required tactical resources to accomplish response objectives
- Establish and disestablish Staging Areas
- Assemble and disassemble Strike Teams and Task Forces
- Assist in the development of the Incident Action Plan

Actions to Take upon Arriving at the Incident as the Oncoming OSC

If you've been notified that you're to assume the responsibilities of the OSC for an incident, there will likely be a lot of pressure for you to take over operations almost immediately, because the Incident Commander (IC) determined that he or she can no longer give operations the time and attention that the incident requires. You're going to have to come up to speed quickly on the incident situation, understand the response objectives and priorities, the operations organization that you're inheriting, and resources committed to the response. This knowledge can be achieved only by having a thorough briefing from the IC. In the heat of battle it's easy to forget to cover important issues and we encourage you to use the checklist in *Figure 6.2* as a way to ensure that the major items are covered.

> **Operations Section Chief Briefing Checklist**
>
> At a minimum the briefing should include:
>
> ☐ Incident situation: magnitude and potential of the incident
> ☐ Political, environmental, and economic constraints
> ☐ Facilities already established (including Staging Areas)
> ☐ Priorities and objectives
> ☐ Command structure (single or unified)
> ☐ Agencies and jurisdictions involved
> ☐ Information on committed resources
> ☐ Resources ordered
> ☐ Incident investigation
> ☐ Resource-requesting process
> ☐ Operations organization
> ☐ Support organization (Planning, Logistics, etc.)
> ☐ Communications schedule with Supervisors and Leaders
> ☐ Meeting schedule, if established
>
> Make sure you request a copy of the ICS-201, Incident Briefing Form. This form helps you manage your responsibilities.

Figure 6.2. Checklist will help you remember what you need to cover during your in-brief.

Hopefully the IC has been keeping an ICS-201, Incident Briefing Form, or similar document that contains a map of the incident, response objectives, and initial actions taken, an organization chart, and list of resources. The ICS-201 is an excellent briefing tool that is short and concise, and provides documentation of what transpired on the incident. If the IC has not started a 201, we recommend that you do (refer to Appendix F).

One of the best things you can do is "ground truth" the tactical operations. Between the briefing and a quick look at field operations you'll be much more in tune with where the response is heading.

Once you have assumed the role of the OSC ask yourself some questions:

- Are operations being conducted safely?
- Is the current operations organization adequate for the situation on the ground?
- Is my span-of-control within acceptable limits?
- Will I need a Deputy to execute operations?
- What can go wrong?
- Do the supervisors and leaders know that I am now the OSC?

Tactical operations are now in your hands. If the incident is brought under control within a fairly short timeframe, the primary documentation of the response will have been captured on the ICS-201. Most of the incidents that we face never go beyond the initial response effort or require documentation and planning beyond the ICS-201, which can be considered a reactive plan, because you take actions and then record what you did.

If, however, multiple operational periods, days, weeks or longer are required to resolve the incident, the IC/UC will shift the response organization to the ICS Planning Process to proactively develop an Incident Action Plan. The IAP is built in advance of the next operational period and details the organization, tactical assignments, and resources that will be used. A sample IAP can be found in Appendix A. As the OSC, you have a major role in the development of the IAP. *Figure 6.3* provides a visual depiction of the ICS Planning Process. The Operational Planning "P," as it's commonly referred to by its users, breaks the ICS Planning Process into *steps* to help illustrate what has to take place in order to successfully move through the planning cycle.

The Operational Planning "P"

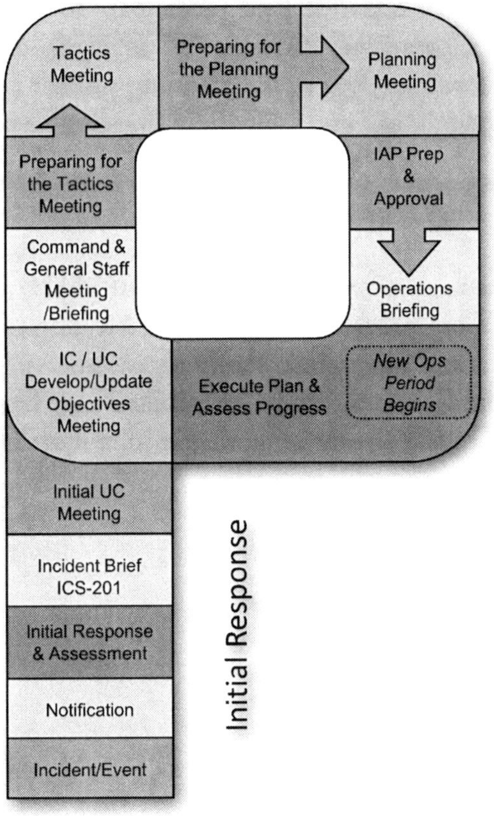

Figure 6.3. The ICS Planning Process is best depicted by the Operational Planning "P" that shows the various steps the command team will go through in the development of the Incident Action Plan.

We're going to spend time in this chapter dissecting your responsibilities in helping to develop an Incident Action Plan.

The OSC Role in the ICS Planning Process

Before we launch into a discussion of your role as the OSC in the ICS Planning Process, there are two things to keep in mind: first, the primary reason for the ICS Planning Process is to develop an Incident Action Plan (IAP) for you, the Operations Section Chief, to use in directing incident resources in the accomplishment of the IC/UC objectives. Second, the participation of the OSC in the process is absolutely critical. If you do not provide the necessary input into the planning process, the IAP cannot be appropriately developed. This would be your quickest way to seek new employment since no IC/UC is going to tolerate an OSC that does not fully support the development of the IAP. It's going to be challenging for you as you attempt to manage current operations (the here and now) as well as take time to plan and prepare for the next operational period, but everyone on the command team is there to support you.

The ICS Planning Process was created by emergency responders, and it's designed to methodically work the command team from the development of incident objectives all the way through to the delivery of an operational brief for field personnel. To help visualize the various *steps* in the planning process the Operational Planning "P" in *Figure 6.3* will be our guide. In the following pages we cover the various *steps* in the planning process and your part as the OSC in the development of a well thought out and logistically supportable tactical plan.

Command and General Staff Meeting

After the IC/UC has determined or updated the incident objectives for the next operational period, the team will have a Command and General Staff Meeting to review those objectives, set priorities, identify limitations and constraints, clarify organizational issues, clarify staff roles, and answer any questions that the Command and General Staff may have. Make sure that you take this opportunity to get clarification on any issues that require more information. Specifically, the OSC should, if necessary:

- Clarify operational issues (objectives)
- Agree on operational priorities
- Clarify organizational issues
- Identify limitations and constraints
- Discuss any interagency issues

Now, in addition to managing current operations, you have received direction on what the priorities and objectives are for the next operational period. Work on the strategies and tactics to achieve these objectives as soon as you receive them. This process helps prepare you for the Tactics Meeting.

Preparing for the Tactics Meeting

After you have met with the IC/UC and received the objectives for the next operational period, you'll quickly find yourself re-immersed into the time-critical decision making of managing ongoing operations. Although the demands on your time will be significant, and, at times overwhelming, you'll need to spend time roughing out how you want to accomplish the objectives. Remember that ICS is about teamwork so lean on the Planning Section Chief for help as you consider the various strategies (how you're going to accomplish the objectives) that are available to you. Once you determine the strategy you want to use you'll be able to choose the tactic (particular action to achieve the strategy) and identify the necessary equipment and personnel that can implement the tactic and, as a result, achieve the IC/UC objectives.

One way to help you think through your strategy options is to refer to the notes you made during the Command and General Staff Meeting where you should have listed the objectives that are your responsibility. Next to each objective and based on your tactical experience list the various strategies that can be implemented to achieve the objectives. Lastly, select the strategy that best meets the constraints of time and resources available and develop a quick tactical plan. You should ensure that the strategies you choose and the tactical plan that you develop are reasonable and achievable. The Planning Section Chief (PSC) and Safety Officer (SOF) can give you a reality check on your penciled-in plan at this stage.

Figure 6.4 illustrates an easy way to capture your planning ideas and is one of the tools you want to bring into the Tactics Meeting. The more you prepare for the Tactics Meeting the better your plan will be. The scenario for the below identification of response strategies is a flood with potential survivors who must be located and taken care of.

Identification of Response Strategies and Tactics

Objectives	Strategies (How)	Tactics (Who, Where, What, When)
Conduct thorough search for survivors	■ Use aerial assets ■ Use on water assets	On water teams are to go house-to-house in search of survivors. Transport the injured to triage center.

Figure 6.4. Take the time to think through the various options to accomplish the IC/UC objectives so you can select the best course of action and apply the right resources. You do this sort of thinking naturally. Spend a few minutes to jot down these ideas.

Tactics Meeting

This meeting is totally focused on the development of the Operations Section Chief's tactical plan for the next operational period. If this is the first Tactics Meeting of the response, it may take time to create the operations organization because the incident situation is dynamic, resources are still arriving on-scene, and objectives and priorities shift. The Tactics Meeting is facilitated by the Planning Section Chief and attended by the Resources Unit Leader (RESL), Safety Officer, Logistics Section Chief (LSC), Communications Unit Leader (COML) and you, the OSC. You have many responsibilities during the meeting so we recommend you use the checklist in *Figure 6.5* to help guide you.

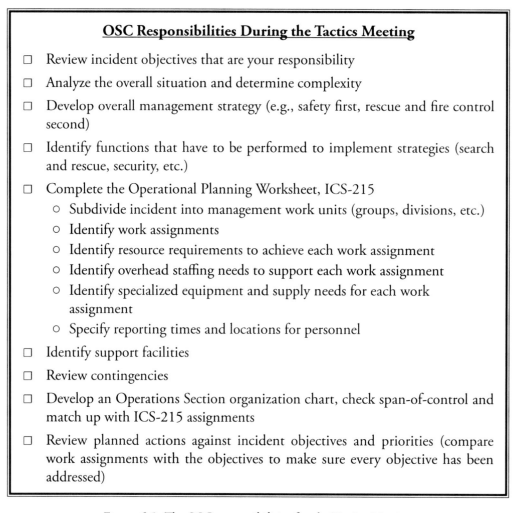

Figure 6.5. The OSC responsibilities for the Tactics Meeting.

When you leave this meeting you will have filled out the ICS-215, Operational Planning Worksheet. The Operational Planning Worksheet is designed to walk you through the development of your tactical plan and when completed you'll have a complete "picture" of what you want to do and how you want to do it. By the time you've completed the ICS-215 you'll have:

- Identified the operations organization you'll need for the next operational period (Groups, Task Forces, etc.)
- Established work assignments to be accomplished for each organizational element (What you want the Groups and Divisions to do)
- Identified the type, kind and number of resources needed to conduct operations
- Identified the number of supervisory personnel required to manage field operations

- Noted any special equipment that may be needed
- Designated the time and location where each organizational element is to meet at the beginning of the operational period

Using the Operational Planning Worksheet (ICS-215)

Prior to completing the ICS-215 you and the Planning Section Chief (PSC) will review the IC/UC objectives for the next operational period. This is critical because the plan you develop must be completely aligned with the objectives or your tactical plan will not be approved. The ICS-215 can be a little intimidating the first time you see it, but it's a simple form despite its appearance. *Figure 6.6* is an example of the ICS-215, Operational Planning Worksheet. The Resources Unit Leader will actually do the writing as you outline your plan because the RESL has to move the information from the ICS-215 to the IAP. Let's walk through the parts of the Operational Planning Worksheet that you are responsible for as the OSC.

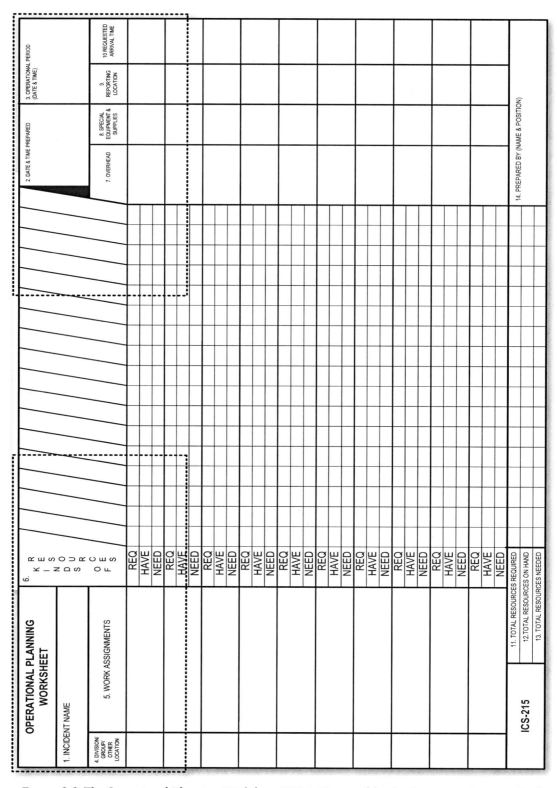

Figure 6.6. The Operational Planning Worksheet, ICS-215, is used by the Operations Section Chief (OSC) to develop the tactical plan for the next operational period. To illustrate how the ICS-215 is filled in by the OSC, two areas of the form within the dashed lines have been enlarged in Figures 6.10 through 6.13.

Step One. List the operational components that you'll be using to divide the incident into manageable units. These units may be Branches, Divisions, Groups, Task Forces, or Strike Teams. Any Staging Areas that are required during the next operational period are also placed on the ICS-215. *Figure 6.7* is a list of some of the things to consider when dividing the incident into manageable work assignments.

> **Considerations When Dividing the Incident**
> - ☐ Incident priorities
> - ☐ Size of effected area
> - ☐ Complexity of the incident and number of tasks
> - ☐ Amount of work to be accomplished
> - ☐ Span-of-control issues
> - ☐ Open water versus shoreline activities
> - ☐ Topography
> - ☐ Logistics requirements
> - ☐ Kind of functions to be accomplished
> - ☐ Contingencies

Figure 6.7. There are many things to take into account when you divide the incident into manageable units.

Unlike the other sections (Planning, Logistics, Finance/Administration), the Operations Section always builds from the bottom up, based on the needs of the incident. Response to an incident begins with the arrival of a few tactical resources and once on-scene, the Incident Commander determines if more resources are required to bring the incident under control or if it can be dealt with by using the current responding assets. As the Incident Commander expands the amount of tactical resources, he or she adds layers of management to divide the incident into manageable units. Generally, Divisions and Groups are added before Branches. This is not a hard-and-fast rule and in the end you have to do what you think is best for managing your tactical responsibilities. At some point, the IC is no longer able to manage tactical operations along with all other aspects of the response, and this is where you come in as the OSC. *Figure 6.8* is a generic operations organization. Use only the organizational elements you need to manage the tactical response. "Over-organizing" can be just as bad as "under-organizing." Experience and the availability of resources will guide your organizational decisions.

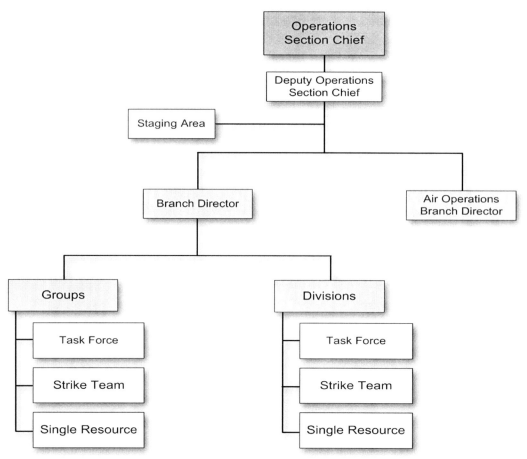

Figure 6.8. The size of the Operations Section organization is determined by the needs of the incident. Add layers of management as necessary, and be sensitive to the fact that over-organizing is as bad as under-organizing.

Branches are primarily used to manage span-of-control and can cover a specified geographic area or be functional and focused on a specific aspect of the response operation such as a search and rescue. When naming Branches use Roman Numerals for geographic Branches (e.g., Branch II) and name functional Branches according to what the Branch is doing for example Search-and-Rescue Branch.

Divisions divide an incident geographically. When you name Divisions use letters of the alphabet. Some of the things to think about when you're trying to determine the size of a Division are:

- Amount of work to be done
- Access
- Terrain
- Numbers of resources assigned (span-of-control issue)

Groups divide an incident by function. Often in the initial response to an incident, Operations is organized by Groups. For example, you may create a Fire Group to manage the fire aspect of the response and a Medical Group to deal with the injured. As you gain a better understanding of the situation, you may elect to include Divisions as part of the Operations structure. Groups are named based on the function they perform, such as Triage Group.

Some expectations you should have for your Division and Group Supervisors are included in *Figure 6.9* below. Make sure that you convey those expectations clearly in your operational briefings.

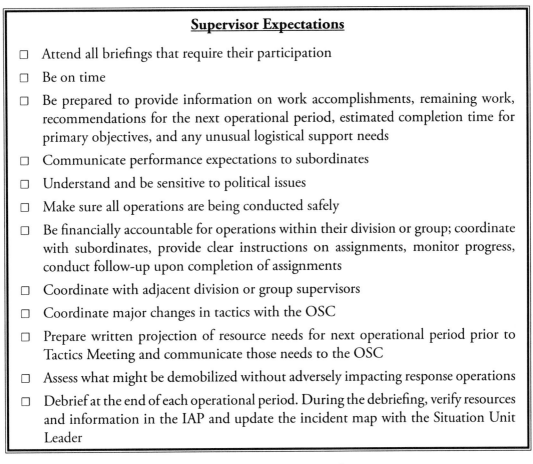

Supervisor Expectations

- ☐ Attend all briefings that require their participation
- ☐ Be on time
- ☐ Be prepared to provide information on work accomplishments, remaining work, recommendations for the next operational period, estimated completion time for primary objectives, and any unusual logistical support needs
- ☐ Communicate performance expectations to subordinates
- ☐ Understand and be sensitive to political issues
- ☐ Make sure all operations are being conducted safely
- ☐ Be financially accountable for operations within their division or group; coordinate with subordinates, provide clear instructions on assignments, monitor progress, conduct follow-up upon completion of assignments
- ☐ Coordinate with adjacent division or group supervisors
- ☐ Coordinate major changes in tactics with the OSC
- ☐ Prepare written projection of resource needs for next operational period prior to Tactics Meeting and communicate those needs to the OSC
- ☐ Assess what might be demobilized without adversely impacting response operations
- ☐ Debrief at the end of each operational period. During the debriefing, verify resources and information in the IAP and update the incident map with the Situation Unit Leader

Figure 6.9. OSC expectations for Division and Group Supervisors.

Staging Areas are used by the OSC for contingencies that require the rapid deployment of resources. Staging Areas are established and disestablished by the OSC and the life span of Staging Areas can be one or multiple operational periods depending on need. Staging Areas are named by their location such as Bayfield Park Staging Area.

Strike Teams are a grouping of resources of similar kind and capability that are supervised by a Leader. The use of Strike Teams helps with span-of-control. Strike Teams are assembled and disassembled by the OSC. An example of a Strike Team can be several Urban Search-and-Rescue Teams that are working under a single leader.

Task Forces are similar to Strike Teams; however, unlike a Strike Team a Task Force is comprised of different kinds and capabilities of resources. Task Forces are an excellent way to manage span-of-control. A Task Force is supervised by a single leader. An example of a Task Force can be the combination of an ambulance, flood rescue boats, and dog teams. These resources working together as a single entity and under a common leader would be established by the OSC to accomplish a particular assignment.

A parting thought on naming your organizational elements: if at all possible, once you have named a Branch, Division, or Group make every attempt to stay with that name. The Resources Unit and Situation Unit build their displays based on the name designations that you establish so if you do decide that the original name has to be changed, immediately notify Planning so that they can continue to support you.

Figure 6.10 is a cutaway of the top left corner of the ICS-215. In this example the Operations Section Chief has created a *Search Group* to address an objective laid out by the IC/UC.

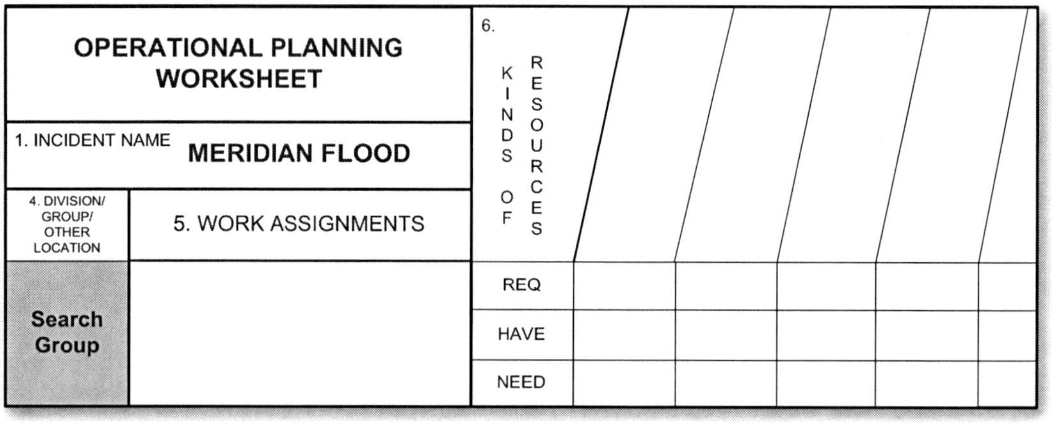

Figure 6.10. This cutaway of the ICS-215 shows where the Operations Section Chief lists the different organizational elements that will be used to divide the incident into manageable pieces. In this example, the Operations Section Chief (OSC) has decided to establish a Group as part of the organization. The OSC has labeled it Search Group.

Step Two. Define the work assignment that you want the specific organizational element to accomplish during the next operational period. You want to make sure that you're clear on your assignment so those who must do the work understand your expectations. *Figure 6.11* shows the work assignment for the Search Group.

OPERATIONAL PLANNING WORKSHEET		6. KINDS OF RESOURCES					
1. INCIDENT NAME	MERIDIAN FLOOD						
4. DIVISION/ GROUP/ OTHER LOCATION	5. WORK ASSIGNMENTS						
Search Group	Conduct house-to-house search for injured persons. Mark each dwelling searched with a red "X" on the front door. Evacuate injured persons to triage center.	REQ					
		HAVE					
		NEED					

Figure 6.11. Next to the organizational element, Search Group, the Operations Section Chief outlines the work assignment that the Search Group supervisor is to accomplish during the next operational period.

Step Three. The OSC lists the resources required for the Search Group to accomplish the work assignment (see *Figure 6.12*). You want to keep in mind that the required (REQ) portion of the ICS-215, which lists the resources that the OSC needs to accomplish the work assignments, is a wish list. Hopefully, the resources that you require are already there or on their way. If not, the Logistics Section Chief will have to order them. If all the resources do not show up in time for the start of the next operational period, you'll need to evaluate whether the strategy that you selected can be accomplished with fewer resources. This is one reason why we recommended jotting down various strategy options early on in the ICS Planning Process (refer to *Figure 6.4*, Identification of Response Strategies) so you can quickly implement another strategy if necessary. If your original strategy will still work, but it will take longer to accomplish unless more resources check in to the incident, make sure that you brief the IC/UC on the expected delay. It's absolutely critical to keep the IC/UC informed.

OPERATIONAL PLANNING WORKSHEET		6. KINDS OF RESOURCES	Ambulance (type 2)	Helicopter	Search Team (six person)	Rescue Boat	
1. INCIDENT NAME	MERIDIAN FLOOD						
4. DIVISION/ GROUP/ OTHER LOCATION	5. WORK ASSIGNMENTS						
Search Group	Conduct house-to-house search for injured persons. Mark each dwelling searched with a red "X" on the front door. Evacuate injured persons to triage center.	REQ	1	1	1	1	
		HAVE					
		NEED					

Figure 6.12. Once the OSC has listed the work assignment for the Search Group, the OSC will determine the kind, type, and number of resources necessary to accomplish the work assignment. In this example, an ambulance, helicopter, search team, and rescue boat are required.

Step Four. Once you list the resources that you want, there are a few more items you need to do before you go to the next organizational element and repeat Steps One through Four. You want to note the overhead (supervisory) personnel you need to oversee field operations. For every Division and Group that you designate you must have a Supervisor assigned. If you designate a Branch, you must have a Branch Director assigned. If you and the Safety Officer determine that an Assistant Safety Officer is needed to support operations in a particular Division or Group, you want to note that on the ICS-215 as well.

Finally, there's a place to note any special equipment that may be needed and a place to note the time and location that each organizational element is to meet at the beginning of the operational period. *Figure 6.13* is an example of the information that you would want to include in Step Four in completing your part of the ICS-215.

Ambulance (type 2)	Helicopter	Search Team (six person)	Rescue Boat	2. DATE & TIME PREPARED 15 Nov 2100		3. OPERATIONAL PERIOD (DATE & TIME) 16 Nov 0600 to 16 Nov 1800	
				7. OVERHEAD	8. SPECIAL EQUIPMENT & SUPPLIES	9. REPORTING LOCATION	10. REQUESTED ARRIVAL TIME
1	1	1	1	Supervisor Assistant Safety Officer	Red Spray Paint	4th Street and County Road 502	0530

Figure 6.13. In addition to the organizational element (Search Group), the work assignment for the Group, and resources required, the ICS-215 also has a place to record any required overhead personnel, special equipment, reporting location, and time to report.

You have just laid out your tactical plan for the next operational period, which will be presented to the IC/UC during the Planning Meeting. As you develop your plan, the PSC, LSC, and SOF are right there with you and their input and suggestions help you create a plan that all support. The last thing you want is to go to the Planning Meeting and not be united in your support of the tactical plan. The IC/UC will not be reassured that the proposed plan is viable if there's general disagreement in the success of the tactical plan.

Use of Deputies

When you develop your organization, do not forget to consider bringing in Deputies. Deputies are an excellent resource especially when the pace of operations and demands on your time makes it difficult for you to manage both current operations and support the ICS Planning Process. For operations running continuously around the clock, you may need to bring in a Deputy to manage operations so that you can get some rest. If the incident is complex and many branches have been established, Deputies can be used to help you manage your span-of-control.

Just remember that Deputies need to be as equally qualified as you.

Air Operations

Aircraft, both fixed-winged and helicopters, are a tremendous asset during response operations, but the management of these high-cost, extremely technical assets requires personnel trained and experienced in management of air operations. We strongly recommend that if you have more than a few aircraft supporting your operation for any length of time that you bring in an Air Operations Branch Director (AOBD). The AOBD ensures that air operations are tactically implemented and integrated with ground operations. The AOBD is in constant contact with the Operations Section Chief to provide critical coordination in the use of air assets. Using an AOBD solves span-of-control problems and provides the Operations Section with the technical knowledge necessary to safely manage aircraft.

Preparing for the Planning Meeting

In preparing for the Planning Meeting, the OSC must continue to manage ongoing operations and update the tentative plan for the next operational period based on changes in current operations — for instance, some work may be running either ahead or behind schedule. Make sure that you keep both the PSC and the LSC up-to-date as they plan for the operations you discussed at the Tactics Meeting.

Planning Meeting

The purpose of the Planning Meeting is to provide the opportunity for the OSC and the PSC to present their proposed plan of action to the IC/UC and Command and General Staff members in response to the Command's direction, objectives, and priorities that they have set for the next operational period. The Planning Meeting is a complete and polished executive-level version of the Tactics Meeting. The PSC will facilitate the meeting and *Figure 6.14* is a checklist of the OSC's responsibilities for that meeting.

> **OSC Responsibilities During the Planning Meeting**
> - ☐ Review operational objectives and Command's decisions (e.g., daylight only operations) and direction
> - ☐ Brief on the overall strategy, tactics, and functions that will be performed to accomplish the IC/UC objectives
> - ☐ Review how the incident will be managed
> - ☐ Review work assignments and resources required
> - ☐ Discuss special needs (communications, facilities, security)
> - ☐ Discuss how operations plans to respond to contingencies
> - ☐ Discuss proposed Operations organization structure
> - ☐ Answer any questions and make changes to plan as needed

Figure 6.14. The OSC responsibilities during the Planning Meeting.

Briefing the Operational Planning Worksheet is one of the primary agenda items for the Planning Meeting. *Figure 6.15* is an example of a completed ICS-215 that the OSC would use to conduct the briefing of their plan for the next operational period.

If you have done your job well, the Unified Command will give you tentative approval of the "plan of action" for the next operational period. This plan of action must now be incorporated into an Incident Action Plan (IAP). You, the OSC, should assist the PSC in developing the Incident Action Plan. Specifically, you should help develop the ICS-204 Assignment Lists or, at a minimum, review them to ensure that they are complete, accurate, and meet your needs.

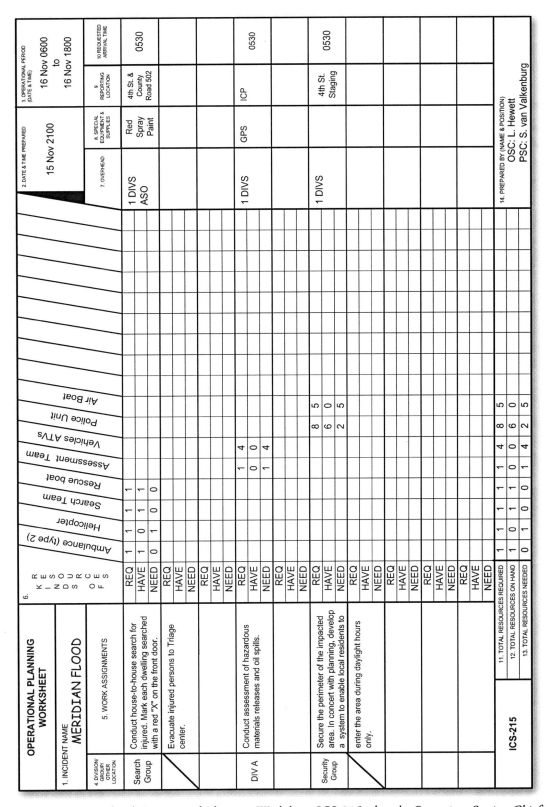

Figure 6.15. A completed Operational Planning Worksheet, ICS-215, that the Operations Section Chief will use to present the tactical plan for the next operational period.

Using the Assignment List (ICS-204)

The ICS-204, Assignment List, is the culmination of all the time and effort that you put into preparing for and participating in the Tactics Meeting. The ICS-204s are the most important part of the Incident Action Plan. These forms provide the Branch Directors, Supervisors, Staging Area Managers, and Leaders who must carry out your tactical direction with the necessary information to accomplish the work that you outlined. Your Branch Directors, Supervisors, Managers, and Leaders have not attended any of the previous meetings and the ICS-204 provides the information on what you want them to do—their assignment. The ICS-204 should include everything they need to know to get their work done for their piece of the operation.

The information that you used in the development of the ICS-215 is used to complete the ICS-204. Let's look at the ICS-204 for the Search Group so that you can see the links between the ICS-215 and the Assignment List. In *Figure 6.16*, the Search Group's Assignment List has been filled in with information from the ICS-215. The Search Group's work assignment, resources, and special instructions have been placed on the form as well as the name of the supervisor.

It is very important that you review all of the ICS-204s before the Incident Action Plan is sent to the IC/UC for final approval and signature.

Assignment List (ICS-204)

1. BRANCH	2. ~~DIVISION~~/GROUP Search	ASSIGNMENT LIST	
3. INCIDENT NAME **MERIDIAN FLOOD**		4. OPERATIONAL PERIOD (Date and Time) 16 Nov 0600 to 16 Nov 1800	

5. OPERATIONS PERSONNEL

OPERATIONS CHIEF __L. Hewett__ ~~DIVISION~~/GROUP SUPERVISOR __P. Robert__

BRANCH DIRECTOR _____ AIR TACTICAL GROUP SUPERVISOR _____

6. RESOURCES ASSIGNED THIS PERIOD

STRIKE TEAM/TASK FORCE RESOURCE DESIGNATOR	EMT	LEADER	NUMBER PERSONS	TRANS NEEDED	DROP OFF POINT/TIME	PICK UP POINT/TIME
MESA #3		B. Riggs	21		0530	
44120		V. Kammer	4		0530	
Ambulance #1		S. Miller	2		0530	
Helicopter 12		T. Troutman	2			

7. ASSIGNMENT

Conduct house-to-house search for injured persons. Mark each dwelling searched with a red "X" on the front door. Evacuate injured persons to triage center.

8. SPECIAL INSTRUCTIONS

Work in teams of two or more at all times. Snakes are suspected to be in the work area, wear snake gators. Personal floatation devices should be worn. Daylight operations only. Conduct regular communications checks. Send hourly updates on the group's progress to the Situation Unit.

9. DIVISION/GROUP COMMUNICATIONS SUMMARY

FUNCTION		FREQ.	SYSTEM	CHAN.	FUNCTION		FREQ.	SYSTEM	CHAN.
COMMAND	LOCAL REPEAT	CDF 1	King	1	SUPPORT	LOCAL REPEAT			
DIV./GROUP TACTICAL		157.4505	King	3	GROUND TO AIR				

PREPARED BY (RESOURCES UNIT LEADER) A. Worth	APPROVED BY (PLANNING SECT. CH.) J. Gafkjen	DATE 16 Nov	TIME 0400

Figure 6.16. The Assignment List ICS-204 is the most critical part of the Incident Action Plan. The ICS-204s lay out the work assignments, resources, and special instructions for each operation's established organizational element.

The ICS-204 identifies the incident name and the Branch and Division or Group by name. It includes the operational period for which the assignment is being made and key Operations personnel—essentially an Operations chain of command. The ICS-204 provides a list of resources that are assigned to the supervisor, including the leader name, transportation needs, and resource designator. The ICS-204 also has a block for you to describe their work assignments and any special instructions. Don't worry if the block looks too small, there are addendums that can be added to this form. Just keep in mind that the assignment must be clearly laid out so that the supervisor understands the expectation. Special instructions are just that, and may include an additional safety message, such as to be sure to pack bug spray, or to make sure you wear your bulletproof vests, or it may be something like a communications check-in schedule — whatever applies to the situation. Last, the ICS-204 includes a communications summary that the Communications Unit Leader will fill in, so the supervisor can see at a glance what radio frequency they should be using to communicate with you. The OSC is responsible for filling out part of the ICS-204s because you know what you want your people to know and it is absolutely essential that you communicate that on their work assignment. You will have the opportunity to review each ICS-204 with the Directors, Supervisors, Managers, and Leaders at the Operations Briefing.

Operations Briefing

After the Incident Action Plan has been built and approved by the Incident Commander/Unified Command, it must be presented to the workers who will carry out or implement the Plan. This is done in an Operations Briefing. The Briefing ensures that all oncoming Operations Section supervisory personnel are made and kept aware of the planned tactical actions. The Operations Briefing serves many purposes. Specifically it helps reduce confusion, improves communications among operations personnel, enhances cooperation, facilitates team interaction, increases overall productivity, and ensures continuity of operations. Remember, some people at the Operations Briefing may be very new to the incident. It is absolutely critical that you present your brief so that everyone is brought up to speed. Everyone needs to be very clear on the status of the situation and what their responsibilities are for the response. Make sure that no one is left in the dark on what has to be done and any special considerations that must be made. The following checklist in *Figure 6.17* includes things to cover in the Operations Briefing:

> **OSC Operations Briefing Checklist**
>
> - ☐ Current situation
> - ☐ Overall strategy and priorities
> - ☐ Short- and long-range predictions
> - ☐ Command issues
> - ☐ Safety and security issues
> - ☐ Medevac procedures
> - ☐ Accident injuries reporting
> - ☐ Decontamination procedures
> - ☐ Expected outputs and accomplishments
> - ☐ Resource ordering and resupply
> - ☐ Resource status changes
> - ☐ Assigned tasks and resources
> - ☐ Chain of command
> - ☐ Transportation issues
> - ☐ Reporting time and location
> - ☐ Performance expectations
> - ☐ Sensitive/critical information reporting
> - ☐ Updating work accomplishments
> - ☐ Reporting any changes in tactics
> - ☐ Technical specialists assigned to Operations
> - ☐ End of shift pickup time and location
> - ☐ Debriefing instructions

Figure 6.17. Information the OSC covers in an Operations Briefing.

Conduct a thorough Operations Briefing to ensure that operations start right and continue to go well, that everyone understands his or her role, and that on-scene operations go smoothly (see Appendix H, OSC Operations Briefing Checklist for the Meridian Flood Incident). It's absolutely critical to make sure that key personnel attend these briefings. Specifically, for a large incident, everyone down to the Division/Group Supervisors and Staging Area Managers should attend. In a medium incident, everyone down to team/crew leaders should attend. In a small incident, everyone should attend.

Execute Plan and Assess Progress

At this step in the ICS Planning Process you have developed a tactical plan (plan of action) that can be logistically supported and has been peer reviewed. Your plan has been approved by the IC/UC. The Branch Directors, Supervisors, Leaders, and Staging Area Managers have been briefed on the plan and your expectations. You have done all you can to prepare for the next operational period, but experience has shown that in a dynamic response environment, the IAP is always a work in progress. As the OSC, you must constantly evaluate the plan against current reality on the ground. Remember that you built the tactical plan hours ago based on predictions and input from others. Do not hesitate to modify the IAP to fit the conditions you face. Just make sure that you notify the IC/UC, Planning Section Chief, and Logistics Section Chief of any significant changes so they can continue to provide the support you need to be successful.

In Appendix B you see a one-page summary of the OSC responsibilities during the various *steps* of the ICS Planning Process. The page contains the Operational Planning "P" and calls out specific Operations activities that we present throughout the chapter.

Conducting Operational Debriefings

Conducting a thorough operational debriefing is extremely important to the response operations. Debriefings ensure that the Incident Management Team is aware of the progress made and the hindrances or problems encountered. Typically the OSC or Situation Unit Leader conducts debriefings. The OSC and PSC should agree on the process used. Information gathered during a debriefing include:

- Ability to accomplish tasks with assigned resources
- Resource allocation recommendations
- Coordination issues
- Accomplishments
- Special logistics requirements
- Performance issues (personnel and equipment)
- Hazards
- Crew rest/work ratio
- Other recommendations to improve operations

Use the feedback gathered in operational debriefings to make constructive changes to operations and to processes that aren't working well.

Daily Self-Evaluation of the Operations Section

This evaluation is internal to the Operations Section to help improve service within the Section and to the entire command team. Completion of this evaluation is the responsibility of the Operations Section Chief.

Operations Section Chief

- Are we meeting the IC/UC expectations?
- Have I provided clear direction with good follow-up?
- Is the Operations Section functioning as a team?
- Is Operations talking to the Situation and Resources Unit Leaders with updates?
- Are we producing the highest level of quality achievable?
- Are we conducting daily Operational Briefs and Debriefs to keep things running smoothly?
- Are we conducting safe operations?
- How well does the plan reflect reality?

Operational Issues

- Do we have good familiarity of the situation on the ground; is span-of-control good?
- Are we aggressively debriefing line personnel after every operational period?
- Are we communicating with the other sections?
    - Are they satisfied with our performance?
- Are tailgate safety meetings occurring?
- Are we effectively utilizing our technical specialists?
- Are we adequately staffed to support incident operations?
- Are we using General Message Forms, ICS-213?
- Do we have adequate supplies?
- Is resource status being kept up-to-date?

Personnel Issues

- Are we providing for our well-being (safety, rest, food, etc.)?
- Are ICS-214s (Unit Logs) being kept up-to-date and submitted to the DOCL?
- Are timesheets being submitted?

Demobilization Issues

- Have we thought about demobilization yet?

- What will be the procedures?
- What will be the priorities?
- Have we discussed demobilization with the other Section Chiefs?

Closing Thoughts

The position of Operations Section Chief on any incident is an exciting and intense job. The entire response organization is there to support you so that you can be successful: because your success is everyone's success. Depending on incident complexity and size, your experience and training as a tactician may be tested. Your ability to lead in a high-stress situation will become immediately apparent.

The Incident Command System principles, processes, and common terminology are absolutely essential in the multi-agency response environment where we typically must operate. You can be the greatest tactician in the world and fail miserably if you cannot work as a member of a team. You want to be as good at using the ICS as you are at managing the tactics.

CHAPTER 7

PLANNING SECTION CHIEF (PSC)

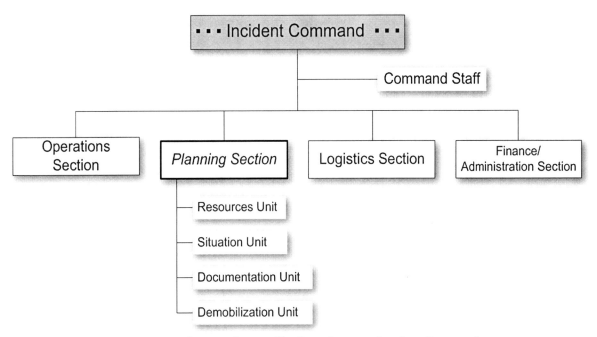

Figure 7.1. Planning Section Chief in relation to Incident Command.

If you've been designated as the Planning Section Chief (*Figure 7.1*) for an incident or event, you've been given an extremely challenging and rewarding assignment. The Planning Section Chief (PSC) plays a critical role in moving an incident from a reactive response to a proactive response. Like the conductor of an orchestra, the PSC ensures the cohesiveness of the incident management team, keeping everyone working together. The PSC must be a strong facilitator and communicator. In addition to having a thorough understanding of the Incident Command System, a PSC needs to be a team builder, enforcer of the Incident Command System (ICS) Planning Process, part diplomat, and have the ability to simultaneously manage current issues as well as plan for future contingencies. Regardless of the initial complexity of the incident, the Planning Section Chief must look far beyond the apparent situation and ask "What if?" The PSC must be aware of immediate challenges and those that lie on the horizon.

When you respond to an incident or support event planning (such as the World Cup Soccer tournament), as the Planning Section Chief, take this chapter with you. The information it contains will help you organize and manage your planning responsibilities. We cover information on all of the Planning Units: Resources, Situation, Documentation, and Demobilization. Remember that in the ICS, unless you bring in personnel to assume the responsibilities of the subordinate positions, you'll have to do the job yourself. For the PSC, this means that if you do not bring in a Resources Unit Leader (RESL), you have to do the work of resource tracking. Chapter 2 in this book is a must read for you: "Incident Command System Planning Process." That chapter is a process guide that everyone will use, and that you reinforce, to make sure that all efforts are focused on meeting the priorities and direction set down by the Incident Commander/Unified Command (IC/UC).

Let's start our discussion of the Planning Section Chief with some of the major responsibilities of that position. PSC responsibilities:

- Provide current, accurate situation status and concise briefings in support of the ICS process meeting schedule and Incident Commander/Unified Command expectations.
- Accurately track all resources through the use of T-cards or another resource tracking system, and aggressive, proactive field observers.
- Establish and maintain resource control through the use of check-in locations/recorders.
- Facilitate the planning process by conducting timely meetings in accordance with the meeting schedule and work closely with the Operations Section Chief, Logistics Section Chief, and Command Staff.
    - Determine the meeting schedule based on the operational period (you can find a helpful guide in Chapter 2)
- Ensure thorough documentation of all key decisions.
- Establish and maintain an "open action items" list of things that must be accomplished. Ensure that each item on the list is assigned to the appropriate ICS command element (e.g., Operations Section, Liaison) for completion.
- Ensure that a complete and thorough Incident Action Plan, along with appropriate support plans, is delivered in support of the operations.
- Use technical specialists in coordination with operations to provide critical information to develop specialized operations and planning efforts to support incident operations. Examples of technical support include: salvage plans, air monitoring, hazardous materials modeling, fire behavior, oil spill trajectory, and intelligence efforts.

Managing the Planning Section Begins Long Before the Incident or Event

As a designated Planning Section Chief for your agency, organization, or company, you should ensure that you're prepared to deploy. If it's early on in an incident, your personal preparedness significantly affects the performance of your section and team. We recommend that you create a personal "go-kit" with the essential items to carry out your responsibilities as the PSC. It's best to assume that no one else will bring the items you'll need. Appendix I, Planning Section Chief Support Kit Checklist, has a list of items for you to have in a deployable kit. It's not all-inclusive, and you should modify it as you see fit.

Actions to Take upon Arriving at the Incident Command Post

If you're the first PSC assigned to the incident and not a relief, the response will most likely be chaotic when you arrive at the incident scene, and it's easy to get quickly pulled in many different directions. Emergency response by its nature is a confusing environment to operate in, especially in the early phase, and although you cannot completely remove this aspect of the incident you can take some steps to help minimize the impact of this chaos on you. Below are some checklists that can help you navigate through the fog of initial response during your first hours on an incident and accomplish your responsibilities as the Planning Section Chief.

Situation Brief

Get a situation brief from the Incident Commander/Unified Command (IC/UC). This initial briefing is important and does a few things: (1) you get a feel for the size and complexity of incident operations, and this helps with staffing decisions and space requirements; (2) you gain essential information to establish effective resources management and develop a situational status; and (3) you assess the initial direction and priorities of the IC/UC. Use the checklist in *Figure 7.2* to remind you of some of the more important issues and items that you should get answered during your in-brief.

Planning Section Chief Briefing Checklist

At a minimum the briefing should include:

- ☐ Incident situation: magnitude and potential of the incident
- ☐ Political, environmental, and economic constraints
- ☐ Response emphasis
- ☐ Facilities already established
- ☐ Priorities, limitations, constraints, and objectives
- ☐ Command structure (single or unified)
- ☐ Operational period and start time
- ☐ Hours of operation
- ☐ Agencies and jurisdictions involved
- ☐ Information on committed resources
- ☐ Resources ordered
- ☐ Media interest
- ☐ Incident investigation
- ☐ Resource requesting process
- ☐ Any established check-in locations

Make sure you request a copy of the ICS-201, Incident Briefing Form, or other documents that may help you begin planning

Figure 7.2. The checklist is a great way to help remember the things you need to get answers to during your in-brief.

Planning Section Chief's Actions and Considerations

Once you receive your briefing it's time to establish your Planning Section. The checklist in *Figure 7.3* covers some of the major actions and considerations that will get you started in the right direction. Depending on the incident, you may only use some of the items on the list. We want you to be successful and have confidence in establishing an effective Planning Section and checklists are great memory joggers. If you have not performed the role of the PSC in a while, the checklist will help.

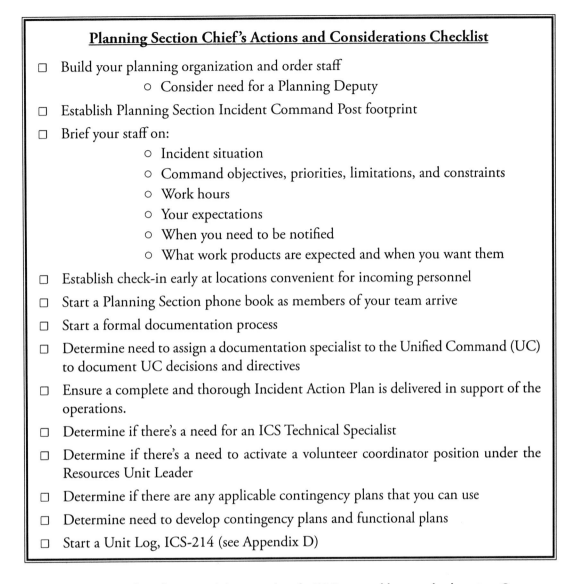

Figure 7.3. *The number of issues and decisions that the PSC must address can be daunting. One way to help you manage is to use a checklist. This list is only a beginning and we recommend you add to it as you gain experience.*

Order Your Staff

For most incidents that you respond to as the PSC, full activation of all of the Planning Section Units (as depicted in *Figure 7.4*) will not be required. For any positions that you do not activate, you'll do the job yourself. So you want to have technical proficiency in the various positions that work for you (excluding the Technical Specialists). The briefing you received from the IC/UC on the current incident status and their view of the incident's future potential will help you make some initial staffing decisions. The actual size of the Planning Section will be based on span-of-control and the needs of the incident. It's our recommendation that if you face a complex incident and/or one that will last several days, do not hesitate to immediately request a Resources Unit Leader (RESL) and a Situation Unit Leader (SITL) to support the planning effort. Ordering just the right amount of staff can be tricky and there's always lag time between when you request the support and when it actually arrives.

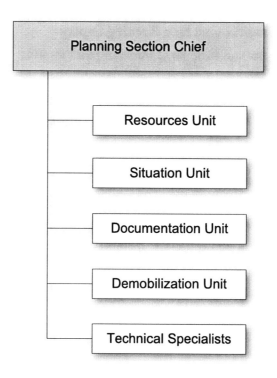

Figure 7.4. Typical Planning Section organization when fully activated. The diagram does not show Deputies, but, if the situation warrants it, Deputies can play an important role in the success of the Planning Section in support of incident operations.

Establishing the Units of the Planning Section

There are a few things to keep in mind as we discuss building your Planning Section organization. One of the Incident Command System's greatest strengths is that it's scaleable. You tailor the size of the organization based on the demands of the incident not on the organizational chart that depicts the variety of positions within the ICS. We mentioned earlier that for the majority of responses there may be no Planning Section Chief assigned or the incident may only require a PSC with no support staff. It all depends on what you face. However, if you need to bring in support, do not hesitate to do so. Below are some thoughts on the various positions that make up your Planning Section.

Deputies

Deputies can be an invaluable asset, and if you're involved in a complex long-term incident, bring in a deputy to help you run the Planning Section. Deputies can be used in many different ways such as to manage the day-to-day responsibilities of the Planning Section as you oversee the planning process; provide nighttime relief for you; or conduct long-range planning.

Resources Unit

Experience has shown that without an efficiently run Resources Unit, an incident will not be able to run effectively. Move quickly and determine whether you're going to order in a Resources Unit Leader (RESL) or do the work yourself. You must get a handle on incident resources or you'll never be able to enter the ICS Planning Process. The RESL job is difficult during the early hours and/or days of a response as the influx of equipment and personnel cascade in to support operations. It's during this large ramp-up of resources that you need to pay particular attention to the Resources Unit and provide the necessary support. It can be difficult to gain cooperation from responding resources to properly check in at the incident and it's equally challenging for the RESL to determine what resources arrived on-scene before check-in was established (unless the initial Incident Commander documented incident resources). Time spent getting the Resources Unit established and providing critical resource information will greatly benefit the response effort.

Directions to the Resources Unit Leader (RESL):

- Ensure that check-in locations are established in the locations where the majority of resources are arriving
- Establish a Resources Status Display
- Conduct field verification of resources that arrived before check-in was set up
- Be prepared to attend the required meetings and briefings
- Ensure the resource information is as accurate as possible
- Provide maximum support to the Operations Section Chief (OSC)
- Work closely with Supply Unit Leader to ensure that the OSC resource requirements are met

Situation Unit

A picture speaks a thousand words, and in an emergency operation the ability to rapidly communicate situational information is paramount. The priorities and objectives established by the IC/UC are grounded in their understanding of the current situation and predicted course of events. The Situation Unit provides that information to ensure that the IC/UC and rest of the command team are all looking at the same problem. Failure to provide timely, accurate situational information will become readily apparent when you start to see situation maps cropping up all over the Incident Command Post (ICP) as people try and fill the vacuum created by the Planning Section. That is not a position you want to put yourself in so make sure that you deliver the best information possible. In a highly dynamic incident this will not be easy.

Directions to the Situation Unit Leader (SITL):

- Establish a master map or chart
- Establish a schedule with field observers to report in

- Ensure information is accurate and posted in a timely manner
- Displays must be neat and legible and information on the displays must be readily understood
- In addition to a master situation board, consider placing status boards at other locations such as:
  - Joint Information Center
  - Unified Command Meeting area
  - Operational Briefing area
- Be prepared to deliver a situational briefing throughout the ICS Planning Process and at other times as directed
- Evaluate accuracy and verify the information you receive. You do not want the IC/UC and the other members of the command team to lose confidence in your work
- If the incident is large, consider setting up a debriefing area for overhead personnel (branch directors, supervisors, etc.) coming off-line

Documentation Unit

Documentation is absolutely critical. Do not let documentation of the incident be a secondary concern. As you read through this chapter, you see time and again reference to documentation. Every member of the command team must have the same appreciation for the importance of documentation as you do.

Directions to the Documentation Unit Leader (DOCL):

- Collect pertinent documents (not all-inclusive list)
  - ICS-201, Incident Briefing Form
  - Incident Action Plans
  - Command Decisions/Directives
  - ICS-211s, Check-in Sheets
  - ICS-215s Operational Planning Worksheets
  - ICS-209s, Incident Status Summaries
  - ICS-214s, Unit Logs
- Ensure that any documentation submitted to the Documentation Unit is accurate and complete. If not, work with the submitting party to correct errors or omissions
- Obtain PSC approval prior to the release of any incident-related documentation or reports
- Establish incident files; ensure that the PSC and the DOCL agree on a particular filing system to use for the incident to ensure continuity and to avoid confusion later on. Filing system options include:
  - File by operational period

- File by calendar date
- File by form number
- Attend meetings
  - Unified Command Objectives Meeting
  - Command and General Staff Meeting
  - Planning Meeting

Demobilization Unit

Inattentiveness to demobilization will come back to haunt you. Resources cost money, and if they sit idle, the wait to be released can run into substantial costs. This is particularly true when the type of incident that you're responding to, such as a hazardous materials incident, oil spill, or biological response, requires that equipment be decontaminated before it can be released. Efficient demobilization can only occur with some forethought and good planning. For a major incident a Demobilization Plan must be written. Here are a few thoughts on demobilization:

- Start planning early
- Recognize the indicators to begin demobilization
  - No new resources ordered
  - End of incident is in sight
  - There are unassigned resources
- Ensure that the Demobilization Unit Leader (DMOB) is getting the support they need from the entire command team

Directions to the Demobilization Unit Leader (DMOB):

- Develop a Demobilization Plan (see Appendix J, Example Demobilization Plan, to familiarize yourself with the content of the plan)
- Work closely with the Command and General Staff to ensure a smooth and orderly release of resources

Technical Specialists (THSP)

Technical Specialists provide expertise that cannot be found within the ICS organization. They provide subject matter expertise and counsel such as developing specialized maps, data analysis, or operating specialized equipment. Technical Specialists bring unique experience and/or knowledge to an incident that the response team relies upon to ensure that operations are conducted in the safest most efficient manner achievable. Technical Specialists can work anywhere their skills are needed on the command team (Operations, Incident Command, Safety, etc.). Some examples of technical specialists include: environmental, family assistance coordinator, salvage, fire behavior, wildlife, safety, public health, and legal.

All Technical Specialists report to the Planning Section, but will be assigned to support the response anywhere on the command team.

Planning Section Incident Command Post Footprint

You need to ensure that the Planning Section has the space it needs to support operations. If the incident you're responding to is large and complex, the Planning Section can swell to 50, 60, or more, personnel. This is where it's important to discuss the incident potential with the IC/UC to help determine your spacing needs. *Figure 7.5* is a generic layout for the Planning Section. The actual configuration will be driven by many variables such as the physical limitations of the facility and your own personal style. Work with logistics to have the section set up and consider these guidelines:

- One entry point
- Status displays near entry
- PSC located near entry
- Situation and Resources Unit located next to each other
- Documentation Unit in close proximity to PSC
- Ensure that the Planning Section space is in close proximity to the Operations Section space

Equipment

- Copier for the documentation unit
- Fax machines
- Enlarger machine to make poster-size displays

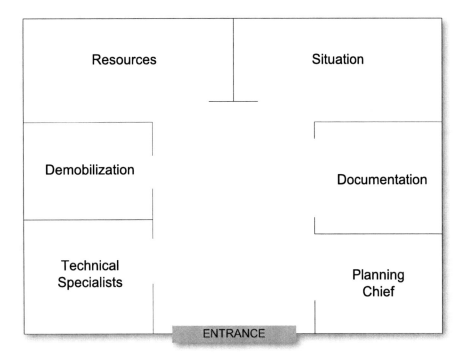

Figure 7.5. For incidents that require a large planning organization it's important to have adequate space. This generic layout provides an example for organizing the Planning Section.

Meetings and Briefings

Meetings and briefings are the primary method for keeping the command team united and moving toward the incident objectives. The PSC sets up and facilitates the meetings and briefings to ensure that they are held on time and that all presenters are prepared. One of your goals as the PSC is to protect the time of the IC/UC and the Operations Section Chief (OSC) and, in order to do that, meetings and briefings have to be conducted in a professional manner. We will briefly cover some of the PSC responsibilities at the various *steps* in the ICS Planning Process depicted in *Figure 7.6*.

The Operational Planning "P"

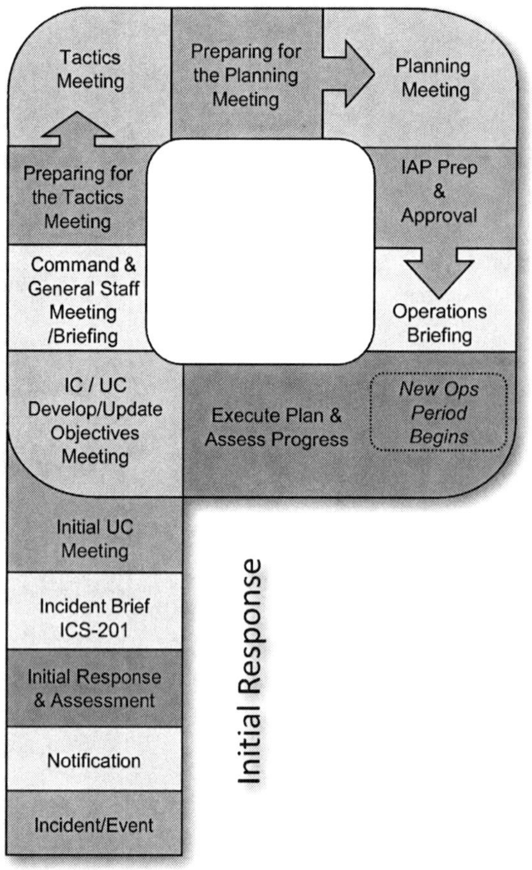

Figure 7.6. The PSC is responsible for ensuring that all meetings and briefings (steps) are conducted on time and that all presenters are prepared.

ICS-201, Incident Briefing

The ICS-201, Incident Briefing, is an important but informal briefing with no established agenda with the exception of following the format of the ICS-201 form. However, the intent of the ICS-201 briefing is to ensure an orderly transfer of command when a more qualified person arrives on-scene to assume command of the incident. The briefing may also occur when a Unified Command is assuming responsibility for the incident. The main player at this briefing is the initial response incident commander who will conduct the briefing to the incoming IC or UC. Some of your specific responsibilities as the PSC are to:

- Facilitate the ICS-201 brief
- If possible, provide a copy of the ICS-201 to attendees
- Document discussion points, concerns, and open action items
- Distribute a copy of the ICS-201 to RESL and SITL

Initial Unified Command Meeting (held only once)

The Initial Unified Command Meeting is held only once and the agenda for the Meeting covers a wide range of issues that the UC will need to come to agreement on, such as which agency will provide the Operations Section Chief, the length of the operational period, and many other issues. At a minimum, what you need from the meeting is:

- A good set of objectives and priorities
- Operational period duration (e.g., 12 hours, 24 hours) and start time (e.g., 6:00 AM, 12:00 PM)
- Unified Command's expectations of the incident management team
- Documentation of the UC decisions, including limitations and constraints

Unified Command Develop/Update Objectives Meeting

At the beginning of every operational period, the Unified Command will meet informally to review current objectives and may add new ones or modify existing ones. This is a time for the UC to assess the incident situation and determine the future direction of the response. As the PSC, you should personally attend this meeting and take notes on the Unified Commanders' decisions and tasking and provide an update on any open action items. It's also important for the Operations Section Chief to attend to give the most current operational update. Following this meeting:

- Brief the Planning Section staff on objectives and decisions
- Provide documentation to the Documentation Unit Leader (DOCL)
- Have the Situation Unit Leader post objectives and decisions
- Prepare for the Command and General Staff Meeting

Command and General Staff Meeting

This meeting is an opportunity for the Incident Commander/Unified Command to meet with their command and general staff and brief them on decisions, assign tasks to individual members, discuss current situation, and resolve any potential problems among staff members that may be impacting teamwork. You need to be flexible with the agenda of this meeting as it might change from meeting to meeting based on the desires of the IC/UC. It's important to capture decisions and directives and display them in the ICP. Some example decisions might be:

- The UC will review and approve all media information dissemination
- 24-hour operational period will be used running from 0600 to 0600
- The OSC will be the local fire department chief

As the PSC, it is your responsibility to:

- Facilitate the meeting
- Document any work tasks and note who is responsible for completing them
- Resolve conflicts and clarify roles and responsibilities before meeting is adjourned
- Brief the Planning Section staff
- Provide documentation to the DOCL

Preparing for the Tactics Meeting

During the time between the Command and General Staff Meeting and the Tactics Meeting, the Operations Section Chief (OSC) should think about and develop a rough draft of their tactical plan in support of the objectives for the next operational period. It's very easy for the OSC to become completely focused on the immediate operational needs of the incident so that they do not spend time to carefully think about future needs. As the PSC, you want to help the OSC put together the plan of action that they will present at the Tactics Meeting. You want to ensure that the Safety Officer also provides input on the chosen tactics. To do that:

- Meet with the OSC to discuss his or her strategies (how he's going to accomplish an objective) and the tactical plan (the equipment and personnel he requires to actually implement the strategy)
- Help the Operations Section Chief fill in the Operational Planning Worksheet, ICS-215 (read Chapter 6 on the OSC to learn more about the Operational Planning Worksheet)
- Ensure that the Safety Officer (SOF) reviews the proposed tactics

Take the time to rough out the future tactical plan and document potential strategies before the meeting to significantly reduce the length of the Tactics Meeting.

Tactics Meeting

Although the Tactics Meeting is driven by an agenda and kept on track by the PSC, it's an informal meeting between the participants to help the OSC develop a workable tactical plan that meets the IC/UC objectives.

As the PSC, it's your job to ensure that this meeting occurs on time and that the following items are completed before the meeting adjourns:

- Completed draft of the Operational Planning Worksheet, ICS- 215
- Operations organization is identified and diagramed
- Any logistical requirements are identified
- The Safety Officer has worked with the OSC to identify and mitigate any safety concerns

- The OSC's tactical plan has been "scrubbed" against the IC/UC objectives and priorities to ensure no objective was overlooked

Failure to Hold a Tactics Meeting Will Result in:

- A forced discussion of strategies and tactics in an open forum
- Promoting tedious and lengthy planning meetings that waste the IC/UC time
- Promoting excessive external influence because the job isn't getting done
- Promoting a perception of disorganization, undermining the confidence of the organization of the IC/UC

Preparing for the Planning Meeting

Time is a limited commodity in an emergency response operation and one way to save time is to ensure that a high degree of preparation is conducted before a meeting or briefing occurs. Time spent in advance of the Planning Meeting will ensure that the IC/UC is presented with a tightly run meeting that focuses on the essential information that the team requires to make informed decisions. Honor the IC/UC time and that of your peers in the command team by facilitating a well-scripted meeting. Make sure that you:

- Talk to the participants who will be at the Planning Meeting to determine if there are any "hot" issues that will come up during the meeting and verify that they're prepared to discuss them
- Have the Operational Planning Worksheet, ICS-215 cleaned up and ready for the OSC's briefing
- Verify the Situation Unit Leader (SITL) is prepared to brief

Planning Meeting

It's appropriate that the Planning Meeting is the final *step* at the top of the Planning "P" before the "P" heads back down. It is symbolic of the fact that this meeting serves as the pivot point in the ICS Planning Process. It's here that the IC/UC and Command and General Staff reconvene in a formal setting after the earlier Command and General Staff Meeting that occurred hours earlier. If you've prepared right, there should never be any surprises during this meeting. You want the IC/UC to see a united team that can support the Operations Section Chief's tactical plan for the next operational period. It's at this meeting that:

- Final touches are put on the ICS-215
- The IC/UC gives tentative approval of the plan
- You obtain concurrence from the Command and General Staff that they can support the plan
- You confirm the availability of resources
- You document decisions and any tasking by the IC/UC

Incident Action Plan Preparation

The Incident Action Plan (IAP) is the culmination of all the effort the command team has expended in going through the ICS Planning Process. The IAP is a written plan that describes the tasks that must be performed during the next operational period and the resources necessary to perform those tasks in order to achieve the Incident Commander/Unified Command objectives. As the PSC, you're responsible for supervising the preparation of the IAP, ensuring all components of the IAP are completed. This requires working with the:

- Safety Officer
- Communications Unit Leader
- Medical Unit Leader
- Situation Unit Leader
- Resources Unit Leader
- Documentation Unit Leader
- Operations Section Chief

After the IAP is complete:

- Review the completed IAP for correctness
- Provide the IAP to IC/UC for review and approval
- Once approved, have copies of the signed IAP made in preparation for the Operations Briefing
- Provide the DOCL with the original signed IAP for inclusion into the permanent incident records

Operations Briefing

The Operations Briefing is the first time those who are responsible for carrying out the assignments outlined in the IAP will hear what it is that they have to do and what resources they'll be given to do it. Remember that up until this time few people in the command team have seen the contents of the IAP. The quality of this briefing is primarily dependent on the OSC, but you as the PSC will help to ensure that the briefing goes smoothly:

- Distribute a copy of the IAP to each branch director, division/group supervisor, staging area manager, and others with field supervisory assignment responsibilities
- Document any changes to the IAP made at the briefing

Execute Plan and Assess Progress

Once the operational period starts, the plan outlined in the IAP must be constantly assessed to ensure that it still meets the actual situation on the ground. Should the OSC adjust the plan, ensure that any critical changes are communicated to those who need to track them. For example:

- The IC/UC for changes in the tactical plan that will delay completing objectives
- The Resources Unit for any changes in a resource's location or status
- The Situation Unit for any observed changes in the situation that were not anticipated

Appendix B has a one-page summary of the PSC responsibilities during the various *steps* of the ICS Planning Process. The page contains the Operational Planning "P" and calls out specific planning activities that we presented above.

Good Meeting Practices

Experience shows that there are some steps you can take to ensure that you conduct crisp, well-executed meetings. Early on in the incident, the meetings that you facilitate are conducted under great stress and anything that you do to help alleviate that pressure is good for the whole team, so keep the meetings on target. To do that, post some basic meeting rules and review them with attendees. For example:

Rules
☐ Cell phones and pagers off or on vibrate
☐ Radios off
☐ No text messaging
☐ Stick to the agenda
☐ No sidebar conversations

Agenda - Post the meeting agenda and stick to it. Each meeting in the ICS Planning Process is conducted for a certain purpose and the agendas vary from meeting to meeting (see Appendix K, for all Meeting and Briefing Agendas in the ICS Planning Process). An example agenda that's used in the Unified Command Develop/Update Objectives Meeting:

> **Unified Command Develop/Update Objectives Meeting Agenda**
>
> 1. PSC brings meeting to order, conducts role call, covers ground rules, and reviews agenda
> 2. Develop or review/select objectives
> 3. Develop tasks for Command and General Staff to accomplish
> 4. Revalidate previous decisions, priorities, and procedures
> 5. Review any open actions from previous meetings
> 6. Prepare for the Command and General Staff Meeting

Open Action Items - We encourage you to establish an Open Action Items list to record any tasking that the IC/UC may assign. Each item on the List should have someone assigned to complete the task. The Open Action Items list does two important things: (1) it lets the IC/UC know that their concerns are being captured and addressed; and (2) the list ensures accountability for the tasking, enabling you as the PSC to monitor progress toward completion. The Open Action Items list is reviewed at subsequent meetings with those items that have been completed, checked off. It's a great tool to help keep track of the myriad issues that must be resolved. An example of some Open Action Items that may come out during a meeting and who is responsible for completing each one:

> **Open Action Items**
> ☐ Establish a Joint Information Center by 0900 (Public Information Officer)
> ☐ Bring in a Critical Incident Stress Management team (PSC)
> ☐ Develop a Disposal Plan (PSC)

Incident Action Plan (IAP)

The IAP is the primary vehicle for providing responders with the direction necessary to ensure that the tasks they're asked to perform are in alignment with the objectives and priorities established by the IC/UC. Remember that management by objectives is one of the principles of the ICS and the IAP is built to ensure that everyone's efforts are supporting the IC/UC.

The IAP is a formal document that becomes part of the incident's permanent record and is good only for the operational period for which it is developed. Whether the operational period is 12 hours or 6 days, at the end of that time another IAP must be ready to go or the incident is deemed over.

The Planning Section is responsible for the development, preparation, and distribution of the plan. The following are things to keep in mind as you develop the IAP:

- The plan must be accurate
- Distribute the IAP in a timely manner
- Prepare an IAP for each operational period
- The first IAP will take the longest to prepare
- Resist information overload – keep the IAP brief
- Determine who needs a copy of the IAP and keep a distribution list
- Maintain initial naming conventions to reduce confusion

Contingency Plans (Asking "What If?")

Be prepared for the unexpected. One of the important responsibilities of the PSC is to look far beyond the current problem and identify potential problems on the horizon. Examples of contingency plans may include:

- Severe Weather Change Plan
- Secondary Explosion Plan

Managing the Planning Team

Way back in paragraph one of this chapter we listed some attributes that a Planning Section Chief should have to be successful. One of those attributes was to be a good team builder and for good reason. The Planning Section relies on teamwork and it's your responsibility to keep the team together. The Planning Section will most likely be comprised of individuals from different agencies and there's a good chance that you'll not know everyone personally. There are steps you can take to build the team and keep them moving toward a common goal:

- Hold daily Section briefs if on a 24-hour operational period or a least one brief a shift. This short meeting will help you get the "pulse" of the planning team
- Ensure each new member is given a brief by their supervisors
- Conduct a daily self-evaluation of the Planning Section
- Set your expectations and communicate them to the planning team
- Make adjustments as necessary

Daily Self-Evaluation of the Planning Section

This evaluation is internal to the Planning Section and serves to help improve service within the Section and to the entire command team. Completion of this evaluation is the responsibility of the Planning Section Chief.

Planning Section Chief

- Is the Planning Section being the leader on this incident in ICS management?
- Have I (PSC) provided clear direction with good follow-up?
- Is the Planning Section functioning as a team?
- Is the Planning space neat and orderly?
- Are we producing the highest level of quality achievable?
- Are we conducting daily Planning Section meetings to discuss internal issues?

Planning Issues

- Do we have good familiarity with the situation on the ground?
- Are reporting deadlines established and being met?
- Are we aggressively debriefing off-going branch directors and division/group supervisors (DIVS) after every operational period?
- Are we communicating with the other Sections?
  - Are they satisfied with our performance?
- Has a meeting schedule been developed?
- Are we effectively utilizing our technical specialists?
- Are we adequately staffed to support incident operations?
- Is there a documentation filing system set up?
- Do we have adequate supplies?
- Is resources status being kept up-to-date?
- Is the Documentation Unit being fully supported by the entire command team?
- Are the situational displays up-to-date?
- Are we delivering a quality IAP?

Personnel Issues

- Are we providing for our well-being (safety, rest, food, etc.)?
- Are ICS-214s (Unit Logs) being kept up-to date?
- Are timesheets being submitted?

Demobilization Issues

Have we thought about demobilization yet?

- What will be the procedures?
- What will be the priorities?
- Have we discussed demobilization with the other Section Chiefs?

Planning Section Daily Work Schedule

The *time a task should be accomplished* in the Daily Work Schedule will be based on the operational period established by the Incident Commander/Unified Command.

Time task should be accomplished (e.g., 9:00 AM)	ICS Position Responsible for the task	Task
	PSC, SITL	Attend daily Operational Briefings
	RESL	Verify T-cards (or other resource tracking system) accuracy for the operational period
	SITL	Assure that a system for daily reporting from field observers is established. Ensure up-to-date meeting schedule posted.
	DOCL	Maintain meeting/activities schedule
	PSC/DOCL	Attend the Command and General Staff Briefing. Initiate Open Action Tracking.
	Entire Planning Department	Daily Planning Section meeting focused on the day's expectations and Planning Section self-evaluation.
	SITL	Debrief off-coming personnel (DIVS, Branch Directors, and Leaders) on accuracy of previous operational period Incident Action Plan, progress made, and any hindrances.
	SITL	Ensure that the most current field reports are submitted to Planning for inclusion into the development of the IAP. Develop predictions on the spread and impact of the incident.
	SITL	Update situational maps and displays with accomplishments. Post a copy of the incident map, Incident Action Plan, and other incident information in a public location to keep responders informed on the overall incident.
	RESL, SITL, DOCL, DMOB	Clean up work areas.
	PSC, RESL, SITL	Prepare for the Tactics Meeting
	PSC	Conduct the Tactics Meeting
	RESL	Coordinate with Finance/Administration Section to verify resources that are on the incident.
	SITL, RESL	Prepare for the Planning Meeting to include wall-size ICS-215 (Operational Planning Worksheet), incident map/chart, situation briefing notes, etc.
	PSC	Conduct the Planning Meeting.
	PSC, SITL, RESL	Following the Planning Meeting assemble the IAP.
	PSC, DOCL	Have the IAP reviewed and signed by the Incident Commander/Unified Command. Make copies of the IAP, labeled and ready for distribution.
	SITL	Prepare the ICS-209 (Incident Status Summary)

Resources Unit Leader

As the Resources Unit Leader (RESL), you're responsible for maintaining the check-in, current status (assigned, available, or out-of-service), and location of all personnel and tactical resources at an incident. The RESL tracks resources from the time they arrive on an incident and check-in, until the time they're properly demobilized from the incident. The definition of resources in the Incident Command System includes both personnel and major items of equipment. If, as the RESL, you're queried as to the availability of ice chests, send the requestor to the Supply Unit and provide them with a dog-eared copy of this book.

- *Assigned Resource:* A resource that has checked into the incident and is currently tasked to support incident operations.
- *Available Resource:* A resource that has checked into the incident and is available for tactical assignment, but is not yet tasked.
- *Out-of-Service Resource:* A resource that has checked into an incident, but is unable to be assigned due to mechanical reasons, resting of personnel, or other reasons that prevent it from engaging in operations.

The Resources Unit Leader position is perhaps one of the most challenging positions within the ICS organization. Without an efficiently run Resources Unit, an Incident will not be able to operate effectively. As the RESL, you must stop unaccounted-for resources or you'll never be able to support the ICS Planning Process.

In your position as the Resources Unit Leader, you're reliant on everyone else involved in the response to support your resource tracking needs. You usually get on-scene well after the initial response personnel and equipment, and from that moment on, you find yourself in a race to catch up. As rapidly as possible, you must get an accurate "picture" of what resources were on the incident before you arrived and what resources are currently arriving. To the uninitiated this may sound easy, but in a large, rapidly evolving response, it requires tenacity, patience, and above all, teamwork.

The degree to which the RESL interacts with other members of the command team varies, but the most important relationship is between the RESL and the Operations Section Chief. The organization chart in *Figure 7.7* highlights the many ICS positions that the RESL will either support, by providing them necessary resource status information to do their job, or rely on, to receive information. The positions with small boxes next to them are the primary positions with whom the RESL interacts. ICS is all about teamwork.

ICS Organization Chart

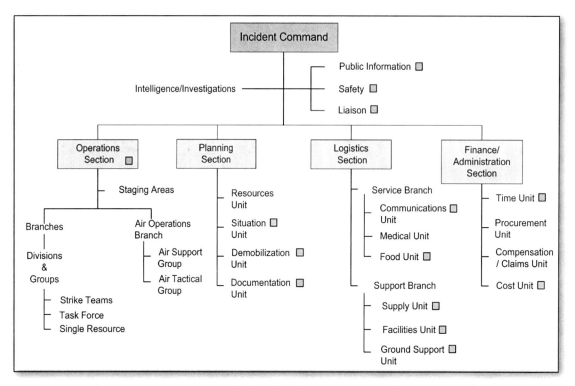

Figure 7.7. The Resources Unit must work closely with many command positions to effectively carry out its responsibilities.

You have just arrived at an incident as the Resources Unit Leader, now what? Either the Incident Commander (IC) or the Planning Section Chief (PSC) will determine if the Resources Unit should be activated. When you arrive at the incident there are some key things that you'll need to do fairly quickly:

- Obtain a briefing from the IC or PSC
- Select a work location
- Determine staffing needs
- Order your staff
- Establish Check-in locations
- Determine what resources are already in the field (field verification)
- Establish a Resources Status Display

Obtain a Briefing

To get a good start on establishing an effective Resources Unit, get as detailed a briefing as possible when you report to the incident (see *Figure 7.8*).

Resources Unit Leader Briefing Checklist

At a minimum the briefing should include:

- ☐ Incident situation: magnitude and potential of the incident
- ☐ Command structure (single or unified)
- ☐ Resources currently working the incident and their assignment (e.g., Search Group)
- ☐ Any established check-in locations
- ☐ Operational period
- ☐ Agencies and jurisdictions involved
- ☐ Resources ordered
- ☐ Resource requesting process

Make sure you request a copy of the ICS-201, Incident Briefing Form (pages 3 and 4) or other documents that can help you begin to develop an accurate resource "picture."

Figure 7.8. Use a checklist to ensure that your briefing covers the information you need to get the Resources Unit up and running.

Ideally, this briefing will provide you with a fairly accurate picture of what resources are currently working the incident along with their incident location, what additional resources have been ordered, and whether any check-in locations have been established. The few minutes spent getting a good brief is time well spent. If you're working for an experienced IC or PSC, he or she will most likely have started documenting the resources already on-scene. Ideally, an ICS-201, Incident Briefing Form, or a similar document, is available (see Appendix F for an example of this form), but the most important thing to do is to get accurate information, any way you can. You do not want to start from zero. Armed with information, you can begin to establish your unit and start to put together an accurate resource "picture."

Select a Work Location

To be efficient, the Resources Unit requires adequate working space in which to display a current and accurate "picture" of all incident resources. In a large and expanding incident there may be hundreds of resources that will have to be tracked and displayed. Your incident-management experience combined with the briefing you receive will help you to anticipate the growth in incident resources and allow you to make an educated guess as to your workspace requirements. If possible, establish your unit in a way that protects the displays from unwanted handling, but allows others on the team to readily view them.

Determine Staffing Needs

Take a few minutes after your briefing to determine the size of the staff that you need to run your unit. There's going to be lag time from when you order your staff to their actual arrival on scene. Without them, you're limited in what you can do.

Experience is the best guide for determining your staffing needs, but if you're new at the game or a bit rusty, the guidelines in *Figure 7.9* provide you with a good reference to help determine the number of personnel necessary to successfully manage all the responsibilities of the Resources Unit. The number of check-in recorders is not fixed because it depends on the number of check-in locations established and the number of check-in recorders needed at each of those locations.

Resources Unit Staffing Guide (per 12-hour period)

Resources Unit Position	Size of the Incident (Number of Divisions/Groups)				
	2	5	10	15	25
Resources Unit Leader	1	1	1	1	1
Status Recorders	1	2	3	3	3
Check-in Recorder	As needed				
Total Staffing	2	3	4	4	4

Figure 7.9. The Resources Unit Staffing Guide.

Order Your Staff

On small incidents the Resources Unit Leader may perform all of the unit's duties with only minimal support; however, in a large incident the Resources Unit Leader requires substantial support especially during the early stages of the response when large quantities of resources are responding to the incident.

It's easy to get caught up in a response that is rapidly changing. As a Resources Unit Leader, you're really a manager responsible for maintaining the status of all resources at the incident. Your staff will be comprised of Check-in Recorders and Status Recorders (*Figure 7.10*). Check-in Recorders and Status Recorders are interchangeable and you can shift them back and forth based on the needs of the incident. Although this book does not cover all the duties of the Check-in and Status Recorders, here are the highlights.

Check-in Recorders, as the name implies, are responsible for ensuring that all resources are properly accounted for (checked in) as they arrive at the incident. Later in this chapter you'll find the information that Check-in Recorders capture from incoming resources. Status Recorders, as the name implies, maintain the resource status display (later in this chapter you'll read about the role of the Status Recorder).

Resources Unit Organization

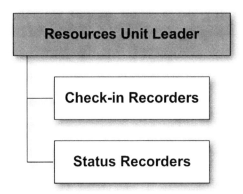

Figure 7.10. The Resources Unit organization.

Establish Check-in

The first step in organizing resources is to make sure that resources are properly checked in to the incident. During the early stages of a response when large numbers of resources are arriving, check-in locations are usually established at many different locations to handle the influx of resources. Check-in locations include the following five ICS support facilities:

- Incident Command Post
- Staging Areas
- Helibase
- Base
- Camps

Once the response shifts from a reactive mode, where resources are first arriving on-scene and then being given an assignment to the proactive response, where operational planning drives the demand for resources, check-in should ideally occur only at one or two different locations such as the incident base or Incident Command Post.

The ICS-211, Check-in Form, is the first of several forms that you'll be introduced to in this chapter. Even though we're in the computer age, our forests are shrinking, and using paper is often viewed as archaic, it's important that you understand the information that needs to be captured when resources first arrive on scene. Armed with the knowledge of necessary check-in information, you can confidently use paperless systems without losing the years of empirical experience that went into developing the ICS-211. *Figure 7.11* is an example of an ICS-211 Check-in Form.

Check-in List ICS-211

CHECK-IN LIST (ICS-211)					1. INCIDENT NAME Meridian Flood		2. CHECK-IN LOCATION 4th Street Staging						3. DATE/TIME 11-15 1230
CHECK-IN INFORMATION													
4. LIST OF PERSONNEL (OVERHEAD) BY AGENCY NAME- OR LIST OF EQUIPMENT BY THE FOLLOWING (see note below)*: S=Supplies H=Helicopter O=Overhead C=Crew E=Equipment D=Dozer A=Aircraft VL=Vessel * If the resource does not fit one of the above categories, make sure that whatever abbreviation is used is documented and used consistently throughout the response (e.g., VL=Vessel)					5. ORDER/ NUMBER	6. DATE/TIME CHECK-IN	7. LEADER'S NAME	8. TOTAL NO. PERSONNEL	9. INCIDENT CONTACT INFORMATION	10. HOME UNIT	11. METHOD OF TRAVEL	12. INCIDENT ASSIGNMENT	13. OTHER QUALIFICATION	14. TIME SENT TO RESTAT
AGENCY	ST/TF	KIND	TYPE	RESOURCE IDENTIFIER										
FRAM		C		MESA #3	C-008	11/15 1230	B. RIGGS	21	555-2310	SPRINGS	BUS	SEARCH GROUP		1330
BFD		O		P. Robert	O-020	11/15 1305		1	555-0909	ARCATA	AIR	DIVS	Staging Manager	1330
USCG		VL		44120	VL-023	11/15 1310	KAMMER	4		EUREKA		SEARCH GROUP		1330
PHL		E	2	AMBULANCE #1	E-025	11/15 1315	S. MILLER	2	555-3891	ARCATA		SEARCH GROUP		1330

Figure 7.11. ICS-211 Check-in Form.

Check-in of resources includes recording:

- ***Agency:*** This is the agency from where the resource came.
- ***ST/TF:*** In this block of the ICS-211, you want to note whether the resource that is checking in is a Strike Team (ST) or a Task Force (TF). For example, for a Strike Team place ST in the block.
- ***Resource Identifier:*** This is the unique identifier of the resource that distinguishes it from other resources. For example, the crew in *Figure 7.11* is called the MESA #3 crew. If the resource is a person, then that name is recorded as shown in *Figure 7.11* with P. Robert.
- ***Kind:*** Is the resource a crew (C); vessel (VL); equipment (E); etc.? There are many kinds of resources.
- ***Type:*** If the resource checking into the incident is *Typed*, record that information in this block. In *Figure 7.11* you can see where the Check-in Recorder placed the number "2" next to the resource identified as Ambulance 1.
- ***Order Request Number:*** When resources are requested, they should receive an order request number. Currently, outside the wildland fire discipline this is not a well-developed system, however, the National Incident Management System recognizes the value of using a resource order number to track resources.
- ***Date/Time Check-in:*** This is the date and time that the resource (personnel or equipment) arrives on-scene at the incident and checks in.
- ***Leader's Name:*** This is the name of the person that's in charge of the resource; e.g., leader of a 20-person crew or the pilot-in-charge of a helicopter.
- ***Total Number of Personnel:*** This is where the number of personnel who come with a resource is recorded; for example, 4 personnel attached to a fire engine or 2 personnel attached to a law enforcement vehicle.
- ***Incident Contact Information:*** This is where you can record the cell phone number that can be given to the Communications Unit Leader.
- ***Home Unit:*** This is the city or town where the resource is permanently assigned.
- ***Method of Travel:*** This is how the resource arrived at the incident; for example, bus, government vehicle, airplane.
- ***Incident Assignment:*** This is the initial assignment of the resource when ordered to the incident.
- ***Other Qualifications:*** This is where other ICS qualifications are recorded. For example, a Planning Section Chief may also be qualified as an Operations Section Chief.
- ***Sent to Resource Status Recorder (RESTAT):*** This is the time at which the Check-in Recorder sends the check-in information to the Resources Unit.

Without accurate resource information you'll be unable to effectively support incident operations, specifically, the ICS Planning Process.

Determine Resources Already in the Field (field verification)

Establishing check-in enables the Resources Unit Leader to record all resources arriving on-scene, but how about the resources that were already working the incident before check-in was established? It's vital that the RESL verify the kind and type of resource (equipment and personnel) and location of those resources that arrived on-scene prior to the establishment of check-in. To do this, the RESL should conduct field verification. This can be done in several ways:

- Use Field Observers to get the information (see the description of Field Observers in the Situation Unit section of this chapter)
- Send Check-in Recorders when available
- Debrief Supervisors (DIVS) coming off-shift

However it gets done, field verification combined with an effective check-in process will give you, the Resources Unit Leader, confidence in the kind, type, numbers, and location of incident resources. Verification should also include nonoperational resources such as the Supply Unit Leader.

T-card System

When the wildland firefighting community developed ICS, they created a visual method to track the location and status of incident resources called the T-card system. This system is simple in design, inexpensive, extremely portable, and robust. Although there are other ways to track resources, T-cards have been used by emergency responders since the 1970s, and have a proven track record even on the most challenging incidents. The T-card system may seem fairly simple and low-tech, but it works!

It's one thing to stand in front of a class and teach the T-card system and a totally different challenge to try to touch on the basics in a few pages of a book. We use a lot of pictures with the hope that, "a picture is worth a thousand words."

The wildland firefighting agencies developed seven color-coded T-cards to track the most common incident resources encountered on wildland fires (see *Figure 7.12*). Fortunately, several of the T-cards have application in a broad array of emergency response operations, ranging from hazardous materials response to search and rescue. On the back cover of this book you'll find a color version of *Figure 7.12*. Helicopter, aircraft, personnel, and the miscellaneous T-cards are generic to all incidents and can be used regardless of the incident to which you respond. Dozers, engines, and crew T-cards are not as versatile, but can still be used for a nonwildland fire incident. If a particular T-card does not fit neatly into your emergency response, you only have to provide a legend so that everyone understands which cards represent which kind of resource.

For example, in the oil spill response community, the rose-color card has been used to display vessels capable of removing oil from the surface of the water. The RESL is responsible for ensuring that everyone assigned to the Resources Unit understands how to decipher the T-card color code.

Seven T-Card Colors

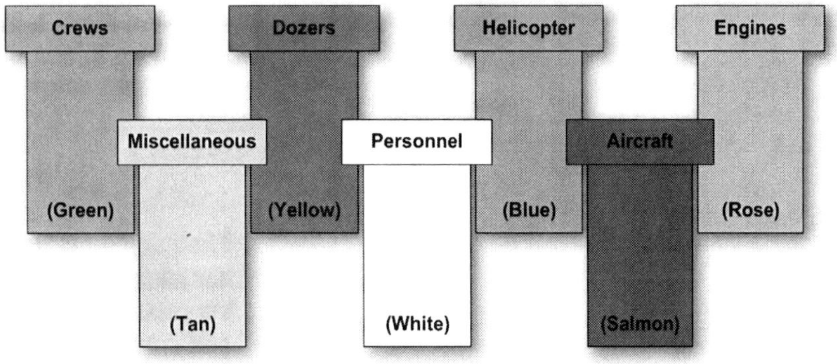

Figure 7.12. The wildland firefighting agencies use seven color-coded T-cards to track incident resources.

Please continue with us through this T-card discussion as it's important. The ICS-211 Check-in form and the ICS T-cards are part of an overall resource tracking system. The resource information captured on the ICS-211 is eventually transferred to the appropriate T-card, which in turn is displayed for all to see. Later in this chapter we'll demonstrate the connection between the Check-in form and T-cards, but first we need to dissect the T-cards so that they do not remain a mystery.

Regardless of the color of the T-card, the information that is placed on them primarily comes from the ICS-211, Check-in form. *Figure 7.13* is an accurate representation of the front side of the rose-color T-card. The layout of each card may differ from the one here, but the differences are minimal.

Front of the Rose-color T-card

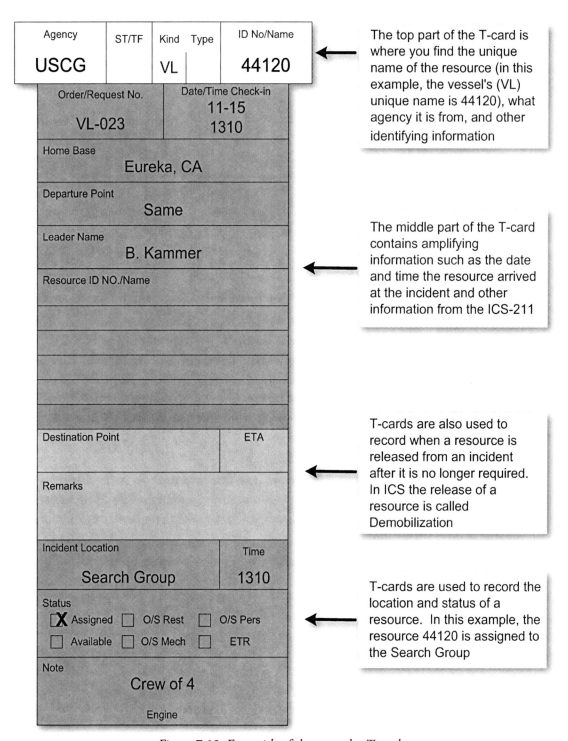

Figure 7.13. Front side of the rose-color T-card.

Every time the status of a resource (Assigned, Available, and Out-of-Service) or its location (Division A, Search Group, etc.) is changed, that change is reflected on the T-card. If the Resources Unit Leader is doing his or her job correctly, a properly maintained T-card will have both the current status and location of the resource as well as a historical record. To avoid confusion, a line is drawn through historical resource status information. *Figure 7.14* shows the backside of the rose-color T-card with both historical information and current information.

Back of the Rose-color T-card

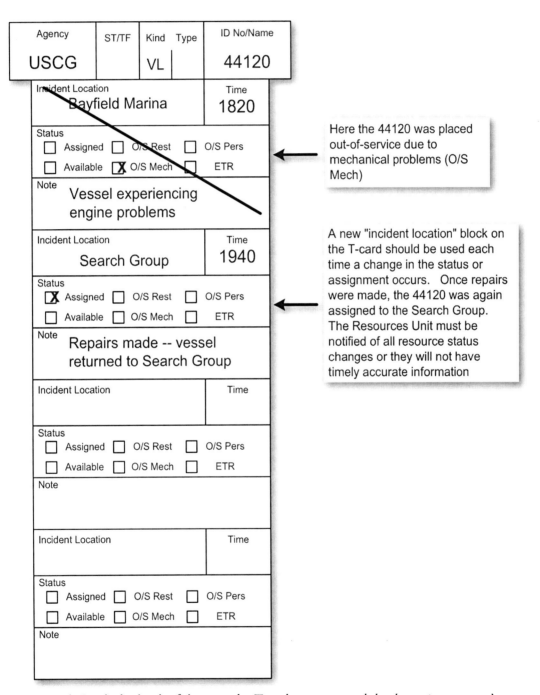

Figure 7.14. On the back side of the rose-color T-card you can record the change in a resource's status or location up to four times before you have to begin with a new card.

Beyond Initial Response   Planning Section Chief

The white-color T-card is used to track what is affectionately, as well as formally, referred to as overhead. Overhead are those responders in management positions such as the Operations Section Chief and those in direct support of management activities such as Check-in Recorders.

In *Figure 7.15* a white-color T-card is used to keep track of P. Robert, a Supervisor. P. Robert checked in to the incident on November 15 at 1305 (1:05 PM) as a supervisor (DIVS). P. Robert was immediately assigned to be the Search Group Supervisor. The T-card also notes that P. Robert is qualified as a Staging Area Manager. Knowing P. Robert has other ICS qualifications enables the RESL to have more flexibility in assigning him on the incident, either as a Supervisor or Staging Area Manager.

Front of the White-color T-card

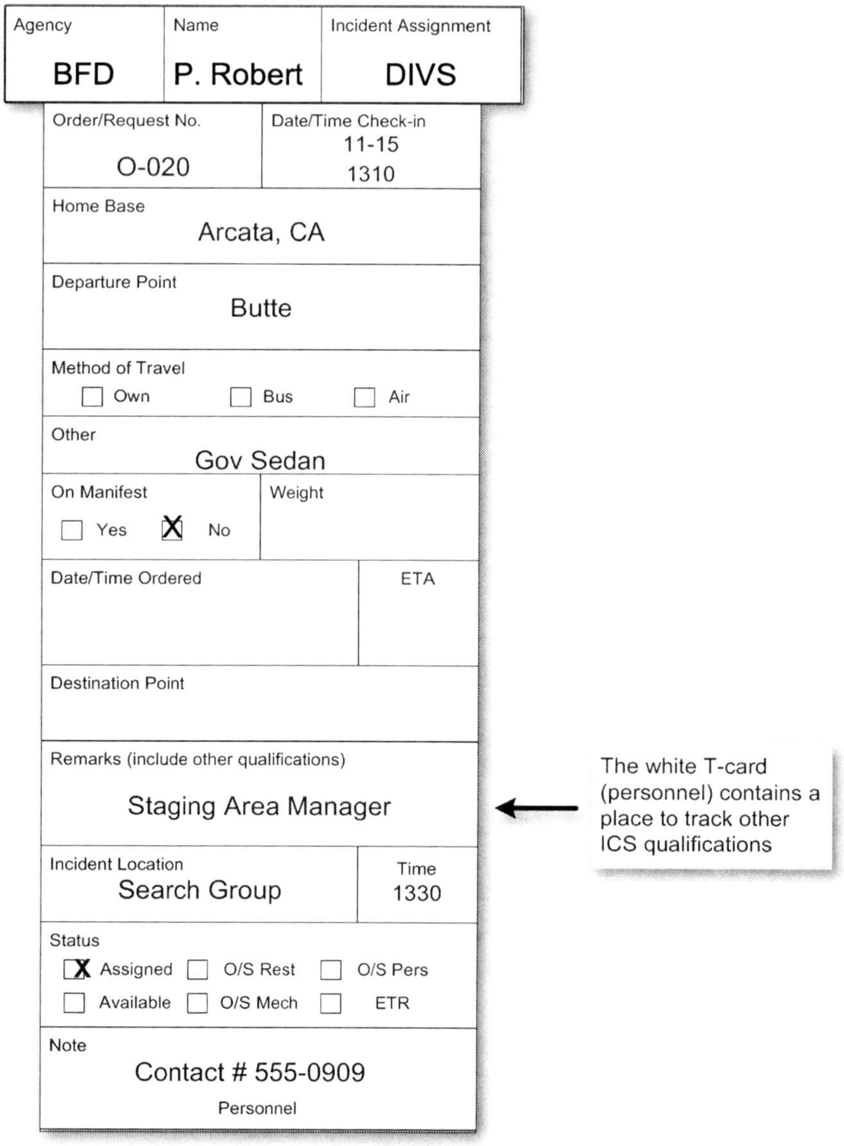

Agency	Name	Incident Assignment
BFD	P. Robert	DIVS

Order/Request No.	Date/Time Check-in
O-020	11-15 1310

Home Base
Arcata, CA

Departure Point
Butte

Method of Travel
☐ Own   ☐ Bus   ☐ Air

Other
Gov Sedan

On Manifest	Weight
☐ Yes ☒ No	

Date/Time Ordered	ETA

Destination Point

Remarks (include other qualifications)
Staging Area Manager

← The white T-card (personnel) contains a place to track other ICS qualifications

Incident Location	Time
Search Group	1330

Status
☒ Assigned   ☐ O/S Rest   ☐ O/S Pers
☐ Available ☐ O/S Mech ☐ ETR

Note
Contact # 555-0909
Personnel

Figure 7.15. The white-color T-card tracks personnel.

As mentioned earlier, T-cards display both current and historical information on a resource's status and location. In *Figure 7.16*, the Search Group Supervisor, P. Robert, came off shift at 2030 (8:30 PM) and was out of service for rest. The following morning, the 16th of November, P. Robert came back on shift at 0500 and was again assigned to the Search Group.

Back of the White-color T-card

Agency	Name	Incident Assignment
BFD	P. Robert	DIVS

Incident Location	Time
Incident Camp	11-15 2030

Status: ☐ Assigned ☒ O/S Rest ☐ O/S Pers ☐ Available ☐ O/S Mech ☐ ETR
Note:

← As soon as P. Robert's new status and location is recorded the old information has a line drawn across it

Incident Location	Time
Search Group	11-16 0500

Status: ☒ Assigned ☐ O/S Rest ☐ O/S Pers ☐ Available ☐ O/S Mech ☐ ETR
Note:

← This is P. Robert's current assignment as of 0500 on 16 November. P. Robert is again assigned to the Search Group

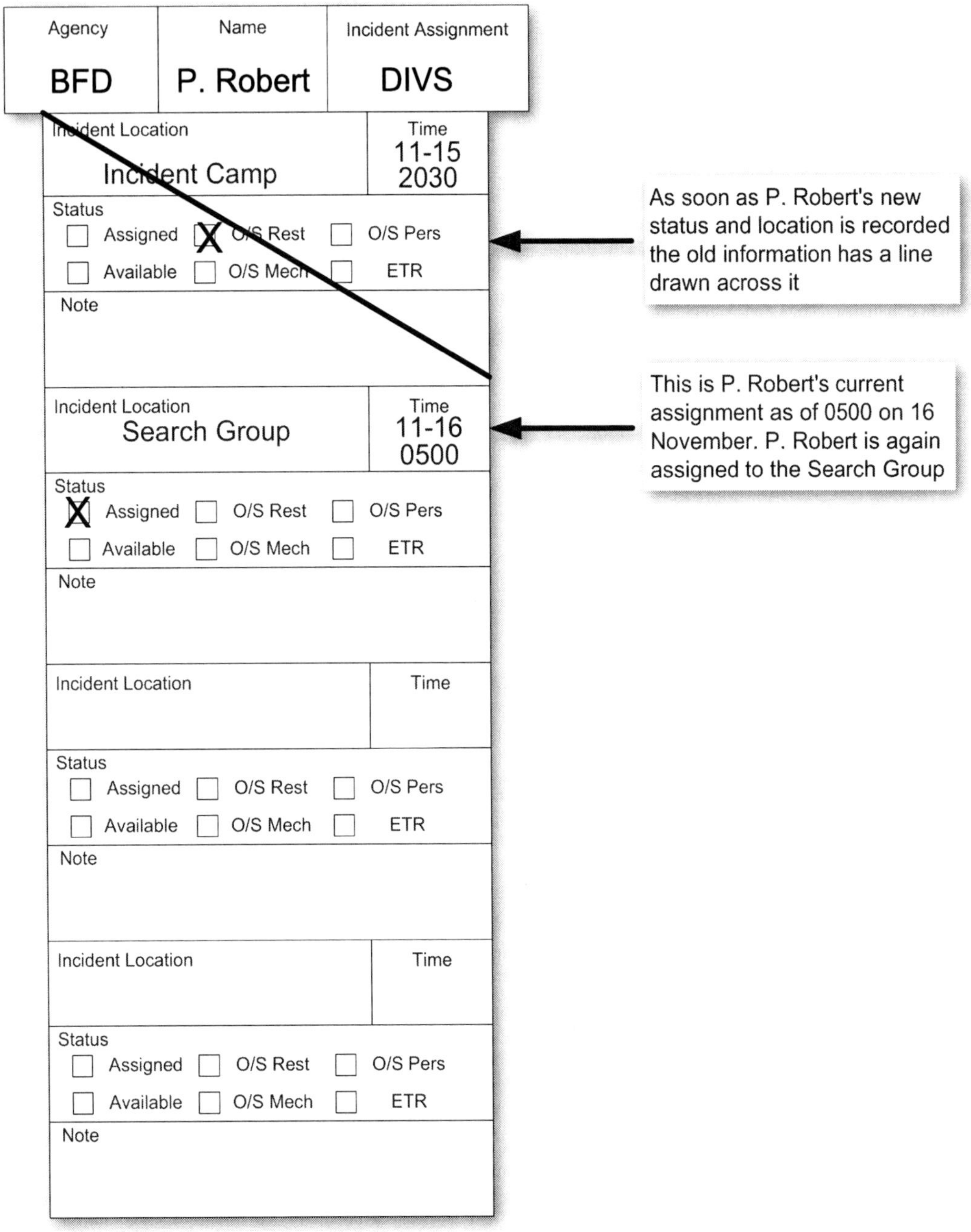

Figure 7.16. Back of the white-color T-card.

T-cards provide both current incident assignments and past assignments. If done properly, you can re-create the history of a resource's assignments or when the resource was out-of-service and not engaged in work. This documentation can be important, particularly in instances such as where financial issues, with regard to when a resource was working, come up.

Although we have only examined two of the seven T-cards used by the wildland firefighting agencies, the information required on each is very similar. The critical challenge is not in filling out T-cards, but that each resource was checked in to the incident and that the information recorded was accurate and complete. This gets back to the importance of teamwork by all of those involved in an incident. Responding resources must fully appreciate their role in the check-in process and the positive impact check-in has on efficient and effective operations.

Transferring Resource Information from the ICS-211 to the T-card

The information recorded on the ICS-211 Check-in form is transferred to the appropriate color T-card. *Figure 7.17* shows where the information that was taken by the Check-in Recorder when P. Robert arrived on-scene is transferred to the corresponding Personnel T-card for P. Robert.

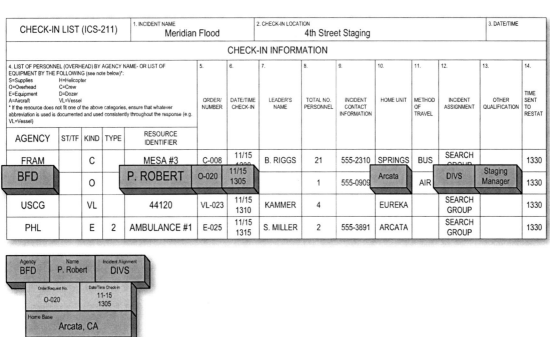

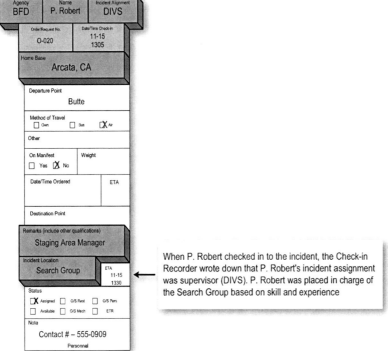

Figure 7.17. Transferring ICS-211 Check-in form information to the appropriate T-card.

Establishing a Resources Status Display

The Resources Status display is the culmination of a process that started with the establishment of check-in at various incident locations to account for any resources arriving on the incident. The quality and thoroughness of the Check-in Recorders in capturing the information on a resource checking in is critical. We have all heard the adage "garbage in equals garbage out," and this is especially true when it comes to check-in information.

The information from the ICS-211 Check-in form is communicated back to the Resources Unit at the Incident Command Post where Status Recorders transfer the information to the appropriate color T-card as illustrated in *Figure 7.17*. Once the T-card is filled out it's placed in the Resource Status Display that shows its location on the incident.

Incident locations are recorded on gray-color cards called Header T-cards or Label T-cards. The naming on Header T-cards can come from a few sources. The Operations Section Chief (OSC) determines how an incident is divided up (read Chapter 6 if you want to know more about the OSC). In *Figure 7.18* the Header cards labeled Division A and Search Group came from the Operations Section Chief. As the Resources Unit Leader, you simply followed the naming provided by the OSC and filled in the Header cards accordingly.

To track overhead personnel in the Incident Command Post the RESL determines how best to display the information, this is a second source for the naming that appears on the Header cards. For example, you might label a Header card *Command*, and place the Incident Commander(s), Liaison Officer, Safety Officer, and Public Information Officer cards under it. For those working in the Planning Section, you may label a Header card Planning and place all overhead personnel working in Planning under that card.

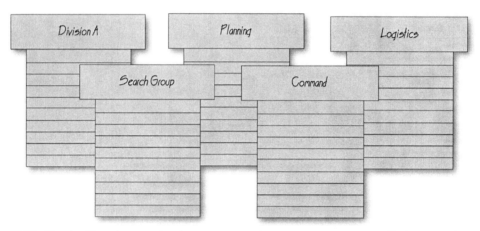

Figure 7.18. Header T-cards are a simple way to organize your resources status display according to the resource's location on the incident. Remember: your job is to maintain an accurate picture of all incident resources, including their location. Others on the team should readily understand your display.

We have come a long way in a few short pages and now we'll pull everything together. In the ICS-211 Check-in form example in *Figure 7.11*, there was an overhead person named P. Robert and a US Coast Guard (USCG) vessel that went by the unique identifier 44120. We have already seen how the information on P. Robert and the USCG 44120 was transferred from the ICS-211 Check-in form to the appropriate T-card. We have discussed Header T-cards and how they work to help you organize your status display.

You have probably noticed that T-cards are shaped like the letter "T." The reason for the distinctive design is so that the T-cards can be slipped into a multipocketed display case leaving the unique resource-identifying information visible (see the top of *Figure 7.13* for an example of unique resource-identifying information). The pockets in the display are similar to those used to hold time cards in businesses that require personnel to clock into and out of work. The displays can be cloth or metal. The remainder of the T-card, which records its status, is hidden from view within the pocket.

On the back cover of this book you'll find a color representation of a partial resources status display. If you look at the status display, you'll find a white T-card with P. Robert's name and incident assignment below the Header card labeled Search Group. You'll also find the Coast Guard vessel 44120 below the same Header card on a rose-color T-card. This matches the ICS-211 check-in information. For this example of a resource status display, the RESL used the yellow T-card to track wheeled vehicles such as ambulances and green T-cards to track work crews. As the RESL, you must provide a legend so that everyone understands which cards represent which kind of resource.

You now have an accurate accounting of incident resources and can support the ICS Planning Process. You, the Resources Unit Leader, are a critical link in the effective and efficient management of the incident.

The resources status display, using the T-card system, is an excellent visual tool for the entire command team. To some, T-cards and T-card displays may be considered an anachronism. We are, after all, in the computer age where paper is often not the medium by which we communicate. Like learning basic math, you have to understand the mechanics before you use a calculator. Our approach in introducing you to the world of resource tracking is based on the same belief. If we can get you to understand the mechanics and nuances associated with resource tracking, then we have given you a firm foundation to intelligently evaluate other systems that you'll encounter in your travels. T-cards have been successfully used for decades to track resources on incidents with thousands of responders.

For the vast majority of incidents in this country a grease pencil on a vehicle windshield may be all that's needed to track incident resources. The use of low-tech T-cards is a quick, inexpensive, and proven way of capturing and displaying resource status and location. It's robust enough to expand and contract as the resource demands of the incident dictate. There are computer programs that have electronic versions of the ICS-211, Check-in form, and that, at the punch of a button, print out a T-card. Other resource-tracking computer programs are totally paperless. Regardless of the system you use, to be effective it should:

- Visually display resources in a manner that is readily understood by the entire team
- Display accurate information for the current operational period
- Be flexible enough to expand and display hundreds of resources

Role of the RESL in Supporting the Operations Section Chief

The Resources Unit Leader and the Operations Section Chief (OSC) must work as a team to ensure that incident resources are accurately accounted for. When the Operations Section Chief reassigns a resource during an operational period (e.g., moves an ambulance from Division A to the Search Group), the OSC notifies the Resources Unit so that the T-card for that particular ambulance can be updated to reflect its new assignment.

The resource flowchart in *Figure 7.19* outlines the process that the OSC follows to fill unanticipated resource requirements. For example, if the OSC recognizes that an additional fire engine is needed to support the tactical operations in the Structure Protection Group, the OSC can:

- Reassign a fire engine from another division or group to the Structure Protection Group
- If a Staging Area(s) has been established, the OSC will check with the Staging Area Manager to see if any fire engines are in staging and reassign them to the Structure Protection Group
- Contact the Resources Unit to see if any fire engines are available and direct the Resources Unit to assign them to the Structure Protection Group. If the Resources Unit does not show any available fire engines on the response, the RESL will order the fire engine for Operations through the Supply Unit in the Logistics Section.

Resource Assignment Flowchart During an Operational Period

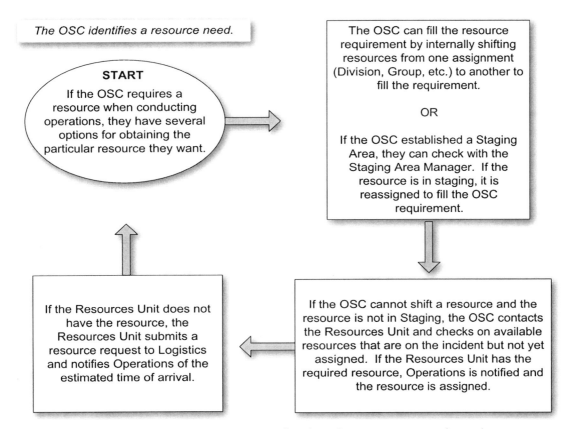

Figure 7.19. Resource assignment flowchart during an operational period.

The vast majority of responses to emergencies are over quickly. A house fire is extinguished with little damage to the structure, two people are safely rescued from their sinking sailboat, the highway is reopened an hour after a severe accident. These emergencies are responded to and dealt with routinely. The complexity and duration of these incidents do not require a separate Resources Unit Leader, Check-in Recorders, or Status Recorders; the person managing the incident performs all of these roles. However, the complexities and duration of an incident can escalate and emergency responders need to be ready to expand their organization to respond. For incidents that are going to take time to bring under control, an organized process for responding is needed to ensure that the response is coordinated, that tactical resources are used effectively, and that operations are being conducted proactively and not reactively. In the Incident Command System this process is called the ICS Planning Process. The Resources Unit has an important role to play in the planning process. In fact, the role directly impacts the effectiveness and efficiency of the response.

The reason you have established a check-in process, conducted a field-verification, and established a Resources Status Display is to support the planning process. As the Resources Unit Leader, you must maintain accurate information on what resources are currently on-scene.

Walking Through the Operational Planning "P"

To walk through your role in the ICS Planning Process, we rely on several pictures. The first picture, seen in *Figure 7.20*, is a visual representation of the ICS Planning Process commonly referred to as the Operational Planning "P." We'll use the Operational Planning "P" as our guide to discuss the various meetings and briefings that the Resources Unit Leader attends and what responsibilities you have during those meetings and briefings.

The Operational Planning "P"

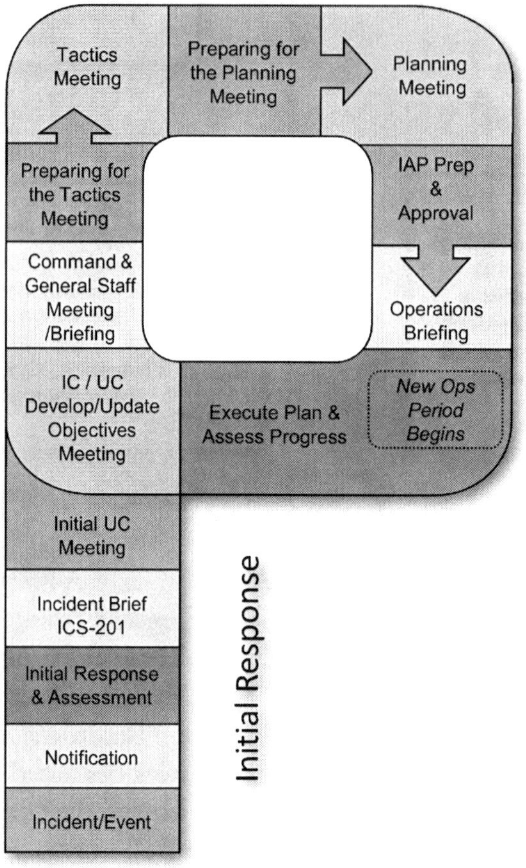

Figure 7.20. The Operational Planning "P" is a visual representation of the ICS Planning Process.

Preparing for the Tactics Meeting

Our first stop on the Operational Planning "P" is preparing for the Tactics Meeting. One of the responsibilities that you have as a Unit Leader is to be prepared when you step into a meeting or briefing. In preparation for the Tactics Meeting make sure that you have verified the accuracy of tactical resources that are on the incident and available for the next operational period.

Tactics Meeting

As the Resources Unit Leader you're an important player in the Tactics Meeting. The way this meeting works is that the Operations Section Chief outlines his or her Operations Section organization (divisions, groups, etc.), and lays out their tactical plan and the resources that they'll require for the next operational period.

The centerpiece of the Tactics Meeting is the completion of a draft ICS-215, Operational Planning Worksheet, which will be finalized at the Planning Meeting, seen in *Figure 7.21*. The Operational Planning Worksheet is unquestionably the "ugly duckling" of the ICS forms, but that belies its elegance as a simple and effective tactical planning tool. The primary ICS positions that attend the Tactics Meeting are: Operations Section Chief, Planning Section Chief, Logistics Section Chief, Safety Officer, Situation Unit Leader, and you the Resources Unit Leader. The RESL is the scribe for the meeting and fills out the ICS-215.

You have been forewarned that the ICS-215 is less than pleasing to the eye, but for the next couple of pages we'll examine the parts of the Worksheet to help you understand how it's used and the role the RESL plays in helping to fill out the form. Along the way we hope you'll appreciate its benefit as a planning tool.

The Worksheet comes in a variety of sizes, but the size most useful for the Tactics Meeting measures 3'x 5' and is affectionately referred to as a horse blanket. This large-size ICS-215 enables all those involved in the Tactics Meeting to easily participate in the discussion and support the Operations Section Chief.

To illustrate how to use the ICS-215, the areas within the dashed lines in *Figure 7.21* have been enlarged in *Figure 7.22* through *Figure 7.26*.

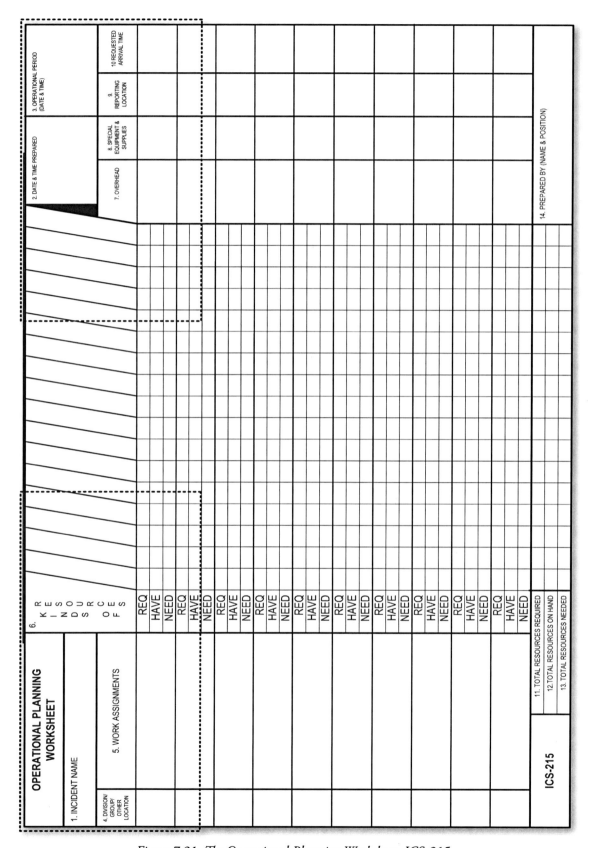

Figure 7.21. The Operational Planning Worksheet, ICS-215.

Beyond Initial Response — Planning Section Chief

To facilitate the explanation of how the ICS-215 works and to illustrate the role of the Resources Unit Leader in filling out the form, only a portion of the ICS-215 will be used. Let's walk through an example as the Operations Section Chief develops a tactical plan for the next operational period in response to a major flood incident.

In *Figure 7.22* the OSC has created a Search Group as part of the operations organization and given the Search Group a work assignment. In addition, the OSC has listed the kinds of resources necessary for the Search Group to accomplish the work assignment. In our example, those resources include: ambulance, helicopter, search team, and rescue boat.

OPERATIONAL PLANNING WORKSHEET		6. KINDS OF RESOURCES	Ambulance (type 2)	Helicopter	Search Team (six person)	Rescue Boat
1. INCIDENT NAME	MERIDIAN FLOOD					
4. DIVISION/ GROUP/ OTHER LOCATION	5. WORK ASSIGNMENTS					
Search Group	Conduct house-to-house search for injured persons. Mark each dwelling searched with a red "X" on the front door. Evacuate injured persons to triage center.	REQ				
		HAVE				
		NEED				

figure 7.22. In the Tactics Meeting, the Operations Section Chief establishes the organization and work assignments for the next operational period.

Once the OSC has determined the work assignment and resources, he or she will list the amount of each resource required (see *Figure 7.23*). In our flood scenario, the OSC requires (REQ) one of each of the resources (ambulance, helicopter, search teams, and rescue boat) as shown in the shaded blocks.

OPERATIONAL PLANNING WORKSHEET		6. KINDS OF RESOURCES	Ambulance (type 2)	Helicopter	Search Team (six person)	Rescue Boat
1. INCIDENT NAME	MERIDIAN FLOOD					
4. DIVISION/ GROUP/ OTHER LOCATION	5. WORK ASSIGNMENTS					
Search Group	Conduct house-to-house search for injured persons. Mark each dwelling searched with a red "X" on the front door. Evacuate injured persons to triage center.	REQ	1	1	1	1
		HAVE				
		NEED				

Figure 7.23. The Operations Section Chief identifies resource requirements for the next operational period. In this example the OSC requires 1 ambulance, 1 helicopter, 1 rescue boat, and 1 search team to accomplish the work assigned to the Search Group.

Once the OSC has laid out the resource requirements for the next operational period, you, as the Resources Unit Leader, will fill in the HAVE block of the ICS-215. The HAVE block contains the number and kind of tactical resources available for the next operational period. In *Figure 7.24* the RESL has listed that there are no helicopters currently on the incident that can be committed to the Search Group for the next operational period (see shaded blocks).

OPERATIONAL PLANNING WORKSHEET		6. KINDS OF RESOURCES	Ambulance (type 2)	Helicopter	Search Team (six person)	Rescue Boat
1. INCIDENT NAME **MERIDIAN FLOOD**						
4. DIVISION/ GROUP/ OTHER LOCATION	5. WORK ASSIGNMENTS					
Search Group	Conduct house-to-house search for injured persons. Mark each dwelling searched with a red "X" on the front door. Evacuate injured persons to triage center.	REQ	1	1	1	1
		HAVE	1	0	1	1
		NEED				

Figure 7.24. The Resources Unit Leader comes to the Tactics Meeting ready to provide resource information to accurately fill in the HAVE blocks on the ICS-215, Operational Planning Worksheet.

The next step we have to cover is to determine what goes into the NEED block on the ICS-215. The NEED block is determined using simple math. It's just the difference between the REQ and the HAVE blocks. In our flood example, the OSC required 1 ambulance, 1 helicopter, 1 search team, and 1 rescue boat and the RESL noted that there were sufficient numbers of ambulances, search teams, and rescue boats, but that there are no helicopters currently on the incident available for the next operational period. This can be seen in the shaded area of *Figure 7.25*. The Logistics Section Chief will attempt to get a helicopter ordered to the incident in time to support the tactical plan that the OSC has laid out on the ICS-215 for the next operational period.

OPERATIONAL PLANNING WORKSHEET		6. KINDS OF RESOURCES	Ambulance (type 2)	Helicopter	Search Team (six person)	Rescue Boat
1. INCIDENT NAME	**MERIDIAN FLOOD**					
4. DIVISION/ GROUP/ OTHER LOCATION	5. WORK ASSIGNMENTS					
Search Group	Conduct house-to-house search for injured persons. Mark each dwelling searched with a red "X" on the front door. Evacuate injured persons to triage center.	REQ	1	1	1	1
		HAVE	1	0	1	1
		NEED	0	1	0	0

Figure 7.25. With the NEED block filled in, the Logistics Section Chief can readily tell the number and kind of resources that have to be ordered to support incident operations.

Finally, the OSC fills in the right side of the ICS-215 (see *Figure 7.26*), where the OSC lists the supervisors and skilled support necessary to conduct the tactical assignment of the Search Group. In our example an Assistant Safety Officer has been specifically assigned to the Search Group. The OSC also notes any special equipment or supplies and a location and time where the Search Group is to be at the beginning of operations.

Ambulance (type 2)	Helicopter	Search Team (six person)	Rescue Boat		2. DATE & TIME PREPARED 15 Nov 2100	3. OPERATIONAL PERIOD (DATE & TIME) 16 Nov 0600 to 16 Nov 1800		
					7. OVERHEAD	8. SPECIAL EQUIPMENT & SUPPLIES	9. REPORTING LOCATION	10. REQUESTED ARRIVAL TIME
1	1	1	1		Supervisor Assistant Safety Officer	Red Spray Paint	4th Street and County Road 502	0530

Figure 7.26. The right side of the ICS-215 is used to record additional information, including specialized equipment and overhead personnel.

With the Tactics Meeting and development of the ICS-215 Operational Planning Worksheet completed, the Resources Unit Leader prepares for the next important meeting in the ICS Planning Process, the Planning Meeting. Prior to the Planning Meeting the RESL must ensure that the Resources Status display is current and that any deficiencies are corrected and, if necessary, that the ICS-215 is updated and made presentable.

Planning Meeting

The Planning Meeting is attended by the Incident Commander/Unified Command and the Command and General Staff. It's at this meeting that the Operations Section Chief presents his or her tactical plan for the next operational period using the ICS-215. The information passed at this meeting must be as accurate as possible and that includes information about the ability of existing resources to support the proposed tactical plan. As the RESL, you have several responsibilities in supporting the Planning Meeting:

- Make changes to the ICS-215 as necessary
- Modify the command objectives as directed
- Confirm the availability of resources to meet the objectives for the next operational period
- Request additional resources for any identified shortfalls

At the end of the Planning Meeting, the Incident Commander/Unified Command will tentatively approve the tactical plan for the next operational period. The Command and General Staff led by the Planning Section Chief then begins putting together an Incident Action Plan. The RESL is central to this effort. Several key parts of the Incident Action Plan are the responsibility of the RESL (see *Figure 7.27*).

Incident Action Plan

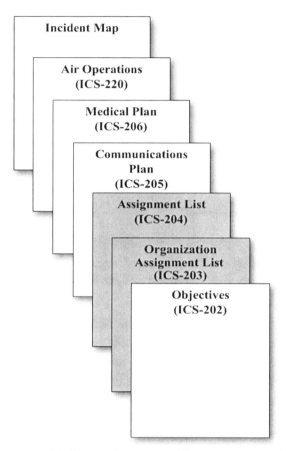

Figure 7.27. The RESL is responsible for completing several key components of the Incident Action Plan.

The Resources Unit Leader is responsible for ensuring that the ICS-204, Assignment List (*Figure 7.28*), and the ICS-203, Organization Assignment List (*Figure 7.30*) are completed. In addition, the RESL collects all parts of the Incident Action Plan and delivers the plan to the Documentation Unit Leader (DOCL) who puts the Plan together for the Planning Section Chief's review. A one-page summary on the Resources Unit Leader's responsibilities during each step in the ICS Planning Process is in Appendix B.

ICS-204, Assignment List

The backbone of the Incident Action Plan is the ICS-204, Assignment List. This form is used by Division and Group Supervisors, Staging Area Managers, and others with supervisory responsibilities in the field, and describes their work assignments (think of the ICS-204 as a type of work order). Remember that the Supervisors (DIVS), Staging Area Managers (STAM), and others were not in the Tactics Meeting or the Planning Meeting so they do not know what was discussed. The Assignment List provides them with:

- Resources assigned to them for the operational period
- Specific instructions on their work assignment
- Who they work for (chain of command)
- Information on any specific safety, environmental, or other concerns
- Communications information (e.g., radio frequencies, cell phone numbers)

Figure 7.28 is a filled-in ICS-204 that's based on the flood example we've been using over the last few pages.

Assignment List (ICS-204)

1. BRANCH	2. ~~DIVISION~~/GROUP Search	ASSIGNMENT LIST
3. INCIDENT NAME **MERIDIAN FLOOD**		4. OPERATIONAL PERIOD (Date and Time) 16 Nov 0600 to 16 Nov 1800

5. OPERATIONS PERSONNEL

OPERATIONS CHIEF _L. Hewett_ ~~DIVISION~~/GROUP SUPERVISOR _P. Robert_

BRANCH DIRECTOR _____ AIR TACTICAL GROUP SUPERVISOR _____

6. RESOURCES ASSIGNED THIS PERIOD

STRIKE TEAM/TASK FORCE RESOURCE DESIGNATOR	EMT	LEADER	NUMBER PERSONS	TRANS NEEDED	DROP OFF POINT/TIME	PICK UP POINT/TIME
MESA #3		B. Riggs	21		0530	
44120		V. Kammer	4		0530	
Ambulance #1		S. Miller	2		0530	
Helicopter 12		T. Troutman	2			

7. ASSIGNMENT

Conduct house-to-house search for injured persons. Mark each dwelling searched with a red "X" on the front door. Evacuate injured persons to triage center.

8. SPECIAL INSTRUCTIONS

Work in teams of two or more at all times. Snakes are suspected to be in the work area, wear snake gators. Personal floatation devices should be worn. Daylight operations only. Conduct regular communications checks. Send hourly updates on the group's progress to the Situation Unit.

9. DIVISION/GROUP COMMUNICATIONS SUMMARY

FUNCTION		FREQ.	SYSTEM	CHAN.	FUNCTION		FREQ.	SYSTEM	CHAN.
COMMAND	LOCAL	CDF 1	King	1	SUPPORT	LOCAL			
	REPEAT					REPEAT			
DIV./GROUP TACTICAL		157.4505	King	3	GROUND TO AIR				

PREPARED BY (RESOURCES UNIT LEADER) A. Worth	APPROVED BY (PLANNING SECT. CH.) J. Gafkjen	DATE 16 Nov	TIME 0400

Figure 7.28. The RESL is responsible for ensuring that the ICS-204s are correctly filled in. The Communications Unit Leader fills in the bottom of the ICS-204, communications summary.

Assignment information on the ICS-204 comes directly from the ICS-215, Operational Planning Worksheet. The IAP will have an ICS-204 for every division, group, and staging area that the OSC lists on the ICS-215. The Resources Unit Leader makes specific resource assignments to meet the requirements dictated by the Operations Section Chief. To help illustrate the connection between the ICS-215 (*Figure 7.29*) and ICS-204 (*Figure 7.28*) we have used various shading. If you move back and forth between these two figures, you'll see that for every resource the Operations Section Chief required, the Resources Unit Leader listed a specific resource to fill the requirement. For example, the OSC required a search team to work in the Search Group. The RESL assigned the search team Mesa #3 to the Search Group to meet the OSC requirement.

OPERATIONAL PLANNING WORKSHEET		6. KINDS OF RESOURCES	Ambulance (type 2)	Helicopter	Search Team (six person)	Rescue Boat
1. INCIDENT NAME	MERIDIAN FLOOD					
4. DIVISION/ GROUP/ OTHER LOCATION	5. WORK ASSIGNMENTS					
Search Group	Conduct house to house search for injured persons. Mark each dwelling searched with a red "X" on the front door. Evacuate injured persons to triage center.	REQ	1	1	1	1
		HAVE	1	0	1	1
		NEED	0	1	0	0

Figure 7.29. The information on the Operational Planning Worksheet is transferred to the ICS-204 Assignment List. The Resources Unit Leader makes specific resource assignments to meet the requirements dictated by the Operations Section Chief.

ICS-203, Organization Assignment List

The ICS-203 serves as a master list of names of personnel who are filling key incident command positions in the Command and General Staff, Unit Leaders in all of the sections, and the supervisory personnel in the Operations Section. In *Figure 7.30* P. Robert has been assigned as the Search Group Supervisor.

Organization Assignment List (ICS-203)

ORGANIZATION ASSIGNMENT LIST		Food Unit	
		Medical Unit	
1. Incident Name **MERIDIAN FLOOD**			
2. Date	3. Time		
		9. Operations Section	
4. Operational Period		Chief	
		Deputy	
Position	**Name**	Staging	
5. Incident Commander and Staff		Staging	
Incident Commander		Staging	
Deputy		**a. Branch I - Divisions/Groups**	
Safety Officer		Branch Director	
Information Officer		Deputy	
Liaison Officer		Search Group	P. Robert
Intelligence Officer		Division/Group	
6. Agency Representatives		Division/Group	
		b. Branch II - Divisions/Groups	
		Branch Director	
		Deputy	
		Division/Group	
7. Planning Section		Division/Group	
Chief		Division/Group	
Deputy		**c. Branch III - Divisions/Groups**	
Resources Unit		Branch Director	
Situation Unit		Deputy	
Documentation Unit		Division/Group	
Demobilization Unit		Division/Group	
Technical Specialists		Division/Group	
Human Resources		**d. Air Operations Branch**	
Training		Air Ops Branch Director	
		Air Attack Supervisor	
		Air Support Supervisor	
8. Logistics Section		Helicopter Coordinator	
Chief		Air Tanker Coordinator	
Deputy		**10. Finance Section**	
Supply Unit		Chief	
Facilities Unit		Deputy	
Ground Support Unit		Time Unit	
Communications Unit		Procurement Unit	
ICS-203	Prepared by Resources Unit		

Figure 7.30. The RESL is responsible for ensuring that the ICS-203 is completed for inclusion into the Incident Action Plan. The ICS-203 is filled out using information from the Resources Status Display (e.g., Search Group Supervisor is P. Robert).

Nonincident Action Plan Documents: Incident Organization Chart, ICS-207

Sharing information with everyone in the command team is part of the Resources Unit Leader's job. The ICS-207 Incident Organization Chart (see *Figure 7.31*) is one way to communicate information. The wall-size version (approximately 3' x 5') is posted on display in the Incident Command Post and contains the names of personnel filling overhead positions such as the Incident Commander, Logistics Section Chief, and Resources Unit Leader. The ICS-207 is not typically part of the Incident Action Plan, but is maintained with up-to-date information.

Resources Unit Leader Involvement in Demobilization

It's equally important that resources properly checked-in and methodically tracked on an incident be released (demobilized) from an incident in the same meticulous and caring manner. Demobilization is an orderly and planned process and the Resources Unit Leader has an important role in ensuring that the process is smooth.

Resources that are scheduled for demobilization are placed under a Header T-card labeled DEMOB. The tracking of resources does not end until the resource has physically departed the incident and is en route to its home (parent) unit or off to another incident. Once the Demobilization Unit Leader has advised the RESL that the resource is released, the T-card is updated with the demobilization information and then it's sent to the Documentation Unit Leader as part of the incident's historical record.

As a Resources Unit Leader, you should always be looking for resources that are checked into the incident but have not been used for several operational periods and discuss demobilizing the resource with the Planning and Operations Section Chiefs. Unused resources cost money.

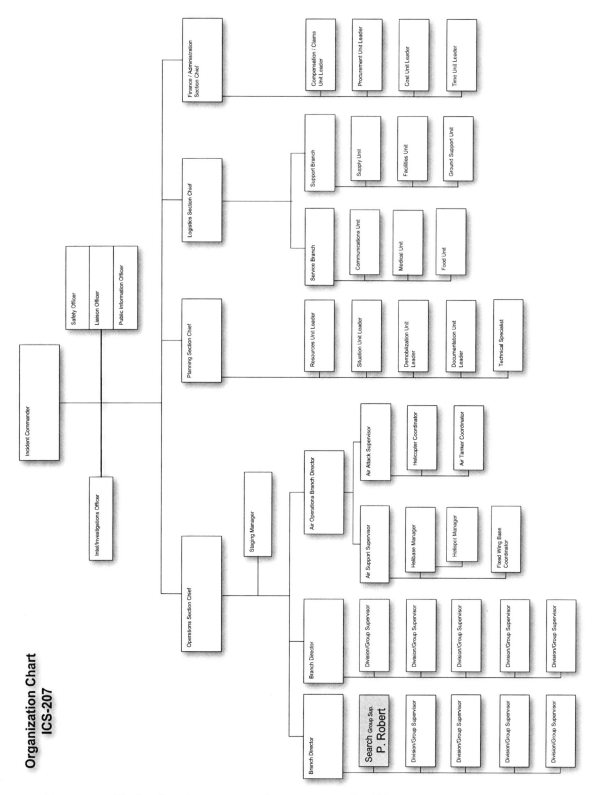

Figure 7.31. The Incident Organization Chart, ICS-207, should be prominently displayed so that personnel entering the Incident Command Post can readily identify the primary management personnel. In this example, you'll notice that the Resources Unit accurately recorded P. Robert as the Search Group Supervisor.

Resources Unit Leader Conclusion

The Resources Unit Leader is a critical position in the ICS. To be successful, the Resources Unit Leader must receive support from others in the team, meaning that if information is not coming to you, you have to go to the information.

No one thinks of the Resources Unit Leader when it comes to the safety of responders, but as the RESL, you have a role in supporting safety as well as tracking resources. Since you assign resources to meet the Operations Section Chief's requirements, you need to be sensitive to responder fatigue associated with difficult or high-tempo operations and make assignments accordingly.

Hopefully, you have a better understanding of how the Resources Unit Leader contributes to the ICS Planning Process, have attained some knowledge of how to actually do the RESL job, and have gained an appreciation for the importance of the Resources Unit Leader.

Situation Unit Leader

Information often drives the small and big decisions that affect our daily lives. Just look at all of the information we gather when purchasing a home: quality of the schools, commuting distance from work, house price, interest rate, condition of the house structure, estimated monthly utilities, and the list goes on. Fortunately, when buying a home we have time to get answers to many of these questions before we commit ourselves to a final decision. In emergency response operations there is no such luxury called "time" and responders are often forced to make critical decisions concerning their course of action with less than ideal information.

If an emergency responder could bring only one thing to an incident, it most likely would be accurate and timely information. Unfortunately, information is not a piece of equipment that can be thrown into a truck and brought on-scene. Information has to be aggressively collected and then properly managed. As the Situation Unit Leader, you are the focal point for incident information collection and dissemination.

ICS knowledge and skills are only part of the equation in making you an effective Situation Unit Leader (SITL). The other parts of the equation are the personal traits you bring with you. As a SITL you should be articulate, a team player, support-oriented, value attention to detail, and be a self-starter who produces high-quality work. You should have excellent communications skills and be a talented briefer. To be an effective SITL you must be able to manage an incredible array of information and distill it into something that can be used by responders. This is especially true of technical information that is often produced by the more scientifically inclined to be used by those at the pointy end of the response spear. The Situation Unit is responsible for gathering incident information and, even more importantly, ensuring that the information is "pushed out" and shared throughout the entire response organization.

As the Situation Unit Leader, you should consider the entire ICS organization as your "informational oyster" and aggressively harvest the information to build an accurate "picture" of the current incident situation. In addition, as the SITL you need to be part soothsayer, predicting what the status of the incident will be in the future based on weather, the impact of mitigation strategies, and other variables. A combination of both current situation and prediction will enable the Incident Commander/Unified Command to develop achievable objectives that guide the direction and priorities of the entire incident management team.

Upon Arriving On-scene

Mr. Chuck Mills, President for Emergency Management Services International, Inc., has more than 40 years' experience in using the Incident Command System and has accumulated a treasure trove of truisms including "What starts right–goes right." Any Situation Unit Leader who takes these five words to heart understands the critical importance of getting it right the first time. When you arrive at the incident there are several tasks you need to accomplish quickly so that you "Start Right:"

- Obtain a situational briefing from the IC or PSC
- Select a work location
- Determine staffing needs
- Establish a Situation Display
    - Incident Information
    - Status Summary
- Conduct formal briefings

Get a Situational Brief

After arriving at the incident and checking-in, you need to find the Planning Section Chief (PSC) or the Incident Commander (IC), if the PSC has not arrived on-scene, and get a briefing. This briefing is critical! Without it you'll be wasting precious time in establishing your unit. The bottom line is to get enough detail so that you can develop an accurate picture of the current situation. Some of the information you need to get from your briefing is provided in the checklist below in *Figure 7.32*.

Situation Unit Leader Briefing Checklist

At a minimum the briefing should include:

- ☐ Current situation
- ☐ Incident potential
- ☐ Location of any incident facilities (e.g., Helispots)
- ☐ Incident objectives
- ☐ Establishment of any Divisions, Groups, Branches
- ☐ Weather information
- ☐ Establish meeting schedule

Make sure you request a copy of the ICS-201, Incident Briefing Form (pages 1 and 2) or other documents that may help you begin to develop an accurate situational picture.

Figure 7.32. The SITL Checklist helps you gather initial information quickly.

If you're working for an experienced IC or PSC, he or she will most likely have started documenting the incident situation, giving you a solid place to start developing the displays necessary to support response operations.

Select a Work Location

The Situation Unit can consume a lot of wall space. Remember that you're in the business of shamelessly sharing information and a big part of that information is shared in the form of maps, charts, and other large displays. So select your work space accordingly. In addition to wall space, the Situation Unit should be near the Resources Unit because the two units work closely with each other. If it's anticipated that the incident size will continue to grow, plan your space needs to meet the increased demands. You do not want to experience the unpleasant feeling of outgrowing your work space and thereby sacrificing the ability to adequately display critical information.

Your Staff

The staff of the Situation Unit is comprised of Display Processor(s), Field Observers (FOBS), and, if necessary, Technical Specialists (*Figure 7.33*).

The Display Processor is the person who keeps all of the situation status board displays up-to-date with incident information. As we move through this chapter, we'll come back to the role of the Display Processor.

The Field Observer is the "eyes and ears" of the Situation Unit. FOBS are sent into the field to gather real-time incident information and report their findings back to the Situation Unit so that they can be analyzed and recorded to help create an accurate picture of the incident.

The Situation Unit may also include Technical Specialists. Technical Specialists have unique skills and experience and can be employed anywhere on an incident. In the case of the Situation Unit, technical specialists might include: Geographic Information Specialists (GIS), weather specialists, or plume modeling specialists.

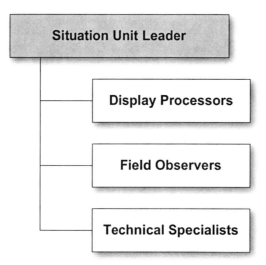

Figure 7.33. The Situation Unit organization is comprised of display processors, field observers and, if necessary, certain types of technical specialists.

Situation Unit Staffing Guidelines (per 12-hour period)

The job of Situation Unit Leader can range from fairly straightforward to immensely complicated, depending on the complexity and magnitude of the incident. One of the first actions you should take upon arriving and checking in to the incident is to determine your staffing needs.

There are no hard-and-fast rules on staffing the Situation Unit. When you try to determine the number of Field Observers you might need to take into account the number of Divisions/Groups where the situation is dynamic and where there is a lack of situational information. If the situation is stable, one Field Observer may be able to cover several Divisions/Groups. As for the number of Display Processors, you need enough of them to keep up with the incoming information. Technical Specialists can range from none to numerous based on the type of incident and its complexity.

We agree that the staffing guidelines for the Situation Unit are not as clear-cut as we would all like, but there are some variables to consider as you make your best determination of what personnel you need to have an effective unit. These variables include:

- Intensity of the operations being conducted
- Size of the incident (is there a large command team in place)
- Complexity of the incident (may require many technical specialists)
- Duration of the incident (need to factor into your staffing needs the ability to manage the Situation Unit 24 hours a day, 7 days a week)
- Number of locations where situation displays are maintained
- Information demand

It is our recommendation that you do not delay the ordering of your staff. Delaying can result in the Situation Unit's inability to support incident operations, resulting in lost opportunities as the command team members struggle to understand the situation they face.

ICS Map Symbology

One of the principles of the Incident Command System is common terminology. Common terminology includes: position titles (e.g., Operations Section Chief), organizational elements (e.g., Officers), facilities (e.g., staging areas); documentation (e.g., ICS-204, Assignment List), and map symbology. We started the discussion of the Situation Unit Leader by telling you that the job of the SITL includes the gathering, analysis, and dissemination of information. One method of communicating incident information is through the symbols used on maps or charts. *Figure 7.34* contains some of the most common ICS symbols used on both small and large incidents.

ICS Map Symbols

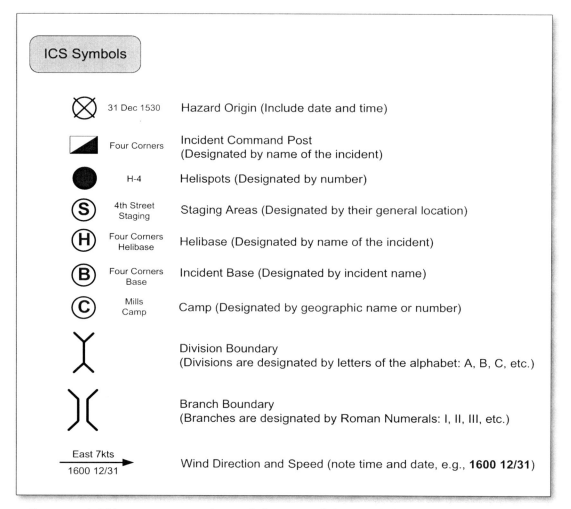

Figure 7.34. ICS common terminology includes map and chart symbols. Some of the most commonly used symbols are displayed along with their naming convention.

A few things to keep in mind with regard to symbols: the Operations Section Chief designates location and naming of Divisions, Staging Areas, Helispots, and Branches. Your job is to ensure that the location and names are displayed on your maps/charts. This is a good place to talk about teamwork. Information such as the number and name of Divisions, Staging Areas, and Branches is equally important to the Resources Unit Leader (RESL). Make sure you take the time to compare notes with the RESL so that you both have the latest information on the Operations Section organization.

You will find that the common ICS symbols do not include symbols for every type of contingency and that your particular response discipline may require additional symbology to better communicate situational status. Your responsibility is to ensure that you clearly display what the symbol represents. For example, in the oil-spill response community the use of physical barriers to prevent oil from spreading, referred to as boom, is an important response strategy. The location of boom is annotated on charts using the symbol in *Figure 7.35*. The SITL did not violate any of the principles of ICS by creating a new symbol specific to his or her response. Rather the SITL recognized that ICS is a flexible system and kept faith with that system by ensuring the symbol for "boom" is prominently displayed on a map legend so that anyone viewing the display knew exactly what the symbol represented.

Figure 7.35. The symbol often used by the oil-spill response community to denote completed boom line. All symbols that are placed on a map/chart must be shown on a legend so that anyone examining the map/chart knows exactly what each symbol means.

Gathering Situational Information

As the Situation Unit Leader (SITL) you must continuously gather accurate and timely information, to ensure that you're providing the response team with the most up-to-date picture of what is happening on the incident. In a fluid response environment it can be a daunting task for you, the SITL, to stay abreast of the most cutting-edge information and you must be aggressive in your quest. There are several avenues that you can use to gain real-time information on the incident: (1) debrief division/group supervisors; (2) talk to technical specialist(s); (3) gather information from meetings and briefings; (4) work with other members of the response team such as the Safety Officer; and, (5) employ your Field Observers (FOBS).

The FOBS are one of your greatest assets for gathering information. They work directly for you and their primary job is to observe and record what is transpiring on-scene and feed this back to the Display Processors. To get the most out of your FOBS we recommend you follow a few simple rules:

- Ensure that the FOBS are knowledgeable in the type of incident for which they are collecting information
- Coordinate the FOBS field activities with the Operations Section Chief (OSC). For safety purposes the OSC must know who is in the field and where they're located
- Ensure that the FOBS are properly outfitted with safety equipment and the tools needed to collect the incident information (maps, radio, transportation, etc.)
- Develop a list of things you would like the FOBS to collect while in the field. For example:
    - Progress of operations
    - Boundaries of the incident
    - Weather
    - Wildlife impacted

o Tactical resources on the incident and their location (work with the Resources Unit Leader to see if they need this information collected ... remember ICS is teamwork)

Make sure to establish a time and method for the FOBS to report their findings. For example, early on in the incident when the situation you're facing is unclear or dynamic you may want information communicated back to you every 30 minutes. The method may be by radio. Don't let the FOBS go into the field until you have set the reporting back expectation.

We recommend that you create written instructions for the FOBS to help guide their actions and to ensure that the FOBS understand what you expect of them. In Appendix L, you can see a sample of written instructions to the FOBS. Much of the information in the sample instruction applies to any type of incident. Make sure to borrow heavily from the sample instructions so that you are not trying to develop a list of instructions from scratch.

Maps and Charts

There is little need to browbeat you on the value of having visual aids as a form of communication. Can you imagine having to build a gas barbecue without the help of the diagrams? (Okay, that's probably a bad example because even with the diagram you can get totally lost, but you get the idea.) The maps and charts you choose to use in displaying incident information must be appropriate for the incident you're facing. For example, you would not want to use a map of the entire state of Colorado for an incident confined to one county located on the western side of the state. The maps/charts must help responders to do their job and the more detailed your displays are for their area of operations, the better. A few thoughts to keep in mind as you establish and maintain your maps and charts:

- Strive for a high-quality presentation
- Ensure accuracy of situational information
- Maintain current information
- Prominently display a map/chart legend
- Establish a method to capture map/chart information for historical purposes
- Date- and time-stamp your map/chart to reflect most recent updates

Figure 7.36 is an example of an incident map. The map contains a legend so that the user can interpret the symbols used to annotate key ICS facilities and other symbols unique to the incident. You'll notice that a non-ICS symbol was incorporated into the legend. This was done because in the oil-spill response community it's important to designate the location of deployed boom on the map display. By placing the symbol for boom on the map legend the SITL ensured that anyone consulting the map can understand the common language used on the map.

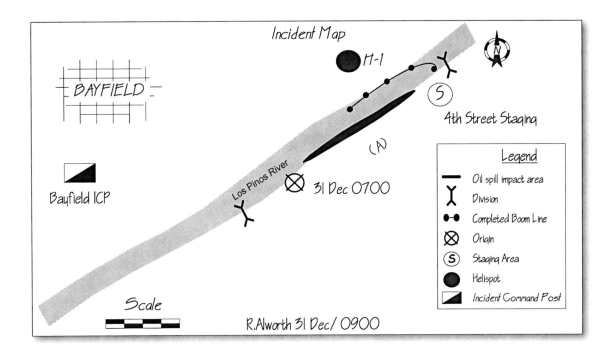

Figure 7.36. The use of common ICS terminology extends to the symbols used on maps and charts. In addition to the common symbols used in ICS, your response discipline may require additional symbols to communicate incident status. Your responsibility as the SITL is to ensure that others on the response are aware of what the symbols mean.

Establishing Situation Displays

All of the effort that you have expended so far is to establish a visual story of what is happening on the incident. This story should include at a minimum:

- The current incident objectives
- Summary of the status of the incident. This includes information on the incident itself (e.g., number of injured, buildings damaged) and information on response resources (e.g., number of ambulances, fire trucks)
- The current situation (e.g., incident boundaries, weather, tides and currents)
- Predictions and potential impacts of what could happen if weather does not cooperate and mitigation strategies do not have the desired outcome
- Schedule of meeting times and locations

Figure 7.37 is an example of how you might consider setting up your displays. The displays should be established in a manner that lets anyone examining them quickly capture the information that they're looking for. That said, access to the displays should be limited to Situation Unit personnel. These displays serve both the responders and are a part of the historical record of what transpired on the incident. The situation display map/chart is used for briefings and meetings and the need for current and accurate information is absolutely essential.

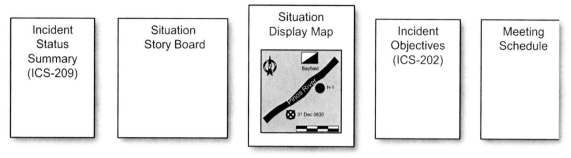

Figure 7.37. As soon as possible, establish a situational display. There is no prescribed manner in how you must do the display, but keep in mind that the information needs to be displayed in a way that benefits the entire response team.

Information Dissemination

The same level of attention and planning that you gave to gathering information needs to be put into disseminating the information you collected and analyzed out to those who need it. Information needs will vary depending on who in the command team needs the information: IC, OSC, RESL, and others. Your job is to "feed" them all.

Use all avenues at your disposal to push information out to the response team. In addition to the large wall displays in *Figure 7.37*, you should use inboxes/outboxes (like those used on a person's work desk) for getting information out to the response team, collecting information, and for placing historical records that the Documentation Unit will require.

Boxes identified for distributing information might contain:

- 8.5 x 11 inch copies of the most recent incident map
- Future weather, tides, and sunrise/sunset
- Extra Incident Action Plans
- Meeting schedule

Boxes identified for collecting information might contain:

- Field Observer reports

Boxes identified for historical records might contain:

- ICS-214, Unit Log
- Annotated maps/charts
- ICS-209, Incident Status Summary
- Photographs

SITL Role in the ICS Planning Process

Every formal meeting or briefing that takes place within the ICS Planning Process (depicted in *Figure 7.38*) starts with a situational brief. This is to ensure that the decisions being made are grounded in the most current information known about the status of the incident. The burden you carry as the SITL is to deliver those situational updates with the greatest accuracy achievable and to assure the highest quality possible. You generally do not have much time to give your updates so you must have everything ready to go: maps, briefing notes, etc. Be a stickler for quality, but realize that you'll be trading time for quality and that somewhere you'll have to settle for the best your unit can provide, based on the number of personnel you have and the dynamics of the incident. Your words and graphics must paint a picture of the current incident status and a glimpse into the future of what the status might be. This is no easy feat on a complex incident. In Appendix M you can see a list of some best briefing practices to help you develop and deliver a high-quality briefing. The list is not exhaustive so add to it as you see fit. However, at a bare minimum your briefs should include:

- The perimeter of the incident
- Operations Section organizational boundaries (e.g., divisions, branches)
- Established support facilities
- Key geographic features
- Wind direction and speed
- Tides and currents (if appropriate)
- Success of mitigation measures
- Projections and predictions of the future incident status

The Operational Planning "P"

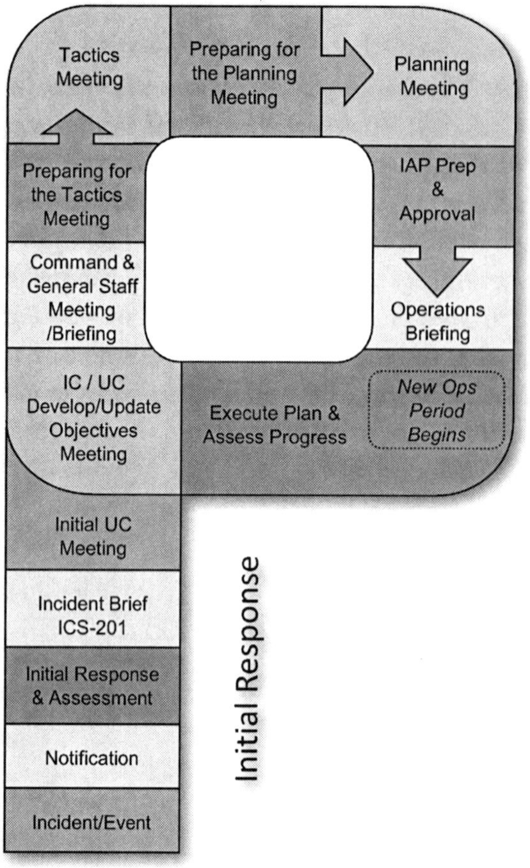

Figure 7.38. The ICS Planning Process is depicted in the Operational Planning "P." The Situation Unit Leader plays a critical role in each meeting and briefing by providing the response team with accurate situational information and a prediction on the future status of the incident.

Incident Action Plan

The ICS Planning Process and Incident Action Plan go hand-in-hand and both are discussed in detail in Chapter 2. The planning process provides a common methodology for responders to work toward the development of a plan of action that is documented and referred to as an Incident Action Plan (IAP). Throughout the planning process you have given the best situational information possible and the IAP is anchored in the current and predicted incident status you provided (no pressure). Your role in the development of the IAP is to provide an *incident map* such as one depicted in *Figure 7.36*. The map provides responders with information on:

- Divisional/branch boundaries
- Incident facilities (e.g., base, ICP, helispots)
- Impacted areas
- Predictions

Measurement of Success

You know you have done a good job as the SITL with your maps, charts, and other displays if responders are coming to you for information. This is especially true if the Operations Section Chief uses your products to outline his or her tactical plans. Take heed if no one seeks the information you're providing; it's a signal that the products you're producing are not valued by the response team. If the OSC begins to maintain a detailed situational display of his or her own, it's because he or she is filling a vacuum left by an inadequate Situation Unit. You have two choices if this occurs: get your act together or pack up and go home.

Situation Unit Leader Conclusion

The job of the Situation Unit Leader is critical to the success of a response. Depending on the size and complexity of the incident, you may be the sole person staffing the Unit or you could have dozens of personnel working for you. Your challenge is to engage the entire command team to support your needs for information and in return provide them with an accurate picture of incident status. ICS is about teamwork and the SITL is central to that team's ability to function.

Documentation Unit Leader

The only thing that remains of the Incident Command team's efforts in the wake of an emergency response operation are the documents left behind that faithfully record the critical decisions, tactical planning, resources committed to the incident, contingency plans developed, situation maps, funds expended, and many other documents that, when taken together, tell their story. It's the Documentation Unit Leader's (DOCL) responsibility to put into place a systematic process to collect critical incident information and to create a filing system that will enable people to access the information years later.

For some types of incidents, the documentation files may have to stand up to scrutiny in a court of law or may be used to determine when key decisions were made during the course of the response operations. In those instances the incident files should be treated as sensitive and need to have the appropriate level of safeguarding. In certain instances the Incident Commander (IC) is required by law to safeguard the incident files.

The complexity and size of the incident will dictate whether the Incident Commander or, if assigned, the Planning Section Chief (PSC) will designate a Documentation Unit Leader. For the majority of incidents that occur, documentation will remain the responsibility of the IC or PSC. However, if you're tasked to assume the duties of the DOCL the information that follows will enable you to focus your energies and create an effective Documentation Unit that supports the needs of the command team as well as others who will, at some point, need to understand how and why the response played out as it did.

The responsibilities of the DOCL are many. The DOCL will:

- Provide incident documentation
- Implement a system to ensure that critical documents pertaining to the response are sent to the Documentation Unit
- Assess the effectiveness of the Documentation Unit's ongoing activities and modify the system, as necessary to ensure proper documentation
- Provide duplication services for the command team
- Establish a comprehensive filing system
- Ensure that any discrepancies and/or missing documents are recorded
- Ensure that any documentation that is submitted to the Documentation Unit is accurate and complete
- Establish a comprehensive archive of files for the response
- Store files for post-incident use
- Document Unit activities on the ICS-214, Unit Log (see Appendix D)

Arriving On-scene as the Documentation Unit Leader

One of the first responsibilities you have after checking-in to the incident is to receive an in-briefing from the PSC. This briefing will give you a sense for how complex the documentation effort is going to be. It's during this briefing that you should:

- Discuss with the PSC his or her expectations for incident documentation
    - Confirm what meetings and briefings the PSC would like you to attend. These are usually the critical meetings conducted during the Incident Command System Planning Process where the Incident Commander/Unified Command is making key decisions on the direction of response operations. These meetings include:
        - Unified Command Develop/Update Objectives Meeting
        - Command and General Staff Meeting
        - Planning Meeting
    - Clarify the DOCL authority to release any incident-related documentation or reports outside the Command and General Staff
    - Determine how the PSC would like the filing system for the incident to be established
        - File by operational period
        - File by calendar date
        - File by form number
- Determine where the Documentation Unit is to be located
- Find out from the PSC what documentation efforts have been undertaken and where collected documents are located (retrieve those documents as soon as you can since they will be your starting place)
- Ask the PSC if there are any additional requirements that the Documentation

Unit is expected to perform. Examples of additional duties may include:

- o Photo documentation of the incident
- o Providing copies of archived documents to outside requesters, responding to Freedom of Information Act (FOIA)

Order Your Staff

Although there are no established guidelines for staffing the Documentation Unit, you can use a table like the example in *Figure 7.39* to help you determine the level of support that you'll require to accomplish your responsibilities. Just grab a piece of paper and list the various tasks that your unit has to perform and the different skills that you'll need to accomplish those tasks. As you develop your staffing you may have to staff the Documentation Unit 24-hours a day during the height of the response. So remember to factor that into your staffing determinations.

Documentation Unit Staffing Chart

(Skills)

Tasking	Administration Assistant	Photo Documentation	Note Taker
Filing	2		
Copy/Fax	1		
Meeting Documentation			1
Photo Management		1	

Figure 7.39. Once you have received your in-briefing, take time to determine your staffing needs. The sooner you order staff, the sooner they will arrive to support you.

Assessing the Status of Incident Documentation

If you're the first Documentation Unit Leader at the incident, you're going to have a big job ahead of you. You'll be in a catch-up mode because the incident-response operations started well before you arrived. While you are waiting for your staff to arrive, conduct an assessment on the status of the incident documentation efforts that are currently underway. Your assessment should at least focus on the following:

- Determine how incident documentation is currently being collected and identify gaps in the collection process
    - o A good example of a time-critical documentation gap is information that is leaving the Incident Command Post (ICP) via e-mail or facsimile. These types of communications cannot typically be re-created and once sent are very difficult to capture.

- Determine what can be done to close any identified gaps immediately and take those steps, even if the solution is only a short-term measure (ensuring that important response documentation is not lost is one of the most crucial tasks you will perform as the DOCL)
- At an absolute minimum, determine if the following documents are being collected, and, if not, start collecting. Always try to get the originals for the incident files.
    - ICS-201, Incident Briefing
    - Incident Action Plans with original signatures (number pages of the IAP: e.g., 1 of 22; 2 of 22; 3 of 22 for continuity)
    - Command decisions/directives
    - Site Safety Plan (SSP) with original signature (number pages of the SSP: e.g., 1 of 22; 2 of 22; 3 of 22 for continuity)
    - Signature sheets of those who reviewed the Site Safety Plan
    - Contingency and other non-IAP plans with original signatures
    - Press releases
    - ICS-211, Check-in sheets
    - ICS-215, Operational Planning Worksheets
    - ICS-209, Incident Status Summaries
    - ICS-214, Unit Logs (everyone's)
- Determine if there are important meetings taking place that are outside the normal planning process and assess if someone from Documentation Unit should attend and record important decisions or discussions
- If there is a Legal Officer attached to the command team, discuss with the officer any documentation requirements that he or she may have

Part of the assessment that you perform should include equipment. You cannot do your job without copy machines, faxes, telephones, and other equipment and supplies.

Some Best Practices

- Introduce yourself to the Command and General Staff and let each of them know what your incident-documentation needs are and ask for their support. Encourage them to emphasize to the members on their individual staffs the importance of thorough documentation
- Encourage everyone in the Command and General Staff to keep records in writing of their critical decisions and issues. The ICS-214 was developed to capture this information, but any pad of paper will do the trick. The bottom line is that the key players on the incident command team need to keep a running log of their activities
- Talk to the PSC about having the IC/UC make incident documentation a discussion issue during the next Command and General Staff Meeting
- Consider having a member of your Documentation Unit staff assigned to

the Unified Command to ensure that formal documentation of decisions are captured and recorded

Capturing Information in the Incident Command Post

At this point in the response, you should have met with the PSC to determine the expectations for documenting the incident, ordered your staff based on incident complexity and documentation requirements, walked around the Incident Command Post and introduced yourself to key individuals, assessed the current documentation effort and identified any gaps in the capture of incident information. It's now time to make your presence felt as an integral part of the incident command team.

Unless you're dealing with a highly trained and disciplined incident management team, documentation of the incident typically has to be "pulled" out of the organization. What this means is that you'll have to go looking for the incident documents instead of the documents being "pushed" to the Documentation Unit. One way to help create an active approach to documentation collection is to have a documentation plan that outlines an effective system for capturing information. Developing the plan and publishing it are only parts of the equation; you'll have to be actively engaged in its implementation. It's important that you regularly reexamine the documentation plan to ensure that all aspects of it are being carried out and that the plan is still current with respect to ongoing response operations.

In addition to developing and implementing a documentation plan, you can institute other methods to increase your capture of information. One method is to centralize all fax machines. A wealth of documentation is coming and going from the ICP in the form of faxes and if the machines are located in or near the Documentation Unit your staff will have access to the information. The trick in centralizing fax machines is to offer fax services to the entire command team. Your personnel would send, receive, and disseminate all faxes. To do this, set up a service window or drop box where people fill out the cover sheet and provide faxing instructions (fax number, recipient, priority, etc). You send the fax for them, make a copy for yourself, and route the original with the confirmation slip back to the originator. This system benefits everyone—people don't have to wait in line or try to learn how to use the latest model fax machine, and the Documentation Unit gets ready access and can copy any documents for the incident record. Use a similar approach with machines set up for incoming faxes and you can capture a copy of those before routing them to the recipients. By doing this, you have just successfully closed an information gap, and in the process provided a service that helped save the time of the command team members.

Photo documentation of the incident is another area where the Documentation Unit can take the lead in helping to develop a chronology of the incident. Bring a couple of photographers on staff and assign to them the responsibility for documenting the response operations in the field. Develop a schedule and have your photographers take pictures in each response area to capture equipment in service, types of operations being conducted, before and after shots, etc. Having a group focused on getting these kinds of shots will ensure that at the end of the response you don't end up with 50 pictures of the local politician who toured the site and no shots of the site itself.

By having the Documentation Unit personnel take the photos you can also ensure that each is labeled and cataloged so you can tell what is in each photo six months later when you're trying to remember where the pictures came from.

Faxing services and photo documentation are just two ways the Documentation Unit actively contributes to capturing incident information.

Documentation and Demobilization

As response operations start to scale back, it's important to determine how documentation efforts will change during demobilization of the incident. This is a critical time period where there is a potential for important documentation to be lost. As the DOCL, there are some actions you can take to help minimize any loss of incident information. These actions include:

- Meet with the Demobilization Unit Leader (DMOB) and ask them to include specific instructions that you as the DOCL would like demobilizing members to follow with regard to any remaining documents they may have in their possession
- If the Documentation Unit, including yourself, is being demobilized, develop procedures for the PSC that will ensure continuity of the documentation process that you put into place

Developing a Response Documentation Archive

The dust has settled in the Incident Command Post, the majority of the active response operations have wrapped up, and now it's time for the paperwork. Organizing the files that have been collected and safeguarded into a comprehensive archive of files for the response is the next task facing the DOCL. Individual agencies will have their own requirements for how these archives are set up, organized, and maintained. However, there are a couple of principles common to all response archives that are addressed below.

Any response archive should be organized in such a manner that will allow easy access and retrieval of response-related documents/information. The best way to ensure the response archive is effective is to develop a set of standard questions or pieces of information that will need to be retrieved from the system and ensure that any properly authorized person can use the system to access that information. Use a standardized approach to enable agencies to produce a high-quality, repeatable product.

The time to conduct the Quality Assurance on the archive is while it's being assembled. If there are missing or inaccurate files this is the time to fix it. The files need to be corrected or there needs to be a note to file placed in the archive describing why there's a gap in the documentation. Remember, people who were not at the response may have to be able to re-create what was done using the response archive. In the event that the archive is used as a legal record of response actions it has to be complete and comprehensive enough to stand up to that type of scrutiny.

For exceptionally large response archives, the DOCL may want to consider the development of a database to capture the entire archive. This approach, while time-consuming on the front end, may save countless hours of searching to respond to legal queries and Freedom of Information Act requests. The type of database used will depend on the preference of the individual agency but it should be something that will enable searches of the archive to assist in the location of specific files/documents.

The final step in building the response archive is to determine how it will be stored, accessed, and maintained after the Documentation Unit has completely demobilized. These kinds of considerations may seem detailed but they are important items that, if not addressed, could mean all the hard work that went into building that comprehensive response archive will have been for nothing if the archive is misplaced or if documentation files/photos are taken and not returned to the archive.

Documentation Unit Leader Conclusion

The true test of the success of your leadership, attention to detail, and organizational skills is in how thoroughly you and your staff were able to capture incident documents that covered all phases of the response operation. In the end, the Incident Commander/Unified Command is responsible for incident documentation. However, they entrusted you with ensuring that the story of the incident can be retold and that the actions they took can be backed up by accurate documentation. If you created a comprehensive response archive that can re-create the response for someone who was not there, you have done an excellent job that set the standard for others to follow.

Demobilization Unit Leader

The orderly release of resources (demobilization) from an incident is an important aspect of incident management, but demobilization does not usually get the level of attention it deserves. There are several reasons for this: first, the vast majority of incidents never rise to the level of complexity that showcase the value of a well-planned and executed incremental drawdown of incident resources; and second, outside of a few disciplines, the Incident Command System demobilization process is a mystery and its intrinsic worth seldom seen.

If you're called to an incident to assume the responsibilities of Demobilization Unit Leader (DMOB), you will be stepping into a very dynamic environment and you must move rapidly once you have checked into the incident to get up to speed and develop an effective demobilization plan. It's your responsibility as DMOB to ensure that all incident resources are released from the incident in a safe and organized manner. You should be sensitive to the fact that a delay in the development and implementation of a demobilization plan is costly in terms of both time and money. For example:

- High dollar assets will remain on the clock even though they are no longer needed for response operations

- Personnel who are no longer needed are unnecessarily delayed in returning to their primary job outside the incident
- Scarce resources that may be required elsewhere will sit idle

It is our hope that by reading this section on Demobilization, you'll have an understanding of the demobilization process and be able to develop and implement a demobilization plan. The Incident Command System is all about teamwork and you will rely heavily on the other members of the team to successfully fulfill your responsibilities. One of the most critical relationships that you must have is with the Resources Unit Leader (RESL). You'll be working closely with the Resources Unit Leader who maintains the latest information on resources that are currently on the incident and those that will be required for future operational periods. The relationship between the Resources and Demobilization Units is so important that we recommend you take the time to read the Resources Section in this chapter to better understand how Resources operates and how it can help you do your job.

In addition to the RESL, the entire Command and General Staff and others are there to support your efforts. The primary command team personnel that you will work most with are shown in *Figure 7.40*.

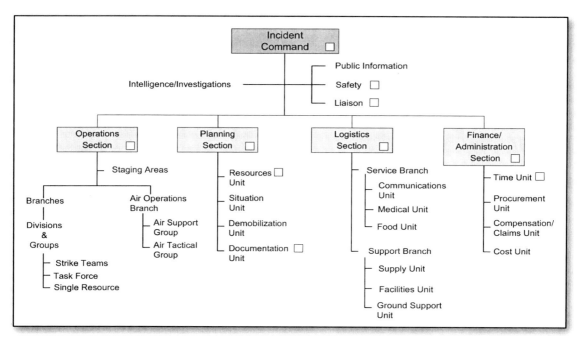

Figure 7.40. Incident Command System organization chart noting the various ICS positions—with blocks next to them—that the Demobilization Unit Leader will interact with during an incident.

Your responsibilities on the incident command team are focused: make sure that all resources (personnel and major items of response and support equipment) are released in a methodical way that maintains the integrity of resource accountability and doesn't impact the continuing response efforts. Controlling the arrival and release of incident resources is absolutely essential to effective incident management. The Resources Unit Leader is responsible for the check-in of arriving resources, and you, as the Demobilization Unit Leader, are responsible for overseeing the release.

Your responsibilities as DMOB include:

- Establishing a Demobilization Plan
- Coordinating and supporting the implementation of the Demobilization Plan
- Preparing Demobilization Checkout forms, ICS-221, for each resource being released
- Keeping the Planning Section Chief apprised of the demobilization progress
- As requested by the Planning Section Chief, attending planning meetings and briefings to provide information on the Demobilization Plan
- Document Unit activities on the ICS-214, Unit Log (see Appendix D)

Arriving On-scene as the Demobilization Unit Leader

Your challenge once you have checked into the incident is to get a feel for the size and complexity of the Demobilization effort. To help accomplish this:

- Obtain a briefing from the Planning Section Chief to get an idea on the scope of the demobilization and to determine how much time you have to establish the Demobilization Unit before resources start demobilizing
- Review incident documents/databases:
    - ICS-201, Incident Briefing Form, (initial response resources may have not been captured during check-in)
    - ICS-211, Check-in Form(s)
    - Incident Action Plans
    - T-cards (what resources are currently assigned and where they are)
    - Computer databases used to track resources
- Meet with the Resources Unit Leader to get his or her thoughts on demobilization

Order Your Staff

Based on the briefing and incident document/database review, order your support staff and establish a workspace. For staff, consider getting multi-agency representation in the Demobilization Unit and work with the Resources Unit Leader to determine if it's feasible to get Resources Status Recorders to help in the demobilization process. Status Recorders are a great source of help to you because they have spent their days and nights keeping track of all incident resources. For work space, try to get near the Resources Unit. You'll be working very closely with Resources throughout the demobilization effort and the proximity will enhance the coordination.

Building a Demobilization Plan

Demobilization Plans generally all follow the same form, but you'll see variances based on who is generating the plan, the type of incident the plan is supporting, and other nuances unique to the particular agencies involved. Regardless, all Demobilization Plans should contain the following to be effective:

I. General Information Section (this section should be short but informative) and include:

    A. Incident Commander/Unified Command expectations

    B. Safety considerations

    C. Directions to the Section Chiefs

    D. Description of the demobilization procedures

II. Responsibilities Section (this section establishes position specific demobilization responsibilities) for example:

    A. Planning Section Chief

        i. Ensures demobilization information is provided to the response organization in sufficient time to conduct an orderly downsizing of incident resources

        ii. Submits proposed release of resources to the Incident Commander (IC)/Unified Command (UC) for approval

        iii. Ensures released resources follow established demobilization procedures

    B. Operations Section Chief

        i. Identifies and communicates excess personnel and equipment available for demobilization to the Planning Section Chief (Operations Section is the biggest customer of demobilization since it has the bulk of the resources)

    C. Logistics Section Chief

        i. Coordinates all personnel and equipment transportation needs to their final destination

        ii. Ensures property accountability for all nonconsumable items (e.g., fire hose, radios)

    D. Finance/Administration Section Chief

        i. Ensures completion of:

            a. Time records (personnel and equipment)

            b. Injury reports

            c. Claim reports

E. Safety Officer

   i. Reviews plan for health and safety issues

      a. Ensures drivers have adequate sleep before driving to their final destination

      b. Verifies personnel tracking system is in place and being used to ensure responders have arrived at their destination safely

F. Liaison Officer

   i. Coordinates with assisting organizations demobilization information of their resources

III. Release priorities

   A. The Incident Commander/Unified Command will determine the release priorities taking into consideration:

      i. Ongoing incident resource requirements

      ii. Personnel welfare (safety and rest)

      iii. Needs of the responding agencies

      iv. Home unit of the resource (out-of-area or local)

      v. Resource cost

IV. Release Procedures

   A. Procedures to be followed for obtaining release

V. Incident Commander/Unified Command approval of the Demobilization Plan

You can find a sample Demobilization Plan in Appendix J.

Distribution of the Demobilization Plan

Ideally, the Demobilization Plan that you have developed and had approved by the Incident Commander/Unified Command should be distributed 24 hours prior to release of the first resource. The reality is that some resources will most likely have been released before you arrived at the incident.

Give your plan wide distribution. Ensure that the following at a minimum have a copy:

- Command and General Staff
- Supporting agencies – work with the Liaison Officer to ensure that they get a copy
- Post on any established information display boards provided to responders

- Documentation Unit (they get the original copy for the incident records)

Steps in the Demobilization Process

Once the Demobilization Plan has been approved by the IC/UC the demobilization process begins. Below are the steps that should be followed in order to ensure the orderly release of resources.

Step 1: All unit leaders in Planning, Logistics, and Finance/Administration identify any surplus resources at least 24 hours in advance of their anticipated demobilization time. The Resources Unit Leader will work with the Operations Section Chief to identify Operation's resources.

Step 2: Lists of identified surplus resources for each Section are given to the Section Chief who will forward the tentative list of surplus resources to the Planning Section Demobilization Unit. *Figure 7.41* is an example of a Tentative Release List Form.

Step 3: The Demobilization Unit will compile a tentative list of surplus resources from all Sections and send them to the Incident Commander/Unified Command via the Planning Section Chief.

Step 4: Incident Commander/Unified Command approves the list of resources to be demobilized.

Step 5: Approved demobilization list is sent to the Resources Unit and to the LSC and any other appropriate Section Chiefs.

Step 6: Section Chiefs notify the resources under their control that they have been approved for demobilization and what procedures they should follow.

Step 7: Demobilization Unit ensures that the checkout process is followed.

Step 8: Demobilization Unit sends completed Demobilization Checkout forms, ICS-221, to the Documentation Unit for the historical record. Each resource that demobilizes from the incident will complete the ICS-221 as shown in *Figure 7.42*.

	TENTATIVE RELEASE LIST	
\multicolumn{3}{l}{From: _____ (Section Chief or Command Staff Officer)}
\multicolumn{3}{l}{The following resources are surplus as of _____ (hours) on _____ (date). At that time, these resources are available for release processing.}
	Name of Individual, Crew, or Equipment in excess	Position on the Incident
1	Patrick Robert	Supervisor
2		
3		
4		
5		
6		
7		
8		
9		
10		
11		
12		
13		
14		
15		
16		
17		
18		
19		
20		

_____
Signature of Section Chief or Command Staff Officer
Date: _____ Time: _____

Figure 7.41. Tentative Release List used by the Command and General Staff to identify recommended resources for demobilization. The Tentative Release List(s) are provided to the Demobilization Unit who compiles a master list for the Incident Commander/Unified Command.

DEMOBILIZATION CHECKOUT		ICS-221
1. INCIDENT NAME/NUMBER Meridian Flood	2. DATE/TIME 27 Nov 1800	3. DEMOB NUMBER

4. UNIT/PERSONNEL RELEASED
Patrick Robert

5. TRANSPORTATION TYPE/NUMBER
Commercial Air (see block 12 for additional information)

6. ACTUAL RELEASE DATE/TIME 28 Nov 0900	7. Manifest YES NO NO
8. DESTINATION Arcata, California	9. AREA/AGENCY/REGION NOTIFIED Region IX notified on 27 Nov 2000

10. UNIT LEADER RESPONSIBLE FOR COLLECTING PERFORMANCE RATING
Not required by Agency

11. UNIT/PERSONNEL YOU AND YOUR RESOURCES HAVE BEEN RELEASED SUBJECT TO SIGNOFF FROM THE FOLLOWING:
(DMOB UNIT LEADER CHECK BOXES THAT APPLY)

<u>LOGISTICS SECTION</u>

- [X] SUPPLY UNIT
- [X] COMMUNICATIONS UNIT
- [ ] FACILITIES UNIT
- [ ] GROUND SUPPORT UNIT

<u>PLANNING SECTION</u>

- [ ] DOCUMENTATION UNIT

<u>FINANCE/ADMINISTRATION SECTION</u>

- [X] TIME UNIT

<u>OTHER</u>

- [X] SAFETY OFFICER

12. Remarks
Frontier Airlines flight 271

Figure 7.42. The Demobilization Checkout Form, ICS-221, is used to ensure the orderly release of resources. Block 6, Actual Release date and time, is normally one of the last items to be filled in.

Developing a Demobilization Tracking Table

To help you keep track of the status of demobilizing resources consider developing a demobilization tracking system like the one in *Figure 7.43*. If you have access to a computer, you can use that, but simple pieces of paper will do the trick (unless, of course, you have hundreds of resources to track). Your goal is to be able to accurately track the movement of resources as they leave the incident. The table you create will be a big help to the Safety Officer because they can use it to ensure that drivers are not leaving the incident before they get adequate rest.

Demobilization Resource Tracking Table

Agency	Name	Order #	Check-In	Last Shift	Sent Home	Home Base	Assignment	Travel
BFD	Patrick Robert	O-020	15-Nov	27-Nov	28-Nov	Arcata, CA	DIVS	Air
BFD	Scott Kelly	O-004	15-Nov	27-Nov	28-Nov	Bayfield, CO	STAM	POV
ANF	Jack Voelker	O-008	15-Nov	27-Nov	28-Nov	Bayfield, CO	THSP	POV
USCG	Mary Yale	O-012	15-Nov	27-Nov	27-Nov	Durango, CO	IC	GOV
FBI	John Murry	O-022	16-Nov	28-Nov	29-Nov	Cortez, CO	PSC	GOV

Figure 7.43. Using a demobilization resource tracking table can greatly facilitate maintaining an accurate and efficient demobilization process.

Demobilization Unit Leader Conclusion

An incident management team that conducts a well-planned and executed demobilization has most likely demonstrated superior leadership throughout the entire response effort. Demobilization is an important part of the response and the attention to detail and quality of that demobilization process begins with you the DMOB.

Closing Thoughts

You probably feel like you've just tried to take a drink from a fully charged fire hose after reading this chapter on the Planning Section Chief. We really just scratched the surface, which is why, if you have the opportunity, you should try to get an ICS position-specific course that is solely dedicated to the PSC. That said, the information in this chapter will prove useful should you deploy to an incident as the PSC. You have the basic tools to get started in the right direction and to have a positive impact on the management of the command team.

There are a few common themes that run through this chapter that we would like to highlight. First, you, as the PSC, are the one person who can have the greatest impact on the cohesiveness of the command team. Leverage your personal skills and professionalism to create that critical team environment that is so necessary to a successful and safe emergency response operation. Second, every function (resource management, situation, documentation, or demobilization) that the Planning Section is responsible for is important. Do not neglect any of them. Do it right regardless of the size of the incident so you're ready when things get really dicey. Remember how you practice is how you play; so practice often.

CHAPTER 8

LOGISTICS SECTION CHIEF (LSC)

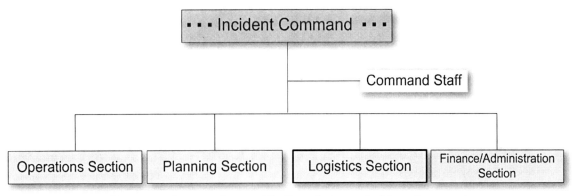

Figure 8.1. Logistics Section Chief in relation to Incident Command

There can be no sustained response operations without logistics. Someone has to order, track, and receive every item necessary to keep a response organization equipped, billeted, and fed. These items run the spectrum from high-dollar assets such as heavy lift cranes to flashlight batteries. The Operations Section Chief may be the tactical wizard of the response, the Planning Section Chief may be the conductor keeping everyone moving in the same direction, and the Finance/Administration Section Chief the keeper of the response organization checkbook, but the Logistics Section Chief (*Figure 8.1*) is the backbone of a strong response organization. Response operations would quickly grind to a halt if the Logistics Section failed to fulfill its responsibilities.

Area Commander, Marc Rounsaville, of the United States Forest Service, noted that the one position in the Incident Command System that can never have a bad day is the Logistics Section Chief (LSC). Operations may not accomplish all of its work assignment for the day and Planning may not have been on top of its game in managing the planning cycle, but if logistics drops the ball everyone takes notice and the impact can be felt across the command team. Imagine if logistics did not order enough fuel or food for a single day! Managing logistics for a dynamic response operation requires a great deal of experience, attention to detail, and a high propensity for customer service and teamwork.

An incident that requires the establishment of a dedicated Logistics Section means two things: (1) the size and complexity of the incident is such that the initial response efforts were unable to successfully bring the incident under control; and, (2) the duration of the incident response, and the amount of logistical support required, have substantially increased. The LSC must be prepared to deliver the necessary material to support tactical operations, provide billeting and food for responders, ensure incident communications are well planned and supported, and take care of the numerous logistical details that are absolutely essential to a successful response. We hope our passion for the importance of logistics is coming through loud and clear. Without logistics, there can be no response.

As the LSC you have many responsibilities. Here's a list of some major ones:

- Anticipate incident potential for growth and plan incident facilities and Logistics Section personnel requirements accordingly
- Develop and implement a resource ordering process
    - Identify agency specifics for requesting and ordering resources
- Ensure daily inventories are conducted
- Implement a "resources ordered" matrix or spreadsheet to track resources
- Ensure an effective communication network is in place to support incident operations (may require need for secure communications)
- Ensure that the following functions are adequately addressed to support incident operations:
    - Transportation of responders on the incident site
    - Medical services for responders
    - Incident facilities
    - Security
    - Food and shelter for responders
- Support development of the Incident Action Plan
- Ensure that Command and General Staff are aware of excessive costs
- Brief incoming Logistics Unit Leaders on:
    - Section operating instructions
    - Chain of command/responsibilities
    - Safety
- Ensure appropriate demobilization of the Logistics Section
    - Account for property and services
    - Return incident facilities to pre-incident condition
    - Properly dispose of surplus supplies
    - Ensure any hazardous materials are properly disposed
    - Make transportation arrangements for incident equipment
    - Submit Section records to Documentation Unit
    - Work closely with Demobilization Unit Leader
- Maintain a Unit Log, ICS-214 (see Appendix D)

Actions to Take upon Arriving at the Incident Command Post

If you're sent to an incident as the Logistics Section Chief, the Incident Commander/Unified Command (IC/UC) has recognized that the situation they're facing will require a dedicated and concentrated logistical effort to help bring the incident under control. The IC/UC and the rest of the response organization are relying on you to provide seamless support. For you to do that, you'll need to get a thorough briefing from the IC/UC. There's going to be a lot going on when you arrive so make sure that you take the checklist in *Figure 8.2* with you as a memory jogger of the questions you need to get answered. We encourage you to add to the checklist as you see fit.

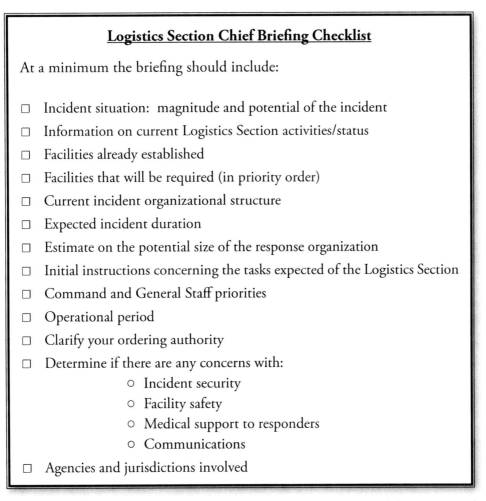

Figure 8.2. Use a checklist to help remember the questions you want answered during your in-brief.

In addition to the IC/UC in-brief, take time to meet the other Sections Chiefs since you'll be working closely with all of them. Each chief will be a great help to you in accomplishing your job.

Logistics Section Chief's Staffing and Space Considerations

After you have received your briefing and have a better understanding of what you are up against, take some time to consider the staffing that you'll need to logistically support the incident and how much work space is needed for your logistics team. The checklist in *Figure 8.3* provides some of the things that you want to consider in staffing the Logistics Section and the space needed for your team to work.

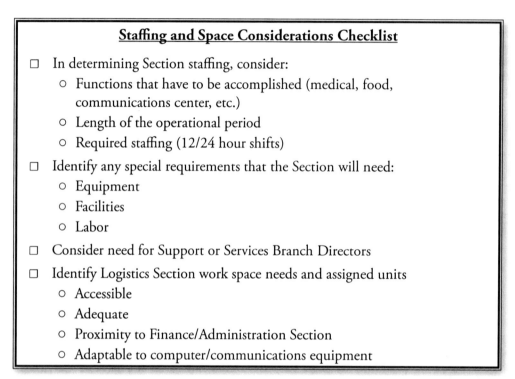

Staffing and Space Considerations Checklist

- In determining Section staffing, consider:
  - Functions that have to be accomplished (medical, food, communications center, etc.)
  - Length of the operational period
  - Required staffing (12/24 hour shifts)
- Identify any special requirements that the Section will need:
  - Equipment
  - Facilities
  - Labor
- Consider need for Support or Services Branch Directors
- Identify Logistics Section work space needs and assigned units
  - Accessible
  - Adequate
  - Proximity to Finance/Administration Section
  - Adaptable to computer/communications equipment

Figure 8.3. There are numerous details that the LSC must attend to in order to successfully establish a highly responsive logistical organization that can attend to the response team's wide range of needs.

Establishing the Units of the Logistics Section

The responsibilities of the Logistics Section are such that you cannot run logistics solo, so you need to order your staff right away. There's going to be a lag time between when you want personnel and when they actually arrive; and with you on-scene, expectations are going to be that Logistics is fully operational even before you're ready. As we have reminded all other Section Chiefs, for any positions that you do not activate you will have to do the job yourself or require someone on your team to take on more than one position. The incident should drive your staffing decisions and not an organization chart. *Figure 8.4* shows the primary positions in the Logistics Section for a large complex incident.

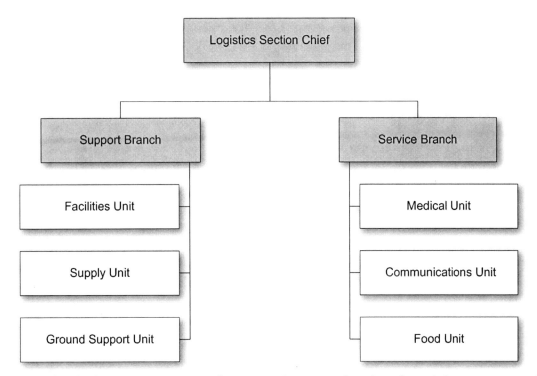

Figure 8.4. Like the Operations Section, the Logistics Section can have branches to help manage span-of-control. The Logistics Section is designed to expand and contract to meet the demands of the incident.

Some of the units that make up the Logistics Section require specialty knowledge and skills to perform, such as the Communications and Medical Unit while others do not, such as the Facilities and Ground Support Unit. As the Logistics Section Chief, you have to decide if some or all of the Logistics Section units will be activated. In addition to designating unit leaders, the scale and pace of the response may require you to bring in a Deputy and Branch Directors. Remember you must provide the necessary support to the incident operation, so staff for success. Over the next several pages we're going to touch on the importance of the various units that you are responsible for and some guidance you should give each unit leader during his or her in-briefing.

Deputies

Your ability to manage incident logistics without the support of a Deputy will hinge on several factors such as:

- The speed of operations
- Length of the operational period (e.g., 12 hours, 24 hours); a shorter operational period is more demanding on your time
- Whether you're able to get adequate rest at night
- Whether you can adequately manage all issues and oversight of the logistics organization without compromising the delivery of services and support

Your consideration of these factors will help you determine the need to bring in a Deputy to support you. A Deputy can help manage the Logistics organization by supporting the planning process, relieving you to get rest, and managing the daily Section functions.

Support Branch Organization

The Support Branch is comprised of the Facilities, Supply, and Ground Support Units (see *Figure 8.5*). Depending on span-of-control needs, you may designate a Support Branch Director to oversee the management of these units or you may manage them yourself.

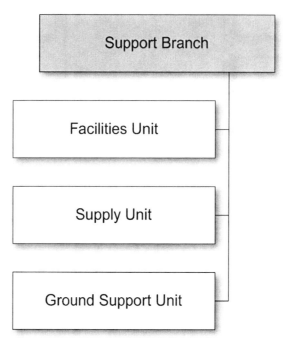

Figure 8.5. The Support Branch is comprised of the Facilities Unit, Supply Unit, and Ground Support Unit

Facilities Unit

The pace of response operations and influx of personnel can quickly gobble up square footage. The Facilities Unit must have the facilities necessary to absorb the increase in personnel or, as with the days of the early West, squatters will quickly stake out their illegitimate claim. Once that happens, removing them can be like removing a tick: bloody and painful. The Facilities Unit is responsible for the layout, activation, and setup of incident facilities and the Facility Unit Leader needs to understand the space requirements for each designated facility. Some facilities may include: Incident Command Post, Base, and Camps.

Directions to the Facility Unit Leader (FACL):

- Conduct a facility needs assessment (see Appendix N for an example Facility Needs Assessment Worksheet)
- Ensure that the ICP is established to meet the needs of the incident (see Appendix O for guidance in establishing an Incident Command Post)
- Ensure that all incident facilities are:
    - Environmentally and structurally sound
    - Located:
        - Upwind and uphill
        - In proximity to area of operations
        - With easy ingress and egress
    - Have adequate:
        - Security and parking
        - Sanitation facilities
        - Electrical power
        - Telephone and broadband capability
    - Clearly marked to enable responders to find them day and night
- Establish Incident Base
- Ensure that sanitation, rest and food facilities are established and maintained
- Provide for the maintenance of incident equipment resources
- Establish incident security for all established facilities
- Demobilize incident facilities as the incident draws down

Relocating the Incident Command Post (ICP)

If the incident requires a larger and more permanent Incident Command Post (ICP), the FACL will have to find an alternate location and develop a relocation plan that is acceptable to the IC/UC. Ideally, you will not have to relocate the ICP once established because it will interrupt operations and impact productivity. However, if you find yourself in a situation that requires shifting the ICP you can minimize the impact on operations by developing a relocation plan. The checklist in *Figure 8.6* is a good guide for what the relocation plan should contain.

> **Relocation Checklist**
>
> Components of a relocation plan:
>
> ☐ Objectives for relocating the ICP
> ☐ Directions and map to the new ICP location
> ☐ Parking information
> ☐ Feeding routine
> ☐ Diagram of ICP showing location of all key ICS functions (e.g., Resources Unit)
> ☐ Facility security
> ☐ List of ICP phone numbers
>
> The transition plan must be signed by the IC/UC and be briefed to the Command and General Staff (see Appendix P for an example ICP Relocation Plan)

Figure 8.6. An ICP Relocation Plan should include, at a minimum, the items in the checklist. The plan must receive approval from the Incident Commander/Unified Command.

Supply Unit

The Supply Unit is at the heart of the logistics machine. If the Supply Unit is not functioning in top form, critical equipment and personnel requirements will not get filled and the ordering, distribution, and accountability for essential supplies that are necessary to keep the response moving forward will remain unaddressed. The Supply Unit has an enormous responsibility in ensuring that the response is successful. A mistake in ordering resources and supplies will ripple through the response effort. Take time to make sure that the Supply Unit Leader is well aware of your expectations.

Directions to the Supply Unit Leader (SPUL):

- Ensure that the Supply Unit is staffed with the appropriate representatives from the primary responding agencies
- Develop an ordering process for Logistics Section Chief's approval. Make sure the process includes:
    - List of who has ordering authority
    - Requirement for a tracking number attached to each ordered item
- Establish an inventory and accountability system for shipping and receiving
- Ensure that all requests for tactical resources are routed through the Resources Unit before acting on them
- Maintain a matrix or spreadsheet to track all resources that have been ordered

- Work closely with the Resources Unit and Procurement Unit
- Manage and inventory nontactical equipment
- Ensure that all invoices are forwarded to the Finance/Administration Section for payment
- Ensure that personnel who requested items are notified in a timely manner of the equipment arrival
- Ensure that storage areas are established for:
  - General supplies and equipment
  - Fuel
  - Hazardous materials
- Maintain a Unit Log, ICS-214

Ground Support Unit

The Ground Support Unit Leader is responsible for all ground support and transportation of personnel, supplies, food, and equipment as well as fueling, service and maintenance of vehicles on the incident, and implementing an incident Traffic Plan.

Directions to the Ground Support Unit Leader (GSUL):

- Work closely with the Facilities Unit Leader to identify and establish equipment service area(s). Services to include:
  - Fueling
  - Maintenance
  - Repair
- Ensure close coordination with Operations, Food Unit, and Supply Unit in providing timely transportation and delivery of:
  - Incident Personnel
  - Supplies
  - Food
  - Equipment
- Work with the Resources Unit Leader to maintain an accurate inventory of all ground support vehicles (e.g., buses)
- Maintain a Unit Log, ICS-214

Service Branch Organization

The Service Branch is comprised of the Medical, Communications, and Food Units (*Figure 8.7*). Establish a Service Branch Director to manage these units if the demands on your time do not allow you to properly oversee and support their work.

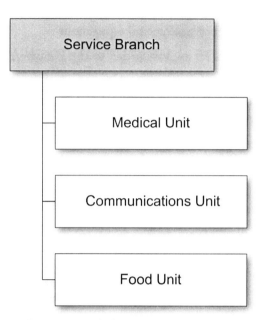

Figure 8.7. The Service Branch is comprised of the Medical Unit, Communications Unit, and Food Unit.

Medical Unit

The number one incident objective for every response is to ensure the safety of response personnel. The Medical Unit has an important role to play in achieving that objective, and, in the event an injury does occur, it's the Medical Unit that must be ready to react to ensure that the injured responder receives rapid attention. Make sure that your Medical Unit Leader has your full support and the equipment and personnel to do the job.

Directions to the Medical Unit Leader (MEDL):

- Ensure that medical aid is in place and ready to meet the needs of the incident
- Ensure that transportation for injured responders is available and well coordinated
- Work closely with the Safety Officer as you develop the incident's evacuation and emergency plans
- Ensure that the Medical Unit has the personnel and equipment to provide for rehabilitation of responders who are suffering from the strenuous work and/or environmental conditions (heat, cold, etc.). Rehabilitation measures should include:
    - Critical Incident Stress Management (CISM) activities
    - A sheltered location for rehabilitating responders where they can get rest, water, and food
- Ensure that the medical portion of the Incident Action Plan is completed on time:

- Develop the Medical Plan, ICS-206
- Ensure that there are adequate medical supplies to respond to any anticipated need
- Maintain a Unit Log, ICS-214

Food Unit

Feeding responders can be a daunting task, and if neglected, can have a negative impact on the response. The delivery of food must be timed to ensure that the field responders are ready to go to work at the start of the operational period and are not unnecessarily delayed due to an unprepared Food Unit. Food plays an important role in safety because hunger can cause inattentiveness and if the work being conducted is physically demanding, the lack of food can speed up fatigue. The message here is that responders must have support to carry out their responsibilities and adequate food and water are part of that support.

Directions to the Food Unit Leader (FDUL):

- Determine the best method for feeding responders
- Determine the amount of food and water needed and ensure that it's distributed to all incident facilities
- Ensure that adequate food supplies, such as potable water and nonperishable food items are ordered to support operations
- Put in place a food-monitoring program that will ensure that the food is maintained and served in accordance with proper food-handling practices
- Monitor food service provider for compliance with proper food-handling practices
- Ensure accountability for all food and water ordered
- Consider need to serve warm meals versus cold "box lunches," especially when response operations are conducted in cold-weather conditions
- Maintain a Unit Log, ICS-214

Communications Unit

The one thing that all emergency responders typically agree on following an actual incident or exercise is that communications did not perform as well as desired. Whether it was incompatible equipment, a poorly designed communications plan, lack of radio discipline, or spotty coverage within the incident area, communications consistently holds the top spot in areas that require improvement.

It is the Communications Unit Leader's responsibility to develop and implement a communications plan that meets the requirements of the incident. This is not an easy task, especially in the multi-agency response environment that we are living in and the added complexity of communicating in a heightened security environment. That said, there are some basics that apply to all incidents that can get communications started in the right direction.

Directions to the Communications Unit Leader (COML):

- Conduct a communications assessment to determine what kinds of equipment (radios, cell phones, computers, telephones) and support are needed for incident operations. Assessment should include:
    - Understanding the topographic features where the incident is located
    - Knowing where the incident is projected to move in the coming hours or days
    - Determining what the future plans are for the incident (e.g., growth in the organization)
    - Determining what communications facilities are in the area currently (e.g., cell towers, repeaters)
    - Knowing whether secure communications are required
- Ensure the communications portions of the Incident Action Plan are completed on time:
    - Develop the Communications Plan, ICS-205; monitor implementation
    - Review and provide input into the Assignment Lists, ICS-204s
- Work closely with the Operations Section Chief to ensure that his or her communications needs are being met
- Provide communication equipment to response personnel and maintain an accountability of equipment that is checked-out
- Maintain a Unit Log, ICS-214

LSC Role in the ICS Planning Process

As a key member of the command team, you'll be attending several scheduled meetings and briefings throughout each operational period. The meetings and briefings are designed to ensure that everyone is working toward the same goals and to provide opportunity to get clarification and direction from the IC/UC. It is the Planning Section Chief's (PSC) responsibility to oversee the various meetings and briefings that make up the ICS Planning Process. It's your responsibility to be on time and prepared to support the process. *Figure 8.8* is a visual representation of the ICS Planning Process known as the Operational Planning "P." As you read on, we'll discuss those meetings and briefings that make up the ICS Planning Process and what applies to you as the Logistics Section Chief. Our goal is to give you a good idea of what is expected of you when you attend these meetings so that you can be prepared to do your part.

The Operational Planning "P"

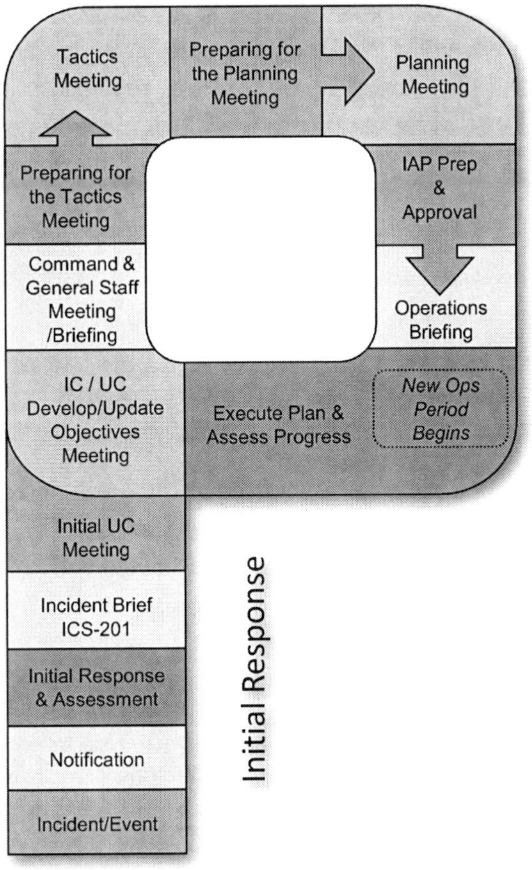

Figure 8.8. As the Logistics Section Chief, you are responsible for supporting the ICS Planning Process by attending required meetings on time and fully prepared.

Command and General Staff Meeting

Early on during each operational period the IC/UC will meet with their Command and General Staff to brief them collectively on how they see the response effort going, discuss incident objectives, provide direction on issues brought up by individual members, and, as necessary, designate Command and General Staff members to undertake specific tasks. This is a good opportunity to brief the IC/UC on logistical concerns and to get clarification on items that involve the Logistics Section. Some areas of discussion may include:

- Clarification on the type and location of support facilities that will be required to support operations
- Identifying logistical limitations that will impact response operations
- Pointing out any problems with internal resource ordering process that are causing unnecessary delays in getting resources to the incident

Preparing for the Tactics Meeting

One of the responsibilities of each member of the Command and General Staff and their subordinates that attend meetings is to be prepared to provide the information necessary to ensure that the objectives for each meeting and briefing are accomplished. Time is one of the most valuable commodities and often the one in the shortest supply; make sure you do your part to keep meetings on time. The Tactics Meeting is extremely important in trying to get ahead of an incident. Prepare for the Tactics Meeting by:

- Understanding what resources are currently ordered to the incident and when those resources are expected to arrive on-scene
- Being prepared to discuss the status of communications and facilities. You need to let the Operations Section Chief (OSC) know the status so that he or she can take advantage of any logistical support
- Evaluating the availability of resources in the local, and, if necessary, regional area (this information will help you better support the OSC as he or she begins to list the resources that he or she would like to have on the incident)

Tactics Meeting

This is perhaps the most critical meeting in the ICS Planning Process. It's here that the OSC will outline the tactical plan for the next operational period. The reason you're attending this meeting is to work with the OSC and Planning Section Chief (PSC) to ensure that the OSC's plan is logistically supportable. Your responsibilities as the LSC during the Tactics Meeting are to:

- Review the proposed tactics
- Identify resource needs and where the OSC wants the resources to report (e.g., Bayfield Street Staging Area)
- Discuss availability of needed resources
- Identify resource shortfalls
- Identify any resource support requirements

The OSC will conduct tactical planning using an ICS-215, Operational Planning Worksheet. Using this form, the OSC will identify the operations organization (Divisions, Groups, Staging Areas, Task Forces, etc.) that will be required for the next operational period, record the work assignments that each organizational element will need to accomplish, and list the resources that are required to carry out each work assignment. It's this last item, the resource requirements, that is of particular interest to you. It will be your responsibility to locate any identified resources that are not already at the incident and get those resources to the incident in time for the next operational period.

We are going to take a few minutes here to go over the ICS-215 so that you can see how the form is to be used and how you, as the LSC, use the form to support operations. In *Figure 8.9*, you'll see a cutaway section of the ICS-215 that has been filled in by the OSC.

The first thing the OSC will do is list an organizational element on the ICS-215 (in our example, the *Search Group*). Next to the organizational element the OSC will outline the work assignment the Search Group is to perform. So far so good, you've mainly been listening to what the OSC has to say up to this point in the development of the ICS-215. The next step for the OSC after finishing the work assignments is to list the various *kinds and type* of equipment that the Search Group will require to accomplish the assignment. In our case, the different kinds of equipment that the OSC listed are: an ambulance, a helicopter, a search team, and a rescue boat. You want to listen carefully here because you're going to get an idea of the range of resources the OSC is seeking. Once the kinds of resources have been written down, the OSC will decide the number of each kind of resource the Search Group will require. On the ICS-215, next to the block labeled REQ (for required), the OSC has placed a **1** below each kind of resource. In our example, the OSC has determined that the *Search Group* will require one of each kind of resource to accomplish the work assignment:

OPERATIONAL PLANNING WORKSHEET		6. KINDS OF RESOURCES	Ambulance (type 2)	Helicopter	Search Team (six person)	Rescue Boat
1. INCIDENT NAME	MERIDIAN FLOOD					
4. DIVISION/ GROUP/ OTHER LOCATION	5. WORK ASSIGNMENTS					
Search Group	Conduct house-to-house search for injured persons. Mark each dwelling searched with a red "X" on the front door. Evacuate injured persons to triage center.	REQ	1	1	1	1
		HAVE				
		NEED				

Figure 8.9. The ICS-215, Operational Planning Worksheet, enables the OSC to lay out the tactical plan including the kind, type and number of resources required.

A constant theme throughout this book is that, in the end, ICS is about teamwork, and the development of the ICS-215 is a prime example. When the OSC has finished listing the required (REQ) resources, the Resources Unit Leader will fill in the HAVE block (*Figure 8.10*). The number of resources in the HAVE block (see shaded area below) are those that are currently on-scene and available to work the next operational period.

OPERATIONAL PLANNING WORKSHEET		6. KINDS OF RESOURCES	Ambulance (type 2)	Helicopter	Search Team (six person)	Rescue Boat	
1. INCIDENT NAME	**MERIDIAN FLOOD**						
4. DIVISION/ GROUP/ OTHER LOCATION	5. WORK ASSIGNMENTS						
Search Group	Conduct house-to-house search for injured persons. Mark each dwelling searched with a red "X" on the front door. Evacuate injured persons to triage center.	REQ	1	1	1	1	
		HAVE	1	0	1	1	
		NEED					

Figure 8.10. The Resources Unit Leader will note the number of resources currently on-scene and available to work the next operational period. This information is placed in the HAVE block (shaded blocks) of the ICS-215.

Once the Resources Unit Leader has filled in the HAVE blocks, there's one more step to take: determine whether there's a need to order more resources to the incident to support the OSC's plan and the number of resources that have to be ordered. You arrive at the number of resources needed with simple math. You subtract what you HAVE from what the OSC requires (REQ). The difference is recorded in the NEED block (shaded gray below) as shown in *Figure 8.11*. As the LSC, you're interested in the NEED block because that is what you'll have to order and get to the incident in time. In our example, everything the OSC requires for the next operational period is already on the incident except for a helicopter. This is the piece of equipment that you'll need to get.

OPERATIONAL PLANNING WORKSHEET		6. KINDS OF RESOURCES	Ambulance (type 2)	Helicopter	Search Team (six person)	Rescue Boat	
1. INCIDENT NAME	**MERIDIAN FLOOD**						
4. DIVISION/ GROUP/ OTHER LOCATION	5. WORK ASSIGNMENTS						
Search Group	Conduct house-to-house search for injured persons. Mark each dwelling searched with a red "X" on the front door. Evacuate injured persons to triage center.	REQ	1	1	1	1	
		HAVE	1	0	1	1	
		NEED	0	1	0	0	

Figure 8.11. The items identified in the NEED block of the ICS-215 are the responsibility of the Logistics Section Chief to order and get to the incident on time to support tactical operations during the next operational period.

We have spent a good deal of time in this chapter explaining the Tactics Meeting and for good reason. The OSC has provided his or her wish list of resources necessary to do the job and accomplish the IC/UC objectives. If you, as the LSC, are unable to locate the resources that have been identified in the NEED block of the 215, you must let the OSC and the PSC know.

The OSC may have to change tactics given resource constraints or the tactical plan that will be used may take more time accomplish the IC/UC objective than anticipated. The OSC is responsible to notify the IC/UC if their objectives will not be reached as originally envisioned.

Preparing for the Planning Meeting

After the Tactics Meeting is complete, you'll have some time before the IC/UC and members of the Command and General Staff will meet to go over the tactical plan that Operations, Planning, Safety, and you have agreed on. Once you step out of the Tactics Meeting you'll likely become extremely busy managing all kinds of logistical issues, but make sure you take the time to properly prepare for the Planning Meeting. Here are some things you need to accomplish before that meeting:

- Meet with the Supply Unit Leader to ensure that the resources identified in the NEED block of the ICS-215 during the Tactics Meeting are ordered
- Keep the OSC and PSC informed on any required resources that you're unable to get for the next operational period
    - Suggest to Operations any alternatives that may be available, such as a smaller crane if a large one cannot be obtained. (It's possible that Operations can use a less-capable resource to do the job, but that the job will take longer. Operations can present this alternative to the IC/UC and set the expectation that more time will be required to accomplish the objective.)

Planning Meeting

You'll attend the Planning Meeting with the other members of the Command and General Staff and it's in this meeting that the IC/UC will give tentative approval of the tactical plan. You're an important player in this meeting even though the Operations Section Chief dominates most of the discussion. The IC/UC is going to be looking for you to verify that the tactical plan is logistically supportable because they know that the OSC's plan is only as good as the resources behind it. The Planning Section Chief will call upon you during this meeting to:

- Verify support for the tactical plan
- Confirm availability of required resources and that they will arrive in time to support the response during the next operational period
- Provide estimates of future service and support requirements

It's possible during the Planning Meeting that the IC/UC will change, add, or delete an objective(s) so you need to be flexible and note any additional resource requirements that were not identified during the Tactics Meeting, and, if necessary, order them following the Planning Meeting.

Incident Action Plan Preparation

With the IC/UC tentative approval of the tactical plan (plan of action), the Planning Section Chief will be overseeing the development of the Incident Action Plan (IAP). The Logistics Section is responsible for several important sections of the IAP and it's your responsibility to ensure that those sections are accurate and submitted to Planning on time. Specifically, Logistics develops the following parts of the IAP:

- Communications Plan, ICS-205 (completed by the Communications Unit Leader)
    - In addition, the Communications Unit Leader fills in the communications portion of the Assignment Lists, ICS-204s
- Medical Plan, ICS-206 (completed by the Medical Unit Leader)

We recommend that you take a few minutes to look at Appendix A, an example Incident Action Plan, to get an idea of the information that Logistics has to provide in the development of an IAP. The sample IAP contains all of the forms that Logistics will be involved with, including the ICS-204, ICS-205, and ICS-206.

Operations Briefing

The Operations Briefing provides the Operations Section Chief with an opportunity to meet face-to-face with the Branch Directors, Division and Group Supervisors, and Staging Area Managers to go over the tactical plan that they'll each be a part of implementing. A portion of the Operations Briefing is set aside for logistical issues, and it provides you with an opportunity to brief the responders on several issues to include:

- Reviewing the Communications Plan
- Reviewing the Medical Plan
- Reviewing the Traffic Plan
- Discussing other logistical issues that may be of concern to the field personnel

Execute Plan and Assess Progress

Once the Operations Briefing is completed, the field personnel will begin implementing their portion of the Incident Action Plan for the operational period. While they're engaged in the tactical operations, you should:

- Meet with the Logistics Section personnel to go over any issues and discuss the section's performance
- Throughout the operational period monitor ongoing logistical support and make adjustments as necessary to deliver the best service possible
- Maintain close interaction with other members of the Command and General Staff

Look in Appendix B to find a one-page summary of the LSC responsibilities during the various *steps* of the ICS Planning Process. The page contains the Operational Planning "P" and calls out specific logistics activities that we presented above.

Briefing the Logistics Section Staff

It's a good habit to get into a rhythm for briefing your staff at least once a shift. Consider conducting your staff briefing following the Command and General Staff Meeting because it occurs early in the ICS Planning Process, and you'll have the latest information from the IC/UC. These internal briefings help keep everyone working as a team. *Figure 8.12* lists some items that you may want to share with your logistics team.

Briefing the Logistics Section Checklist

Consider briefing your staff on the following items:

- ☐ Changes to the incident situation
- ☐ IC/UC expectations
- ☐ Anticipated Logistics Section activities for the operational period
- ☐ Safety and security issues
- ☐ Changes to specific unit tasking
- ☐ Anticipated changes to section staffing
- ☐ Possible changes to operational period and work hours
- ☐ Motivational remarks

Figure 8.12. Conducting Logistics Section briefings following the Command and General Staff Meeting is an excellent way to maintain team cohesiveness.

Demobilization

It's important that, as the incident winds down, the release of resources is conducted in an organized manner. Effective demobilization requires the entire support of the command team. As the Logistics Section Chief, you have four primary responsibilities to support the demobilization effort:

- Demobilize nontactical equipment and Logistics Section personnel in accordance with the Demobilization Plan
- Ensure that all equipment that was checked out, such as radios, computers, and tools are returned and accounted for
- Make any necessary transportation arrangements for incident equipment and personnel
- Ensure that all incident facilities are returned to their original state or better before the incident is over

Closing Thoughts

The importance of a well-oiled logistics machine is best appreciated in the middle of a complex and operationally intense incident response and not in your easy chair reading this chapter of the book. At any time while reclining in your chair, you can go to the refrigerator for something to eat, or to your kitchen sink for a drink of water and stop by the bathroom on the way. Everything is very convenient, but in an emergency response operation this is rarely the case. The basics that we take for granted such as food, water, and sanitation facilities all have to be brought in and maintained throughout the response effort.

The goal of this chapter has been primarily focused on three areas: (1) establishing the Logistics Section; (2) knowing your role in the ICS Planning Process; and, (3) providing direction to your unit leaders so they can meet your expectations and provide the best service and support possible. Above all, we want to reinforce the fact that without logistics there can be no effective, sustained response after the first few hours of an incident.

CHAPTER 9
FINANCE / ADMINISTRATION SECTION CHIEF (FSC)

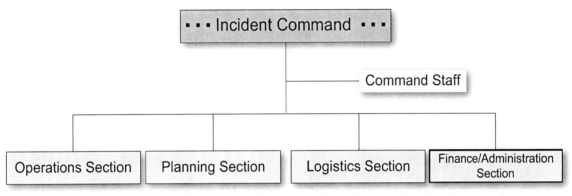

Figure 9.1. Finance/Administration Section Chief in relation to Incident Command.

When responding to an incident, the cost of the response actions is the farthest thing from anyone's mind, especially when lives are at stake; but eventually, when the smoke clears, the water recedes, and the hazardous materials are cleaned up, the expenditures will be scrutinized. On the small incidents that we respond to on a daily basis, funding for the incident response usually comes from the responding agency's operating budget. However, on large complex incidents where there are multi-agencies involved, different sources of funding are being used and response costs are high; financing the response can be complicated. Incident Commanders/Unified Command will have neither the time, nor, most likely, the expertise to manage all of the financial details that must be addressed in support of response operations. This is where you, as the Finance/Administration Section Chief (see *Figure 9.1*), play a key role as a member of the incident management team.

The decision to establish a Finance/Administration Section is generally based on whether the IC/UC determines that it's necessary to have the financial management of the incident done on-site. Some of the factors that influence the IC/UC decision include:

- The financial complexity of the response
- When a Unified Command is established or during a complex interagency operation
- The number of tactical assets deployed (usually measured by the number of divisions/groups established or likely to be established)

As the Finance/Administration Section Chief (FSC), you keep the Incident Commander/Unified Command out of trouble by ensuring that funds are expended appropriately, that contracts are negotiated in accordance with established rules, and that personnel and equipment usage time is accurately recorded. Out of all of the Section Chiefs, your job is the most unique from a management perspective. That is because of all the ICS Sections, yours is the least integrated and your processes the least standardized. *Figure 9.2* illustrates this point by showing the Finance/Administration Section the farthest from the center of the response command team. The closer to the center the ICS function is, the more integrated it is and has to be. As you move away from the center, the integration of the function is not as tight. This doesn't mean that there is not a team effort with clear lines of authority. However, when you get to the Finance/Administration function each organization has its own rules for contracting services, managing cost ceilings, documenting expenditures, and all of the other issues that fall under the Finance/Administration Section.

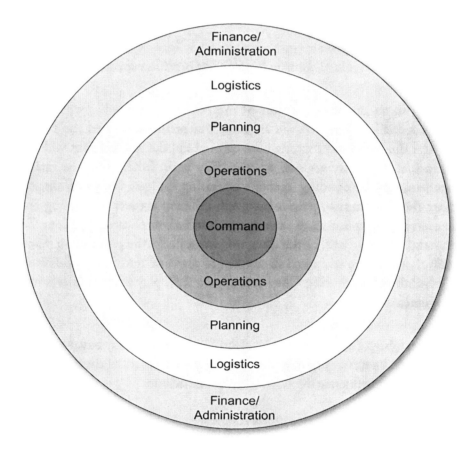

Figure 9.2. The Finance/Administration Section is the least integrated section on the incident management team.

The management challenge you face as the FSC is how to blend all of the various organization requirements for financial protocols while providing top-notch support to the response effort.

As the FSC you have many responsibilities; some of the major ones include:

- Coordinate and ensure the proper completion of response cost-accounting documentation
- Coordinate and manage response ceilings, budgets, and cost estimates
- Provide financial support for contracting services, purchases, and payments
- Identify additional financial service resources or logistics support as needed
- Well before a financial funding ceiling is actually reached, project the "burn rate" and advise the IC/UC when a ceiling must be increased
- With the IC/UC approval, increase various funding ceilings
- Maintain a daily inventory of all accountable equipment purchases by the purchasing agency
- Forward all approved contractor invoices to the appropriate agency processing center for payment, keeping copies for the IC/UC records with the Documentation Unit
- Implement and manage the Finance/Administration Section units needed to accomplish the Section's responsibilities
- As directed by the IC/UC, establish and manage an incident commissary
- Maintain a Unit Log, ICS-214 (see Appendix D)

Actions to Take upon Arriving at the Incident Command Post

You may often find yourself to be the last of the Section Chiefs to arrive on the incident. This has both disadvantages and advantages. The negative side is that you have a lot of catching up to do and depending on how well the initial documentation of finance decisions was made, the climb out of the hole may be steep. On the positive side, the incident brief you receive should be accurate and more detailed than those who arrived during the very chaotic first hours, and you should feel comfortable that you have accurate information to start ramping up your Section. *Figure 9.3* is a checklist of items that you will want to get resolved during your initial briefing.

> **Finance/Administration Section Chief Briefing Checklist**
>
> *At a minimum the briefing should include:*
>
> ☐ Incident situation: magnitude and potential of the incident
> ☐ Information on current Finance/Administration Section activities/status
> ☐ Fiscal limitations or constraints
> ☐ Established incident support facilities
> ☐ Status of any claims
> ☐ Agencies and private sector organizations that are on the incident
> ☐ Funding source(s)
> ☐ Current incident organizational structure
> ☐ Expected incident duration
> ☐ Estimate on the potential size of the response organization
> ☐ Initial instructions concerning the tasks expected of the Finance/Administration Section
> ☐ Command and General Staff priorities, limitations, and constraints and incident objectives
> ☐ Operational period
> ☐ Any accidents or injuries
> ☐ What has been purchased and how procured (e.g., procurement mechanism)
> ☐ Determine if there are any concerns with:
>   o Funding
>   o Contracts
>   o Personnel and equipment time accounting

Figure 9.3. Use this checklist to guide your initial in-briefing.

Finance/Administration Staffing

As you receive your briefing, think about the workload that you face and the staff you need to accomplish all of your responsibilities. Regardless of the agency or organization that you come from, the burden for accomplishing all of the finance and administrative responsibilities should be shared with the agencies and private sector companies that have a financial responsibility or a funding source for the response. As the Finance/Administration Section Chief, you rely heavily on each agency/organization to have representation in your section, so that you can ensure you're meeting the requirements of each.

Establishing the Finance/Administration Section Units

The four units that make up your Section (see *Figure 9.4*) will require personnel who have day-to-day experience in doing that type of work, and for those in the Procurement Unit, the authority to obligate funds.

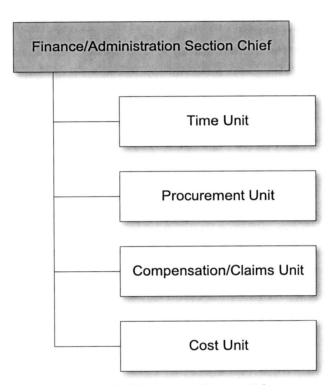

Figure 9.4. This is the typical makeup of a fully activated Finance/Administration Section. The actual size of the Finance/Administration Section will be based on the needs of the incident.

Time Unit

Without accurate tracking of personnel and equipment time on an incident, the cost of the response cannot be computed with any confidence and individual responders may not be compensated for their efforts. If the Time Unit does not do its job correctly, the errors will be compounded. The Cost Unit will take the information on personnel and equipment hours and use it to determine the daily and total costs of the incident operation. There's a high likelihood that the Incident Commander/Unified Command will provide their superiors, the media, and the public with the response costs, so that information has to be accurate. Not to mention that people need to be paid on time!

Directions to the Time Unit Leader (TIME):

- Determine the personnel and equipment time-reporting requirements for each agency and/or organization involved in the response effort
- Ensure that all responders know of the time and method for submitting the daily time sheets
- Maintain a separate log to track overtime expenditures
- Work with the Cost Unit to agree on when the Cost Unit would like to receive the daily totals of personnel and equipment time reports and in what format
- Provide accurate time accounting to any Agency Representatives
- Track personnel and equipment hours against anticipated "burn-rate," update "burn-rate" if necessary
- Determine need, and if approved, establish a commissary for the incident
- Provide completed time records to agencies that are demobilizing from the incident
- Maintain a Unit Log, ICS-214

Procurement Unit

No matter how good the Logistics Section is at either locating the equipment that the Operations Section Chief requires to conduct tactical operations, or finding good incident-support facilities that only require some improvements to be usable, it will be for nothing if the Procurement Unit and Logistics are not working as a team. Timely contracting for services and supplies is critical to the response effort.

Directions to the Procurement Unit Leader (PROC):

- Negotiate all contracts
- Establish local sources for equipment, supplies, and services and notify LSC
- Manage and account for all procurement orders
- Manage and account for all payments
- Document all contracts, procurement orders, and payments
- Report on the status of all contracts
- Administer all financial matters pertaining to vendor contracts
- Maintain a Unit Log, ICS-214

Compensation/Claims Unit

One measure of how successfully a response is being managed is the attention that is paid to claims involving property that individuals make regarding damage from the response activities. The Compensation/Claims Unit needs to act quickly on these claims so that there is no perception that the command team is not taking action. Documentation of responder injuries must also be timely and done with a high standard of care.

> Directions to the Compensation/Claims Unit Leader:
>
> - Investigate all claims involving property associated with or involving the incident
> - Investigate all incident accidents (e.g., vehicle accidents)
> - Ensure that Unit personnel working on injury compensations are coordinating closely with the Medical Unit and the Safety Officer
> - Develop and advertise incident claim process
> - Maintain all files on injuries and illnesses associated with the incident
> - Maintain thorough documentation on all claims (witness statements, photos, etc.)
> - Report on the status of claims processing
> - Maintain a Unit Log, ICS-214

Cost Unit

The importance of the Cost Unit may not be appreciated in the early days of a response, but as things wind down the work the Cost Unit performs in support of incident operations will become critical. Incident Commanders/Unified Command are given wide latitude in determining how best to bring an incident under control. The strategies and tactics that they employ early on may not be as scrutinized, but once a response shifts to more of a project phase the strategies that are used will have to be fiscally defensible. The Cost Unit can be a big help as they collect and analyze where the money is going. To be successful, the Cost Unit must work closely with the Procurement Unit and the Time Unit.

> Directions to the Cost Unit Leader (COST):
>
> - Ensure that personnel and equipment that will receive payment are properly identified
> - Work with the Time and Procurement Units to get all cost data
> - Conduct an analysis of costs and prepare estimates of incident costs
> - Report on documented response costs and projected response costs
> - Maintain accurate information on the actual cost of all assigned resources
> - Identify and distribute the appropriate cost documentation forms

- Monitor direct costs and anticipated costs and track the obligations against various ceilings on a daily basis
- Add up obligations from all sources (contractor, government, etc.) against each fund ceiling (for this reason, it will be important to understand fully the IC/UC decisions about which actions/contracts are directed to be made against which funding source)
- Maintain a Unit Log, ICS-214

FSC Role in the ICS Planning Process

There are several meetings and briefings that you'll attend during each operational period as part of the ICS Planning Process. These meetings and briefs are shown in *Figure 9.5*, the Operational Planning "P." The ones that you attend are called the Command and General Staff Meeting, Planning Meeting, and Operations Briefing.

The Operational Planning "P"

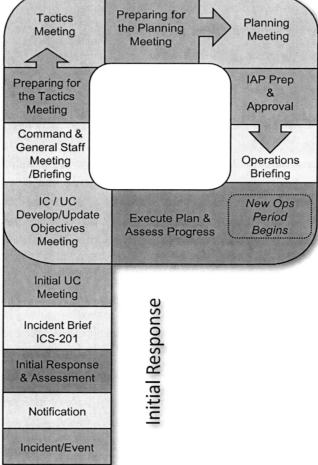

Figure 9.5. The Operational Planning "P" illustrates the various steps in the ICS Planning Process.

Command and General Staff Meeting

The Command and General Staff meeting will provide you with a lot of information on what is going on with regard to the direction of the response effort. The Incident Commander/Unified Command will discuss their priorities, constraints and limitations, and incident objectives, and other items they want to communicate with you and the other members of the Command and General Staff. You'll have the opportunity in this meeting to get clarification on any issues and provide a report on your Section's activities. Some items that you may want to bring up in this meeting include:

- Total incident costs
- Update on any claims
- Any injuries that require follow on medical treatment
- Cost per day (burn-rate)
- Any contracting difficulties
- Any issues with receiving personnel and equipment hours

Planning Meeting

During the Planning Meeting, the Operations Section Chief (OSC) will brief the Incident Commander/Unified Command on their proposed tactical plan for the next operational period. What needs to come out of this meeting is that the IC/UC agrees to the OSC's plan and that you, as the FSC, as well as the other members of the Command and General Staff agree that you can support the plan. While in this meeting there are some things you want to do:

- Obtain information on resource requirements for cost considerations
- Identify high-cost operational items
- Identify actions that may lead to potential claims
- Provide information on anticipated and known fiscal constraints

Operations Briefing

The Operations Briefing is conducted to brief oncoming operations personnel on their assignment for the operational period. This is your best opportunity to pass along to the field personnel any cost or claims issues that they need to be aware of as they go about their work. For example:

- Potential for liability claims
- What to do about problems with a contractor's performance
- How and when the FSC requires accounting of equipment and personnel hours
- Where you want claims referred to (e.g., Compensation/Claims Unit)

Closing Thoughts

A good Finance/Administration Section Chief is critical to the success of the incident management team. The FSC brings specialized expertise that enables the Incident Commander/Unified Command to carry out their jurisdictional responsibilities while providing a cost-effective response effort. An effective FSC can operate in a multi-agency and multi-organization environment and skillfully ensure that the unique fiscal requirements of each entity are met. The FSC must understand the Incident Command System and their role in helping the team to move the response forward. In the months and years following the incident response, the Finance/Administration Section's work products will be highly scrutinized. The products will be important to the settlement of claims and allocation of costs.

CHAPTER 10
AREA COMMAND

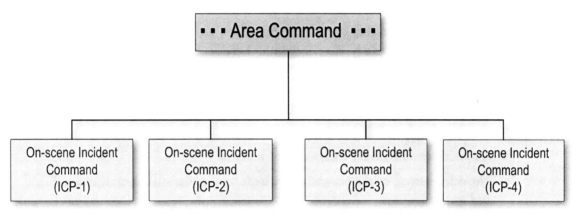

Figure 10.1. Area Command in relationship to on-scene Incident Command.

At 7:25 AM a large magnitude earthquake occurred on the New Madrid fault (the fault lies within the central Mississippi Valley). The earthquake could not have come at a worse time as the roads were clogged with commuter traffic. Initial reports across the city of Saint Louis, Missouri indicate a significant loss of life, hundreds of trapped and injured victims, and substantial infrastructure and property damage. As a result of the earthquake, four on-scene incident management teams have been established around the City to conduct emergency response operations. Due to the complexity of the response effort and demand for response resources, local officials collectively agreed to establish an Area Command organization to oversee the four incident management teams. The purpose of the Area Command is to provide oversight of the incident management teams, focusing primarily on strategic assistance and direction and resolving competition for *critical* response resources. Area Command is an expansion of the Incident Command System and is specifically designed and developed to manage multiple on-scene incidents management teams.

> A critical resource is simply when there is more demand for a particular resource than there are available resources.

At 8:30 AM you are contacted by your Agency Executive (your boss) and told to prepare to deploy and establish an Area Command. As an Area Commander, you will be providing strategic assistance and direction and will be allocating critical resources among the four on-scene incident command teams.

Assuming that you have worked under an Area Command as an Incident Commander (IC/UC), you are sensitive to the fact that your job as the Area Commander is not to replace the individual Incident Commander/Unified Command authority or responsibility, but to provide strategic direction. Execution of tactical operations remains the responsibility of the Incident Commander/Unified Command.

Determination to Activate an Area Command

An Agency Executive/Administrator can determine when an incident is of such magnitude, complexity, or operational intensity that it would benefit from the activation of an Area Command. Factors to consider when deciding to activate an Area Command include:

- Complex incidents overwhelming local and regional government assets
- Overlapping jurisdictional boundaries
- An incident that crosses international borders
- The existence of, or the potential for, high level national/international political and media interest
- Significant threat or impact to the public health and welfare, natural environment, property, or economy over a broad geographic area
- Difficulty with inter-incident resource allocation and coordination
- Major response activities occurring in multiple disciplines or with multiple incident management teams such as search and rescue, fire fighting, and environmental response

Responsibilities of Area Command

Area Command has responsibility for overall strategic management of the incident and will:

- Establish Area Command strategic objectives
- Establish overall response priorities
- Communicate strategic direction to the incident commanders and make timely decisions
- Allocate and track critical resources based on overall response priorities
- Ensure that the incident(s) is properly managed
- Ensure that the on-scene incident objectives are met
- Help to resolve conflict with supporting agencies' priorities
- Communicate, at the commensurate level, with affected parties, stakeholders, and the public
- Ensure that Agency Executive/Administrator direction is implemented
- Ensure that appropriate incident information reporting requirements are met

- Coordinate acquisition of off-incident, unassigned resources such as federal, state, local, and international resources, as appropriate. This coordination may involve other federal, state, and local agencies within the affected areas
- Develop and disseminate strategic guidance and planning information
- Provide direction, guidance, and assistance to the support elements of the incident management teams

Area Command Operating Cycle

As with the incident management teams that are conducting tactical operations, the Area Command team follows a prescribed process for developing and implementing strategic direction to ensure that the incident management teams under its direction are well supported and coordinated. The process is called the Area Command Operating Cycle and is depicted below in *Figure 10.2.*

Area Command Operating Cycle

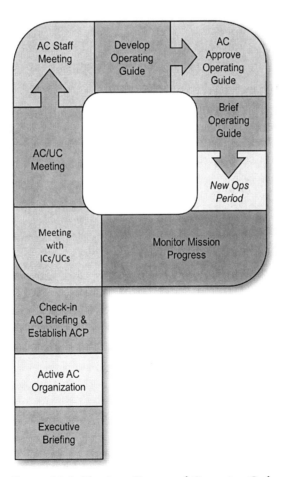

Figure 10.2. The Area Command Operating Cycle

We are going to go through all of the *steps* in the Operating Cycle so you will get a good idea of what occurs at each *step* in the process. We will start with the Executive Briefing and move our way up the stem of the "P" and then around the Area Command Operating Cycle "P." In Appendix Q, you see a one-page summary of your responsibilities as an Area Commander during the various *steps* in the Area Command Operating Cycle.

Executive Briefing

When you receive a call from your Agency Executive to establish an Area Command, there are some important issues to discuss before you deploy. If you receive a good briefing at this stage in the process, it will help avoid confusion later on when you are in the "thick" of the response effort. Items that you want to cover with your Agency Executive include:

- Scope of your authority
- Reporting relationships and responsibilities
- Your organization's/agency's priorities
- Cost constraints
- Critical information reporting thresholds (e.g., serious accidents and/or injuries, loss of life, major operational accomplishments or impediments)
- When you are expected to be operational
- Briefing schedule and content for Agency Executive updates
- Any political, social, economic, and/or environmental concerns
- If predetermined, location of the Area Command Post
- If you will be working in a Unified Area Command, who are the other agencies with Area Commanders that you will be working with
- Other support facilities that are being activated (e.g., Emergency Operations Center [EOC], Joint Field Office [JFO])
- Names of assigned Incident Commanders and locations of Incident Command Posts that you will be managing

If at all possible, try to get a written Delegation of Authority/Delineation of Responsibilities (see Appendix R) that neatly lays out your authority and the limitations and constraints that you will be operating under.

Activate the Area Command Organization

Once you have received your Agency Executive briefing you will need to take some time to determine the initial level of staffing required to establish an effective Area Command organization. If you are working in a Unified Area Command (where there is more than one Area Commander) decisions on staffing and what agencies will fill which Area Command positions will have to be agreed upon. In addition to staffing, you will have to determine where to establish your Area Command Post (ACP) unless your Agency Executive has already decided that in advance.

There are some guidelines that you should try to follow when you select the location of the ACP. To help you make the best choice for locating the ACP, consider the following:

- The ACP should be strategically positioned close to the incidents that you are overseeing.
- You do not want to collocate your ACP with one of the incident management teams. The reasons for this are:
    - It is difficult to maintain a clear line on what the Area Command is responsible for and what the Incident Commanders are responsible for
    - Whether right or wrong, there may be a perception of favoritism since you are collocated with one incident command team and not the others
    - It may become difficult to separate the day-to-day activities that are occurring within both the Area Command and the Incident Command; this invites confusion
- The location that you select for the Area Command Post should have:
    - Sufficient space for your team with the ability to expand should it become necessary
    - Adequate communications capability to enable you to communicate with the Incident Command/Unified Commands, Emergency Operations Centers, Agency Executives, Joint Information Center, and Joint Field Office
    - Adequate backup power to enable 24/7 operations
    - Adequate secure parking
    - Safe and secure work environment
    - Close proximity to a heliport

Check-in, Area Command Briefing, and Establish the Area Command Post

At this *step* in the Area Command Operating Cycle, your initial Area Command team members begin to check-in and it is important for you to give them a good situational briefing. Make sure your briefing includes:

- The current situation
- Area Command's roles and responsibilities in supporting incident operations
- Your expectations
- Initial direction and guidance that you want your team to follow
- Clarification on the overall scope of the assignment, including interaction with the Incident Command/Unified Commands, Joint Field Office (JFO), and Emergency Operations Centers (EOC)

Make sure that you establish some interim operating procedures such as:

- Critical resource ordering and tracking
- Critical information reporting
- Hours of operation and command post staffing
- Meeting and briefing schedule

Once the Area Command team reports that they are prepared to support incident operations you want to notify the Agency Executive, Incident Commanders/Unified Command, EOC, and JFO that your Area Command is operational.

Initial Meeting with the IC/UC

With your Area Command Post established and your team prepared to support the incidents, you are ready for perhaps one of the most important *steps* in the Area Command Operating Cycle—meeting the Incident Commanders. If at all possible, try to have your first meeting with the Incident Commanders face-to-face (follow-on meetings can be by telephone or video conference) and with all of the Incident Commanders present. The first meeting is a golden opportunity to establish a positive working relationship between your Area Command team and the on-scene commands.

You want to walk out of this meeting with a good situational picture for each of the incidents along with a list of critical resources. These two items (incident situation and critical resources) will enable you to start making some hard decisions as to which of the on-scene incident management teams will have priority for receiving the resources that have been identified as critical.

As an Area Commander, one of the most important responsibilities that you have is deciding how best to allocate the critical resources. When you meet with the Incident Commanders/Unified Commands consider having your Assistant Area Command Planning Chief (from now on referred to as AC Planning Chief) begin filling in an Area Command Resource Allocation and Prioritization Worksheet, ICS AC-215, shown in *Figure 10.3*. The AC Planning Chief should be documenting on the ICS AC-215 those resources that the Incident Commanders have identified as essential to their mission and being in short supply and high demand—critical resources.

For example, let's look at our large magnitude earthquake that occurred in Saint Louis this morning. As a result of the earthquake, four on-scene incident management teams have been established around the city to conduct emergency response operations. Due to the complexity of the response effort and demand for similar resources, an Area Command has been established to oversee the management of the four incidents.

As the Area Commander, you want to get a handle on which resources need to be specifically allocated based on some type of incident priority since there are not enough of the critical resources to go around. During your initial meeting with the Incident Commanders, you find out that Federal Emergency Management Agency Urban Search and Rescue (US&R) task forces are a critical resource as they are in high demand and limited supply.

Based on the situational briefings that you received from each on-scene Incident Commander/Unified Command and the consequences that each of the IC/UCs have laid out if they do not receive adequate US&R resources, you are going to have to determine who gets the limited US&R task forces and who does not. Whatever decision you make, it is safe to say that you need to ensure that the criteria you use to allocate the US&R task forces must be well documented. Your decision to provide a critical resource to one incident management team and not another may cost lives. You need to remember that you will eventually have to justify your reasoning so have it documented while it is fresh in your mind.

Using the ICS AC-215, Area Command Resource Allocation and Prioritization Worksheet

The ICS AC-215 is an excellent tool designed to enable your staff to record any identified critical resource requirements and to determine the number of critical resources that are needed to meet the requirements of the on-scene incident management teams. *Figure 10.3* is an example of a partially completed ICS AC-215.

AREA COMMAND RESOURCE ALLOCATION AND PRIORITIZATION WORKSHEET	6. RESOURCE KINDS/TOTALS		US&R task force	Ambulance Type I													2. DATE & TIME PREPARED 12 October 0900	3. OPERATING CYCLE (DATE & TIME) 12 Oct 1800 To 13 Oct 1800
1. AREA COMMAND IDENTIFIER St. Louis Area Command																		
5. INCIDENTS																	7. COMMENTS	
4. INCIDENT PRIORITY																		
Arch		REQ	8	15													Hundreds of injured are reported throughout the incident area. Hazards in collapsed building structures will significantly impede search-and-rescue operations and present a high degree of danger to rescue efforts.	
		HAVE	1	5														
		NEED	7	10														
		RESOURCE PRIORITY	2	1														
University City		REQ	10	13													Several reinforced concrete high-rise apartment buildings are heavily damaged with a high number of trapped and injured victims. Initial assessment indicates a high degree of success if US&R task forces are committed to the incident.	
		HAVE	0	4														
		NEED	10	9														
		RESOURCE PRIORITY	1	3														
Kirkwood		REQ																
		HAVE																
		NEED																
		RESOURCE PRIORITY																
Clifton Heights		REQ																
		HAVE																
		NEED																
		RESOURCE PRIORITY																
		REQ																
		HAVE																
		NEED																
		RESOURCE PRIORITY																
ICS AC-215 RESOURCE ALLOCATION AND PRIORITIZATION WORKSHEET	8. TOTAL RESOURCES REQUIRED																11. PREPARED BY (NAME & POSITION) P. Bakersky AC Planning Chief	
	9. TOTAL RESOURCES ON HAND																	
	10. TOTAL RESOURCES NEEDED																	

Figure 10.3. Area Command Resource Allocation and Prioritization Worksheet, ICS AC-215.

To help you understand how the Area Command Resource Allocation and Prioritization Worksheet is used, we will zoom in on sections of the completed form that you saw in *Figure 10.3*. First, let's look at the upper left section, shown in *Figure 10.4*.

Upper Left Section of the ICS AC-215

AREA COMMAND RESOURCE ALLOCATION AND PRIORITIZATION WORKSHEET		6. KINDS OF	CRITICAL	RESOURCES	US&R task force	Ambulance Type I	
1. AREA COMMAND IDENTIFIER St. Louis Area Command							
4. INCIDENT PRIORITY	5. INCIDENTS						
	Arch			REQ	3	10	
				HAVE	1	5	
				NEED	2	5	
				RESOURCE PRIORITY	2	1	
	University City			REQ	2	7	
				HAVE	0	4	
				NEED	2	3	
				RESOURCE PRIORITY	1	3	

Figure 10.4. A filled-in portion of the upper left side of the ICS AC-215.

Block 1 – Area Command Identifier: This is where you record the name of the Area Command. In our example that would be St. Louis Area Command.

Block 4 – Incident Priority: This block is seldom used outside of the wildland firefighting discipline. In the event that Area Command designates the various incidents under their command based on priority, Block 4 is where you note which of the incidents was priority one, two, three, and so on. The majority of the time Area Command set priorities for which incident is first in line to receive a particular critical resource and not designate an entire incident as having priority over another.

Block 5 – Incidents: In this block you list the names of the various incidents that fall under the responsibility of your Area Command (e.g., Arch).

Block 6 – Kinds of Critical Resources: This is where you record the kind of resources that are determined to be critical. In our earthquake scenario one kind of critical resource is FEMA Urban Search and Rescue task forces. Once the critical resources are identified, the Incident Commanders will tell you how many of those critical resources that he or she will require to accomplish his or her objectives. Your staff should record that information on the ICS AC-215 in the block labeled REQ.

- REQ: For each critical resource that is identified, you want to record the number of US&R task forces that each of the incidents *require* (REQ). For the Arch Incident, eight US&R task forces are required. One thing to remember is that although you may have four incidents working under you, as in our earthquake example, not all of them may require US&R task forces.
- HAVE: Once you know the required number of US&R task forces, you want to know how many US&R task forces are currently assigned to each of the incidents. For the Arch Incident, the number of US&R task forces that the incident has assigned to it is 1.
- NEED: The NEED is simply the difference between the number of US&R task forces that the Arch Incident requires and the number of US&R task forces that the Arch Incident already has checked into the incident.
- Resource Priority: This is where you earn your pay as the Area Commander. The resource priority block is used to record the Area Command's decision as to how he or she will prioritize the incidents that will have precedence in receiving a critical resource. To ensure that you understand how prioritizing incidents works let's walk through an example.

In our earthquake scenario, the Arch Incident and University City Incident both require US&R task forces and ambulances (Type 1) and both of these resources have already been determined to be critical. As an Area Commander, you have made the decision that the University City Incident would have first priority for any US&R task forces, but be third priority for type 1 ambulances (remember that in *Figure 10.3* we showed that there were four incidents under your Area Command). For the Arch Incident, you determined that it was priority 2 for US&R task forces, but was priority 1 for type I ambulances. In *Figure 10.4*, you can see where your staff has placed the incident priority numbers by each incident and under each kind of critical resource.

If you are wondering how you go about making these priority decisions, the answer is that you base them on some established criteria.

Example of Using Criteria to Help You Determine Resource Priority

We mentioned early on in this chapter that it is important to document the criteria that you used to help guide your decision as to why one incident is given a higher priority for the US&R task forces over another incident. You may have received criteria from your Agency Administrator/Executive during your briefing or you may develop some of your own like those listed below:

- Safety of life
- Number of lives at risk
- Risk to responders versus benefit
- Time sensitivity (limited window of opportunity)
- Higher probability of success

Now that we have discussed what is happening on the left side of the ICS AC-215, we are going to move over to the right side of the form and see what information goes there. *Figure 10.5* is a completed section of the upper right side of the ICS AC-215.

Upper Right Section of the ICS AC-215

US&R task force	Ambulance Type I	2. DATE & TIME PREPARED	3. OPERATING CYCLE (DATE & TIME)
		12 October 0900	12 Oct 1800 To 13 Oct 1800
		7. COMMENTS	
3	10	Hundreds of injured are reported throughout the incident area. Hazards in collapsed building structures will significantly impede search-and-rescue operations and present a high degree of danger to rescue efforts.	
1	5		
2	5		
2	1		
2	7	Several reinforced concrete high-rise apartment buildings are heavily damaged with a high number of trapped victims. Initial assessment indicates a high degree of success if US&R task forces are committed to the incident.	
0	4		
2	3		
1	3		

Figure 10.5. Filled in portion of the upper right section of the ICS-AC-215.

Block 7 – Comments: Record briefly some of the relevant information that Area Command will use to determine where the greatest benefit will be gained when assigning critical resources.

What Happens When an Incident No Longer Requires a Critical Resource?

When an incident no longer requires a critical resource (such as the US&R task forces in our Saint Louis earthquake scenario), Area Command is to be notified. This will enable Area Command to re-direct the critical resource to another incident management team that has been waiting for a US&R task force. If there is no longer a need for the task force, Area Command will direct the incident that has the excess US&R task force to demobilize the resource. The Area Command needs to ensure that the Incident Commanders are briefed on how excess critical resources will be demobilized.

Best Practices for your Initial Meeting with the IC/UC

We all know that first impressions are important, so here are some best practices to consider when meeting with the Incident Commanders/Unified Commands for the first time:

- Try to make the first meeting with the Incident Commanders a face-to-face meeting
- Encourage the Incident Commanders to bring their Planning Section Chiefs to the meeting so they hear firsthand what is being discussed and so that they can meet the AC Planning Chief with whom they will be working
- Have the AC Planning Chief start developing a list of potentially critical resources
- Make sure that your staff is prepared to receive and document the situational briefs from each of the Incident Commanders, along with any concerns that they are facing

Responsibility of the Incident Commander/Unified Command

For Area Command to be effective, both the Area Command team and the on-scene incident command teams must work together in a cooperative environment. To facilitate cooperation the Incident Commanders should provide at least the following to the Area Command:

- "Paint" a good situational picture of what his or her team is facing on their incident
- Incident objectives
- Include a listing of resources and any other constraints that are limiting the ability to accomplish the objectives
- An assessment of what the impact to the response effort will be if adequate resources are not received (increased suffering and further loss of life)
- Other areas where the AC could reduce impacts on the incident command teams (VIP visits, national media impacts, political pressures, etc.)
- Hours of operation
- Incident priorities
- Agreed upon operating procedures at the local level
- Progress updates along with hindrances
- Political, social, economic, and environmental impacts

- A copy of the Incident Command Post contact directory
- Long-term projections and/or incident potential
- Copy of any maps/charts/building plans of the incident
- Copy of meeting schedule
- Copy of ICS-201, Incident Briefing Form and/or the Incident Action Plans

Area Command Unified Command Meeting

If you are operating as a sole Area Commander (this would be extremely rare in an all-hazards response), then this *step* in the Operating Cycle is a time for you to set the objectives, priorities, and direction under which the Area Command team and the incident management teams are to operate. However, the reality is that you will most likely be operating under a Unified Area Command and decisions will be made with input from all of the involved Area Commanders. Consider the decisions that the Unified Area Command will have to make:

- Determine critical resources and prioritize which incidents will receive the critical resources and how many resources each incident will receive
- Reach agreement on what work can be shifted from the Incident Commanders to the AC organization
- Develop strategic objectives and priorities
- Establish AC operating procedures
- Identify tasks that the Area Command team will have to perform
- Agree on the division of work among the Area Commanders

How successful the Unified Area Command is in achieving its delegated responsibilities is determined by each individual Area Commander and his or her willingness to work as a team. If the Unified Area Command does not operate with the spirit of cooperation, then the entire Area Command team will suffer.

Area Command Staff Meeting

Let's take stock of where we are in the Area Command Operating Cycle process. You received a briefing from your Agency Executive, assembled your Area Command team, met with the Incident Commanders/Unified Commands that will be working for you, set an initial course that identified strategic objectives, and made some initial critical resource decisions based on what the Incident Commanders told you. Now it is time to gather your Area Command team to make sure that everyone understands the direction you determined. Your briefing should include:

- Any decisions that you have made (e.g., prioritizing the assignment of critical resources, AC Post hours of operation)
- The assignment of any tasks you want accomplished (e.g., immediately initiate

the critical resource ordering process)
- The internal and external reporting process for the AC
- A media and stakeholder intervention strategy
- The AC meeting and activity schedule
- Overall impacts on AC organization (agreed upon shift of work from Incidents to AC)
- Your expectations for documention
- Review of the AC Operating Guide
- Review of AC staffing to ensure it is robust enough to provide appropriate incident support
- Review of AC decisions and direction
- Discussion of potential issues (e.g., potential push back from community leaders who disagree with resource allocation process)
- Identification of any limitations or constraints that Area Command is operating under (e.g., time constraints for obtaining enough life-saving resources within the possible window of opportunity)

The effort you put into providing clear direction and expectations with your staff at the outset of the response will payoff and reduce confusion and frustration for all.

Develop Operating Guide

During this *step* in the Area Command Operating Cycle your staff will be working on putting together the Area Command Operating Guide. This is an excellent opportunity for you to catch up on any issues that you have not had time to address. In addition to any items you have identified, we have a few that you might want to consider:

- Network with counterparts
- Survey the Area Command Post operations
- Talk to staff on assigned task issues and progress
- Liaise with external entities (e.g., EOC, JFO, key stakeholders)
- Communicate with on-scene commanders
- Brief Agency Executive on current direction AC is taking
- Obtain firsthand view of operational areas (over flight or other means)

Area Command Operating Guide

The Area Command Operating Guide is a guidance document that should help to ensure that all players remain focused on what the Area Command roles and responsibilities are and how they will be implemented.

The Area Command Operating Guide is similar to the Incident Action Plan that is produced by on-scene incident management teams. The Operating Guide is designed to provide direction and support to the incident commanders and to ensure that the Area Command organization's activities are coordinated with the incidents under its command. A sample Operating Guide can be found in Appendix S. The actual contents of an Operating Guide will vary depending on the situation, the Area Commander's preference, and external influences such as Emergency Operations Centers and a Joint Field Office. *Figure 10.6* is an example of what may be included in an Operating Guide.

Example Area Command Operating Guide Contents

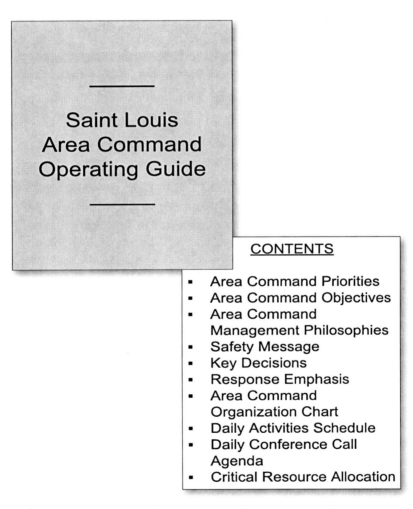

Figure 10.6. Example of Area Command Operating Guide Contents.

Area Command Approves Operating Guide

Once the Operating Guide has been put together, the AC Planning Chief will present it to you (command) for approval. Take time to review the Operating Guide to ensure that it represents the direction and expectations that you established. The Guide should provide information that helps ensure close coordination between Area Command and the incident management teams. Remember that the Operating Guide is part of the incident documentation that will survive well into the future, so make sure that when you place your signature on it, that it meets your standards.

Briefing the Operating Guide

We are almost through the Area Command Operating Cycle and this *step* in the process is where the approved Operating Guide is briefed to the Area Command team, incident management teams, Agency Executive(s), Emergency Operations Center(s), and Joint Field Office.

The AC Planning Chief will make sure that everyone has copies of the Guide and will facilitate the briefing. Your responsibility is to provide any motivational remarks and be prepared to provide clarification on the contents of the Guide.

Monitor Mission Progress

Although we show this as the last *step* in the Area Command Operating Cycle, the reality is that monitoring mission progress is continuous, and an ongoing and essential part of any effective incident management. As the approved Operating Guide is implemented, your Area Command Team will:

- Monitor progress and respond to the needs of the incident management teams
- Develop and use strategic planning tools such as projections, models, forecasts, and other similar information
- Review effectiveness of the Operating Guide and make appropriate changes
- Maintain liaison with entities supporting or coordinating with the AC
- Conduct meetings and briefings
- Ensure situation reports are timely
- Maintain document control system
- Identify critical information issues
- Review AC team staffing
- Maintain status on critical resources
- Maintain close coordination with their counterparts on the incident management teams
- Review and revise operating procedures as needed

Area Command Coordination

As an Area Commander, you may have to work with many entities to effectively support the Incident Commanders/Unified Commands under your command. *Figure 10.7* shows some of the entities that, if established, Area Command will coordinate with.

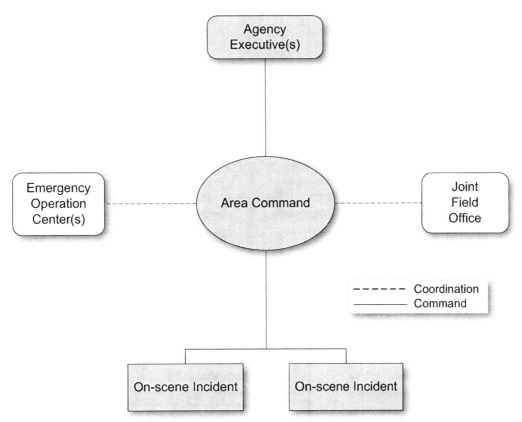

Figure 10.7. Area Command coordinates with several entities such as Emergency Operation Center(s) and the Joint Field Office to support the on-scene incident management teams.

When working with Emergency Operations Centers (EOC) and a Joint Field Office (JFO), Area Command will provide:

- The current status of the situation of each incident under its command which provides a common operating picture
- Any critical resources that the incident(s) needs but that cannot be filled

The coordination relationship between Area Command and the EOC and JFO is a two-way street. EOCs and the JFO provide the needed resources and technical expertise to the Area Command in support of the Incident Commanders.

Area Command Organization

The Area Command organization when fully staffed looks very similar to the ICS organization at the incident level (*Figure 10.8*). However, you may notice two major differences. First, there are far fewer positions in an Area Command organization; and second, there is no Operations function. The role of Area Command in overseeing individual incidents does not require a large organization and Area Command should never be involved in tactical operations; therefore an Operations Section is not required. The last thing that Incident Commanders/Unified Command would want is for Area Command to get involved in the direct use of resources.

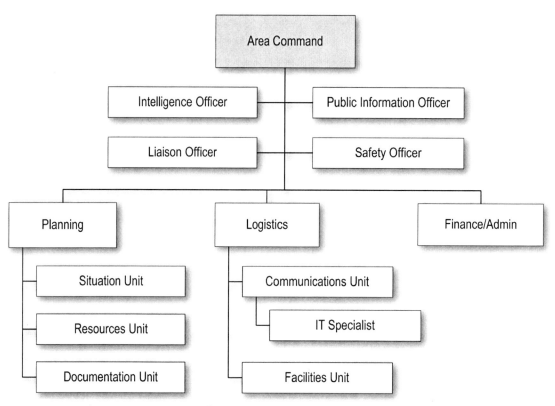

Figure 10.8. The ICS Area Command supports on scene incident operations, but does not, in any way, supplant on-scene incident operations or functions.

Remember that *Figure 10.8* represents a generic Area Command organization. The organization you build will need to meet the demands of the incident(s) that you are facing to ensure appropriate support.

Unified Area Command

Although most of our discussion on Area Command has focused on you as a sole Area Commander, the reality is that you will most likely be operating with other Area Commanders under a Unified Area Command. The agencies that make up a Unified Area Command may closely mirror the agencies that make up the Unified Command on-scene. For example, if the Unified Command on scene is made up of the fire department, law enforcement, public works, and emergency medical services, the Unified Area Command may have a similar makeup. This is important because when Area Command is making critical resource allocation decisions that are directly impacting on-scene operations you want senior officials from those agencies to be involved in the decision making. *Figure 10.9* shows a Unified Area Command organization. Notice that the only difference from the Area Command diagram in *Figure 10.8* is the top of the organization.

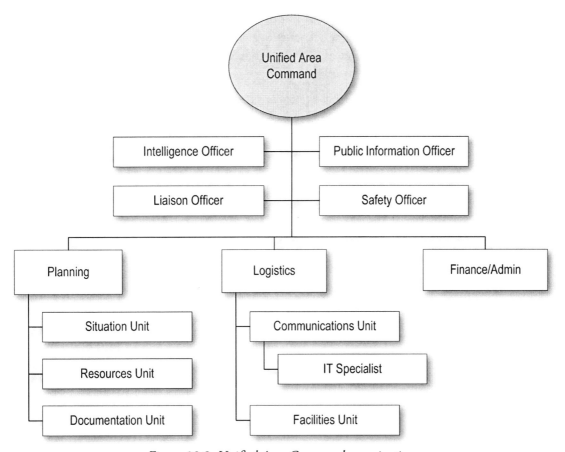

Figure 10.9. Unified Area Command organization.

Best Practices for a Successful Unified Area Command

- Each Area Commander must have the authority to speak for his or her agency and to commit resources and funds
- Area Commanders must collectively agree on an overall direction, priorities, objectives, and decisions
- Each Area Commander's jurisdictional authority is not compromised or neglected
- Each Area Commander should stay focused on providing strategic direction and avoid getting down into the weeds on issues

Closing Thoughts

As an Area Commander, you have an enormous responsibility that can have far-reaching impacts on life, property, and the environment. Even though you are not directly involved in tactically managing the incidents, the direction, coordination, and support that you and your team provide will directly influence the overall outcome of response operations under your authority. You have to use all of your leadership, interpersonal, and organizational skills to effectively provide the level of management that your Agency Executive expects of you.

Success and/or failure depends on how well you are able to meet your Agency Executive's expectations, your team's ability to effectively communicate and coordinate with both the Incident Commanders/Unified Commands under your direct authority and those entities (e.g., Emergency Operations Centers, Joint Field Office) that provide essential assistance needed to support the overall response effort.

CHAPTER 11
MULTI-AGENCY COORDINATION

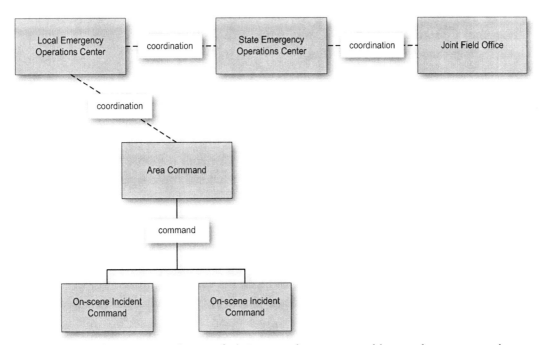

Figure 11.1. Incident Commanders/Unified Commands are supported by a multi-agency coordination system that can be activated based on the complexity, duration, and the resource demands of the response effort.

We have used several different scenarios throughout this book to help explain the Incident Command System (ICS) and the specific responsibilities of several key ICS positions. There was the Meridian flood incident that required an aggressive search-and-rescue operation; the Deer Park incident that involved a private plane crash in a residential neighborhood, resulting in injuries and property damage; and, the Animas incident that involved a forest fire and an oil spill. Each of these incidents was managed under the authority of an Incident Commander (IC) or a Unified Command (UC) that had direct responsibility for conducting a safe, effective, and efficient response. In addition, we also had the St. Louis earthquake scenario that resulted in a tremendous loss of life and infrastructure damage that was managed by several incident management teams operating under an Area Command.

What we did not talk about, but what is a critical component of the emergency management system in the United States is the multi-agency coordination that provides support to the Incident Commander, Unified Command, and Area Command. This chapter is designed to provide a brief introduction on the role of the Emergency Operations Center and Joint Field Office during response operations and their role and relationship to the IC/UC and Area Command.

Some general definitions to keep in mind as we move through this chapter:

> *Incident Commander/Unified Command* — Performs primary tactical-level, on-scene incident command functions. The Incident Commander is located at an Incident Command Post at the incident scene.
>
> *Emergency Operations Center* — Coordinates information and resources to support local incident management activities.
>
> *Area Command* — Oversees the management of multiple incidents. Area Command may be unified, and works directly with Incident Commanders.
>
> *Joint Field Office* — Coordinates the Federal government's response efforts in support of the impacted state. The JFO is a temporary federal multi-agency coordination facility.

To help describe the relationships and interaction among all of these entities we want to use the St. Louis earthquake scenario that was described in Chapter 10, Area Command. The earthquake caused catastrophic damage all across the city of St. Louis with a significant loss of life, hundreds of trapped and injured victims, substantial infrastructure and property damage, and levee failures that inundated low-lying areas.

With the earthquake incident as the backdrop, let's look at how these different entities come together to provide a coordinated response. To make our discussion easier to follow we will use a "building block" approach in building the entire organization that is shown in *Figure 11.1*.

Although we are using a "building block" method to show you the multi-agency coordination, the reality is that the response organization does not necessarily build from the bottom up. In most cases an Emergency Operations Center (EOC) is established and no Area Command or Joint Field Office is ever activated. The actual makeup of the multi-agency coordination system is incident specific and established with only those entities that are necessary to bring the incident to a safe and rapid conclusion.

We are going to look at three particular "building blocks":

- On-scene Incident Command and the EOC
- On-scene Incident Command, Area Command, and the EOC
- On-scene Incident Command, Area Command, the EOC, and the Joint Field Office

On-scene Incident Command and the EOC

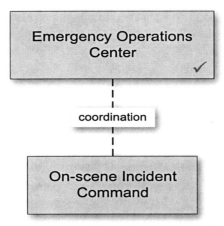

Figure 11.2. Emergency Operations Center in relationship to on-scene Incident Command when there is no Area Command established.

Emergency Operations Center (EOC)

EOCs are predesignated facilities established to provide a location for the various agency representatives who are involved in a response to coordinate their agency's activities. Unlike the organizational makeup at the incident level, an EOC is normally staffed with a broader range of community-based agencies and organizations that are needed to support the overall response effort. Examples of the agencies/organizations that you might see in an EOC include:

- Public officials
- Emergency management staff
- Public utilities
- Public health and welfare (hospitals)
- Military organizations (e.g., National Guard)
- Public works
- Nongovernment organizations (e.g., Red Cross, Salvation Army)
- Public safety (law enforcement, fire, search and rescue)
- Humane Society

EOCs can be found at all levels of government (local, regional, state, and federal) and within private industry. An EOC does not manage the on-scene operations—that is left to the Incident Commander/Unified Command, but the EOC is an integral part of response operations, providing coordination, resource management, and communications. In addition, EOCs collect, analyze, and disseminate information used to provide a common situational picture for decision makers and other entities that require incident status. *Figure 11.2* shows the relationship between on-scene incident command and an Emergency Operations Center.

Although all EOCs perform the same function, they come in all sizes and stages of readiness. A large city like Los Angeles, California, has a well-organized EOC that is staffed 24-hours-a-day 7-days-a-week and is designed to readily expand to accommodate a large influx of agency representatives. Other jurisdictions may only have an EOC that is used on an as needed basis and has only limited space.

EOC Coordination with On-scene Incident Command

One of the most important coordination relationships that an EOC conducts is its relationship with the on-scene incident management team. To ensure that there is no confusion on how the EOC and incident management team will support each other and not duplicate efforts, EOC managers work with the on-scene command team to ensure that their activities are coordinated. It is essential that the EOC team and incident management team work seamlessly to ensure that the impacted population is provided the most professional service possible. Some of the issues that need to be discussed and agreed upon between the EOC and incident command include:

- Information reporting requirements
- Critical information reporting threshold
- Resource ordering process
- Incident documentation process
- Incident closeout procedure
- Areas of responsibility
- Media management issues
- Community outreach issues
- Coordination of VIP visits
- Coordination procedures when dealing with other response entities, including the state EOC and the Joint Field Office (JFO).

Many incidents are routinely responded to around the country without any involvement of an EOC. EOCs generally become involved when the complexity and duration of an incident require a higher degree of coordination among responding agencies. Depending on the incident complexity, there may be many jurisdictions such as the Coast Guard and/or functions such as law enforcement involved and it is the coordination role of the EOC during these types of incidents that will play an important role in ensuring that the responders on-scene are properly supported.

Whether an EOC is coordinating directly with the on-scene incident command or through an Area Command some of the potential areas for interaction include:

- Information reporting requirements, along with timetable for communicating that information
- External situational awareness (what is going on in other areas that could adversely affect support to the on-scene incident commanders)
- Locating and dispatching resources within and external to the EOC's area of responsibility. Resources can include: response equipment, supplies, and personnel
- Interaction with a broad range of community organizations (e.g., utilities, public works, public health and welfare, community action groups, local officials and political entities, nongovernment organizations)
- Liaison between the local and state EOC
- Maintain awareness of the public and local political perceptions of the response effort
- Media feedback
- Manage local mutual aid agreements
- Special interest groups support or adversarial relationships
- Community outreach focus areas
- Key contacts within the community
- Local support contracts for billeting, fuel, and food service

The St. Louis earthquake scenario...

Immediately following the earthquake, the St. Louis Emergency Operations Center (EOC) activates its plan for full activation of the EOC that includes staffing the EOC with representatives from many organizations including the Coast Guard. The Coast Guard, due to its jurisdictional authority on the Mississippi River and its resource capability for on-water search and rescue, is one of the many agencies that the EOC had pre-identified to provide representation at the EOC during an earthquake event of this magnitude. Upon request by the EOC, an agency representative from the Coast Guard, Bob Ward, is sent to the EOC to maintain a Coast Guard presence.

The Coast Guard is just one of many agencies that have been requested to support the EOC throughout the response to the earthquake, and similar to many agencies, the Coast Guard has a pre-identified location within the EOC from which Bob will coordinate Coast Guard operations. Bob's role during this type of incident is twofold: first, Bob will provide situational information to the EOC on Coast Guard activities so that those activities become part of the overall common operating picture. For example, Bob will work closely with the EOC to ensure that time-critical information is shared such as when the Coast Guard closes the Mississippi River to all vessel traffic—the Mississippi River is critical to the economy of the area.

In addition to situational information, Bob is the liaison to the EOC for Coast Guard resources. If the EOC receives a request from an Incident Commander/Unified Command for more rescue boats and crews, Bob may be asked if the Coast Guard can support that resource request. If the Coast Guard can fill the rescue boat request, Bob will coordinate with the Coast Guard to ensure that the resources are dispatched.

The ability of EOCs to bring in multiple agencies to help coordinate response efforts is what makes an EOC a vital link to the on-scene incident commands in support of their response efforts.

On-scene Incident Command, Area Command, and the EOC

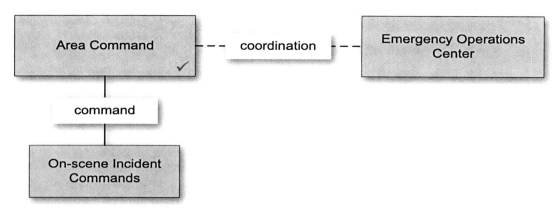

Figure 11.3. Relationship between the on-scene Incident Command and the Emergency Operations Center when Area Command is established.

Unlike an EOC, an Area Command has direct authority over the activities of those incidents that are operating under it (*Figure 11.3*). Some of the most critical responsibilities of Area Command are to:

- Establish overall response priorities
- Allocate critical resources based on overall incident priorities
- Ensure that the incident(s) is properly managed
- Communicate, at the commensurate level, with affected parties, stakeholders, and the public
- Ensure that Agency Executive/Administrator direction is implemented
- Ensure that appropriate incident information reporting requirements are met
- Coordinate acquisition of off-incident, unassigned resources. This could include federal, state, local, and international resources as appropriate. This coordination may involve other federal agencies and the governor(s) of the affected state(s)

Once an Area Command has been established, it will work closely with the EOC to ensure that critical resources required by the on-scene incident management teams are filled. In addition, the Area Command ensures that the EOC is provided with situational information to support its ability to create and disseminate a common operating picture of events taking place on-scene.

Figure 11.4 depicts the relationship among the on-scene incident commands, an Area Command, and an EOC. As you can see in the figure, those on-scene command teams that are not under an Area Command coordinate directly with the EOC. Where Area Command is established it assumes the responsibility for EOC coordination relieving some of the coordination burden on the on-scene teams.

Relationship of the On-scene Command, Area Command and the EOC

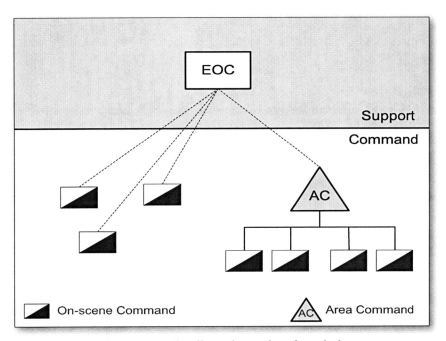

Figure 11.4. On-scene incident command will coordinate directly with the emergency operations center unless an Area Command has been established to oversee the on-scene activities.

The St. Louis earthquake scenario...

Hours after the earthquake occurred, an Area Command was established to oversee the operations of four on-scene incident management teams (Arch Incident, University City Incident, Kirkwood Incident, Clifton Heights Incident). Once the Area Command became operational it took over the coordination activities between the on-scene teams and the EOC. One of the very important roles of the Area Command is to determine the assignment of critical resources. Critical resources are those that are in high-demand but short supply. One kind of resource that met the definition of a critical resource in response to this earthquake was type 1 ambulances (Type 1 refers to the capability of a resource and in this case a type 1 ambulance is the most capable and provides advanced life support). Each of the incident management teams needed type 1 ambulances in order to effectively respond to the tremendous demands involving life safety. Unfortunately, there were not enough type 1 ambulances within the St. Louis area to meet the demand for this type of ambulance.

The Area Command worked directly with the EOC to fill the requests for ambulances. Using its established relationships and procedures to obtain resources outside the impacted area, the St. Louis EOC was able to bring in type 1 ambulances from other jurisdictions to help fill the need identified by each of the incident management teams.

By working with the Area Command and coordinating with other EOCs and organizations, the St. Louis EOC played a vital role in locating and obtaining critical response resources that enabled the incident management teams to accomplish their objectives. Resource management is just one of the important functions that the EOC is responsible for in support of response operations.

If an Area Command had not been established, the EOC would work directly with the on-scene teams.

On-scene Incident Command, Area Command, the EOC, and the Joint Field Office

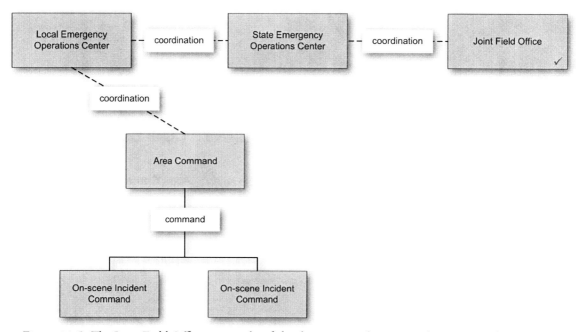

Figure 11.5. The Joint Field Office ensures that federal support to the impacted state is coordinated with broad representation from numerous agencies and private industry.

In the event that a disaster occurs and overwhelms a state's ability to respond (such as in our St. Louis earthquake scenario), the governor of the impacted state may request that the president declare the incident a major disaster. A presidential disaster declaration makes available to the state a tremendous amount of federal resources along with significant financial support.

To ensure that the federal support to the state is well coordinated, the Federal Emergency Management Agency (FEMA) will establish a Joint Field Office (JFO). In *Figure 11.5* you can see that a JFO plays a coordinating role and works directly with the state emergency operations center. The JFO has representation from the state and a wide range of federal agencies such as Health and Human Services, Environmental Protection Agency, Army Corps of Engineers, and many others.

The needs of the impacted state will determine which federal agencies will become part of the JFO organization. In addition to federal agencies, the state will have representation in the JFO; and, as needed, nongovernmental organizations (e.g., American Red Cross), tribal representatives, and private sector organizations. The combination of federal agencies, state representation, and others in the JFO ensures that all activities of the Joint Field Office are closely coordinated.

The St. Louis earthquake scenario…

As the Arch Incident Unified Command gained a better understanding of the situation to which it is responding, the Unified Command recognized that the need for highly capable Urban Search and Rescue (US&R) task forces was critical. The number of buildings that collapsed and potential for trapped victims overwhelm the limited number of search-and-rescue resources that were available. The Unified Command notified Area Command of the number of US&R task forces that would be necessary to search damaged buildings and extract potential victims. It was the Area Command's responsibility to locate the additional US&R task force resources.

After receiving the request, the Area Command determined that it could not fill the US&R needs of the Arch Incident and contacted the St. Louis Emergency Operations Center and asked the EOC to fill the Arch Incident's US&R request.

Upon receiving the request for US&R task forces, the St. Louis EOC staff knew that there were no available task forces and immediately forwarded the resource request to the state EOC. The Missouri State Emergency Operations Center also had no available US&R resources and contacted the Joint Field Office requesting federal support for US&R resources.

The Joint Field Office had five US&R task forces available and assigned two of the task forces (the Salt Lake City and Colorado task forces) to support the US&R resource request from the Arch Incident Unified Command.

It is through the effective coordination and well-established protocols and procedures of the multi-agency coordination system that the Unified Command for the Arch Incident was able to receive the critical urban search-and-rescue task forces that were needed to save lives.

Closing Thoughts

Whether you are an Incident Commander working directly within the disaster area, or an EOC manager working at the state level, understanding how the multi-agency coordination system works is critical. The information in this chapter provides only a glimpse of the system and how the main components (on-scene incident command, Area Command, local EOC, state EOC and Joint Field Office) work together to ensure the most effective response.

The needs of an incident will determine which of the various components of the multi-agency coordination system will be activated to support the response effort. The vast majority of incidents will never rise to the level of complexity or severity to require anything beyond the activities of the on-scene incident commander; however, for those rare instances that require extraordinary coordination from all levels of government and private industry the United States has a proven system in place.

Understanding the system through training and exercises is the best way to ensure that, when it is needed, the components of the system can be put into motion with minimal delay and with clear expectations of what each component is doing.

For more information on multi-agency coordination, we recommend that you take some of the independent study courses offered by the Federal Emergency Management Agency's Emergency Management Institute (EMI). You can find the address for EMI's Web site in the Conclusion section of this book.

CHAPTER 12

EMERGING COMMUNICATIONS

"It's not a big deal to simply use new technology to do old things better. What's revolutionary is using new technology to do new and better things." (Castro and Atkinson)

It has been approximately 14 hours since the large magnitude earthquake occurred, unleashing tremendous destruction across St. Louis and the surrounding communities. As a Battalion Chief with the Consolidated Metro Fire Department, you are serving as an Incident Commander on the Arch Incident Unified Command that is responsible for response operations in the east portion of the city.

Your incident management team has been struggling to obtain accurate and timely information on the number and location of trapped victims, extent of damage to critical infrastructure, status of the numerous uncontrolled fires, and a good count on the displaced population and number of those that were killed. This information is critical to the Unified Command's ability to prioritize the limited resources at its disposal and to be able to communicate the extent of the disaster to superiors and the public. The ability to fill these essential information gaps solely using traditional forms of communications and members of the incident management team will likely take days or longer. This is an agonizing timeframe that will hamper the Unified Command's ability to conduct efficient and effective response operations during those critical hours where lives can be saved. It will also significantly hinder the Unified Command's ability to complete critical off-site reporting requirements to external entities such as the Emergency Operations Center, Joint Field Office and responding agencies.

What if there were other avenues available to the Unified Command and the incident management team to gain information on what is happening on-scene and where it is happening? What if you had at your disposal hundreds of eyes and ears out in the disaster area that could feed the command team with vital information? Would you be interested in leveraging that capability? If so, read on. The chapter Emerging Communications references the world of New Media (a term to indicate the emergence of digital, computer, network and communication technologies) that has the potential to assist in reducing the information gaps that can frustrate responders.

One of the great strengths of the Incident Command System (ICS) is its flexibility to be applied to any type or size of incident. As an Incident Commander (IC), you never know for certain exactly what the incident is going to throw at you and the adaptability of ICS enables you to quickly and efficiently adjust to changing situations. Throughout the previous chapters of this book we have discussed the ICS organizational structure in detail, provided you with time-tested action checklists, and dissected the ICS Planning Process. You need this knowledge to perform your responsibilities as the IC and to use the Incident Command System to its full potential. However, this chapter titled Emerging Communications is unique from the others in that it has not yet been incorporated into the Incident Command System, but it is a topic that we thought was important to include before going to publication to raise awareness among emergency responders

It is because ICS is such a flexible and proven system that we can introduce you to the world of New Media and point out how it might be integrated. Even as we write the words on this page, emergency management professionals across the nation are trying to figure out how to integrate the numerous and ever changing forms of communications into the Incident Command System. Our goal in this chapter is to introduce you, as the Incident Commander, to the forms of New Media, why it is important to you, the challenges that you will face in making New Media part of your response "tool box," and where the future of New Media lies, in supporting incident management teams.

What is New Media?

New Media is the use of various websites combined with social interaction to engage and participate in conversation and communication – but instead of one-to-one conversation, it is a one-to-many conversation. New Media goes beyond static web pages to include tools that let people collaborate and share information.

You might have seen the headlines – More than 270 million Americans, or about 87% of the population, own a wireless device, and about 90% of those people are within three feet of their mobile devices 24 hours-a-day. This availability of communications provides both tremendous opportunities and challenges for you, as the Incident Commander and your command team.

To give you an appreciation for the power of New Media consider that in 2009:

- There were more than 32 million users of Twitter™
- There were over 275 million unique monthly visitors to Facebook©
- YouTube™ added more than 65,000 new videos each day
- MySpace™ had more than 300 million users
- 5.5 million subscribers in North America pay for OnStar global positioning service.

What makes all of these conversations and communications possible is flexible technology that can move digital messages from one person to another, or even to thousands of people, in less than a minute.

One example of the power of New Media was the 2007 Southern California wildfires, where emergency services discovered a new and very effective communications tool to alert those in harm's way; emergency alerting systems that delivered more than six million messages including nearly 3 million voice messages and about 1.2 million email and text messages to almost 400,000 people. More than 96% of these messages were reported to be successfully delivered.

New Media allows people to organize without organizations, using technologies such as blogs, Facebook, Twitter. People are incredibly quick to share what they know, and raise questions about what they want to know. These types of sites provide the user 24-hour access to people around the world and have proven to be a powerful addition to pre-disaster communications such as in the Southern California example above as well as post-disaster communications.

New Media has its own language and to the uninitiated it can be a bit overwhelming. We have listed some of the current components that make up New Media along with an explanation of each. You have probably heard of some of these and even used a few.

Text Messaging

Text messaging or texting refers to the exchange of brief written messages between mobile phones, over cellular networks. Text messages are particularly useful in situations where the participants are mobile or are using mobile or wireless communications devices such as cell phones. It is sending short messages from one person to another and moves instantly. Text messaging has proven to be more resilient than cell phones during disasters due to the use of a different technology which transmits packets of digital information very quickly.

Blogs

Blogs, from the phrase "Web log" is a type of website, usually maintained by an individual providing regular entries of commentary, descriptions of events, or other material such as graphics or video. Entries are commonly displayed in reverse-chronological order.

Many blogs provide commentary or news on a particular subject; others function as more personal online diaries. A typical blog combines text, images, and links to other blogs, Web pages, and other media related to its topic. The ability for readers to leave comments in an interactive format is an important part of many blogs. Most blogs are primarily textual, although some focus on art artlog photographs (photoblogs), sketches (sketchblogs), videos (vlogs), music (mp3blogs), and audio (podcasting).

Technorati™ is an online tool to search for relevant blogs (www.technorati.com) and Wordpress is a specific software site, among many others, that can be used to create a blog (www.wordpress.com).

Twitter (micro-blogging)

Twitter is a short messaging tool from a free social networking site that keeps people connected with one another and with sources of information. With Twitter, group members can follow someone's activities or allow their activities to be followed by others. Post what you are doing up to any moment and others can keep tabs on you.

Twitterers, those who use Twitter can simply follow another's messages by finding a person's username and selecting the "follow" option. Limited to 140 characters for each "tweet," messages are short and to the point. Twitter can be a great resource for fast information exchange. (www.twitter.com)

Facebook and MySpace

These sites allow users to create or join groups, chat with friends, and post photos and videos. Most importantly, trusted friends connect (or 'friend') their friends creating a network of loose and weak ties both locally and distributed. In a disaster existing networks can share critical information with each other or a page could be easily created for a specific incident. Interested participants could join as friends to exchange ideas, concerns, ask questions, post photos or videos and provide status updates (www.facebook.com; www.myspace.com).

Wikis

A wiki is a collaborative web page where participants can post information that relates to a specific topic. The participants see the information instantly and can comment on it, add to it or correct it. The best known is Wikipedia ®, the online encyclopedia. With internal organization wikis, access is limited to those people who are trusted and who are given access.

Baby boomers have always relied on face-to-face meetings or conference calls, but for more and more users, wikis can provide greater flexibility for participants and have the added advantage of transparency (www.wikipedia.com).

Flickr® (and other photo & image sites)

Flickr ® is an image and video hosting website and online community platform. In addition to being a popular website for users to share personal photographs, the service is widely used as an online photo repository. Flickr ® claims to host more than 3.6 billion images. Flickr ® is useful for posting and organizing pictures taken from a disaster scene showing the destruction, response, and recovery activities (www.flickr.com).

YouTube

YouTube is a website for uploading and sharing videos ranging from personal home videos to businesses posting product demonstrations and information. Content is very broad, including user generated movie clips, TV clips and music videos, For both personal and business use, the videos can then be linked to other web pages, either personal or government/corporate. Watchers can also rank the videos, provide comments, and trade thoughts about the clips. YouTube can be very useful for preparedness campaigns, training purposes, and demonstrating appropriate protective actions attached to notifications and warnings. In addition, YouTube is useful for sharing video from a response or showing recovery activities (www.youtube.com).

Ning ©

Ning is a tool that allows users to search existing online communities and create networks around a similar topic by bringing together and collating blogs, photographs, videos, maps and news feeds. Users can request information from the broader network, or provide their own expertise to the subject. Networks can be established around a specific topic such as air tanker operations, or water purification, by bringing together individuals who can comment on and share information relevant to the topic. (www.ning.com)

In addition to understanding the Incident Command System, you and other members on the incident management team such as the Public Information Officer and Situation Unit Leader should spend time learning about these new communication tools and understanding their capabilities since these technologies can be a powerful new resources for you.

Why New Media is Important to you the Incident Commander

Accurate and timely information during emergency response operations is often difficult to obtain, which is ironic given the tremendous growth in the ways in which we can rapidly communicate with one another. From VHF radios to Twitter, you as the Incident Commander have more communications capability than ever before, but you have to know how to tap into that capability. Unleashing the benefits of New Media will help you get answers to critical questions that in the earlier hours and days of a response can leave you frustrated if left unanswered, but you have to know how to use New Media to your advantage.

New Media will help on the ground. It will create significantly improved situational awareness, increase the speed and accuracy of communications, improve the response, allow for better resource allocation and increase abilities to manage the ever-increasingly complex incidents and disasters that we are called upon to fight.

Challenges of New Media for the incident management team

The tremendous benefits from New Media do not come without effort. A commitment to using it has its challenges, some include:

- Relying on information that is coming from unknown sources and where information has not been vetted for accuracy
- The volume of information can quickly overwhelm a management team
- Incident personnel are unfamiliar with New Media and how it can be used during a response to support incident operations
- There is a lack of equipment in the Incident Command Post to support the use of New Media
- Agency and industry policies prohibit robust use of social networking sites

Benefits of New Media to an incident management team

With the challenges that New Media poses clearly understood, here are some of the benefits:

- Access to streams of information that can provide clarity on the incident situation
- Leverage the eyes and ears of the public (human sensors) whose information gathering abilities far exceed that of even the best incident management team
- Access to the public to get the word out on not only what situation is but what you as the Incident Commander want the public to know
- Awareness and participation in New Media will allow your Public Information Officer to:
    - Track misinformation and rumors and to insert corrected information; to direct individuals to trusted sources of information
    - Provide audio-and video-based information about available resources or protective actions
    - Determine public sentiment toward the overall response.

Actual uses of New Media during Response Operations

Virginia Tech School shooting 2007

Research on information sharing during the immediate aftermath of the Virginia Tech school shooting revealed that the social networking technology, Facebook, was instrumental for safety and welfare checks, for connecting with friends, and for posting messages of support. Facebook also served as a mechanism for network groups to coordinate and track information about the injured and deceased. Decentralized and distributed participants used their Facebook connections to accurately compile lists of the deceased, developing norms for turn-taking and validating the veracity of shared information.

In the Virginia Tech incident, New Media served as (or had the potential to serve as) the mechanism for communication among persons affected by a disaster, crisis, or large scale event. It was the means to connect individuals through distributed networks into affinity groups; it was a channel to obtain key life-saving information; it was a mechanism to engage others who had information or needed information.

Red River Flood 2009

During the Red River floods in North Dakota, community members battled the rising river by filling sandbags and shoring up dykes and levees. Local online government-run coordination points served as one source of information about volunteer efforts. Parallel volunteer coordination activities were self-initiated by members of the public through a Facebook fan group, mobilizing youth and young adults who were already connected through this online technology. Twitter users also posted continual updates about the rising flood waters, need for additional personnel resources, and locations at which volunteers could assemble to help. Twitter also served as a secondary channel for relaying evacuation orders, disseminated through a local reverse-911 system. These 'code red' posts were broadcast to the larger Twitter audience following the flood related activities, potentially alerting additional persons at risk.

The value of New Media during the Red River flooding demonstrates the usefulness of network technologies for in disaster response. Technologies were quickly utilized to reach a broad audience while individual volunteers contributed to the creation and distribution of information. Volunteer efforts, however, were not limited to those who resided within the impacted community or to those who were directly affected by the hazardous event. The efforts of grass-roots volunteers include not only those who are within a local community. Now the world-wide online community can also assist and participate in information seeking and sharing about the disaster.

Local and state, and federal government agencies can also use New Media to effectively, fairly, and legitimately respond in disaster. Public officials can use New Media, such as Twitter, as an additional channel to disseminate alerts and direct people to complete warning messages online. Warning messages can be sent via podcast, through audio or video, in real time using an enterprise system such as www.dailysplice.com to provide instantaneous updates to persons seeking information from official sources. Additional podcasts can be used throughout the response to ensure continuous communication from an authoritative source.

Public officials can also use Web 2.0 to monitor public reaction and sentiment to a disaster response as well as identify hot spots or problem areas that are discussed across various media online. Web 2.0 refers to the continued evolution of the internet and its networks, user interfaces and applications all supporting a global community that interacts, collaborates and shares information through text, audio and video. While informal information exchange is a common occurrence in any disaster and the activities of verifying to confirm information and make decisions about protective actions are routine, the opportunity to observe network communications about these collective activities has never before been available.

Closing Thoughts

The world around us is changing rapidly with regard to how we communicate. This chapter on Emerging Communications is recognition that, as responders, we are not shielded from the tremendous growth in the way people communicate and its impact on you, as the Incident Commander, and your command team.

These few pages that you have read on New Media only scratch the surface of an emerging technology that provides enhanced communications capability. In addition to the technology itself, New Media brings with it the world's citizens who are an untapped resource in the transfer of information that can benefit the efforts of your incident management team.

How to best integrate New Media in an Incident Command System will take time to figure out. Policies, processes, and protocols will have to be written and tested through exercises and real life events. Perhaps one day we will see on the ICS-207, Organization Chart a new unit under the Planning Section titled Technical Unit whose job it is to capture, validate, and share information coming from the various forms of New Media. Maybe we will see an entire new Section established, if the incident is complex enough to require that level of organizational support to manage the inputs coming from New Media sources. Since the relationship between ICS and New Media has not been defined, these are only speculations, but we know the technology exists and that the Incident Command System is flexible enough to accommodate new ways of doing business.

By understanding New Media and its benefits and challenges you will enhance the time-tested processes of the Incident Command System far beyond that dreamed by its developers. Timely information is a highly sought after commodity during a response operation and embracing the benefits of New Media will help close the critical information gaps, both coming into and going out of the Incident Command Post.

The impact of New Media on response operations is not limited to on-scene command. Those working in Area Command, Emergency Operations Centers, and Joint Field Offices face the same benefits and challenges.

The function of New Media in emergency response is no longer just a possibility or even a probability, it is a certainty and it is already underway. How you as the Incident Commander and your incident management team understand, adopt, and integrate New Media into your ICS processes is a question with the answer still a work-in-progress.

"It's the framework which changes with each new technology, and not just the picture within the frame." (Marshall McLuhan)

CONCLUSION

As Incident Command System (ICS) practitioners, instructors, and advocates, we have witnessed the dismal implementation of ICS during emergency response operations, as well as seen responders use ICS to its full potential, enabling those responders to efficiently and effectively manage complex incidents. One of the major factors in the failure of responders to put into motion a successful ICS organization and process is the lack of training in how to perform their particular ICS position.

This book provides you a good foundation on many of the key ICS positions and how each of these positions supports the ICS planning process. If your agency or organization has identified you for a particular role in the ICS organization such as Operations Section Chief or the Situation Unit Leader, the information and checklists in this book will enable you to have more confidence in how to perform your specific duties and support the other members of the response team.

Beyond Initial Response, second edition, is a great resource, but you need to use the information that we have given you whenever you have the opportunity to ensure that you maintain proficiency. Opportunities to use ICS can be during exercises, planned events, and actual response.

As authors, we hope that we have provided you the best information in a manner that enables you to use ICS as it was designed—as a flexible system that can be tailored to the complexity of the incident to which you are responding.

Below, we provide a list of some of the many resources available for ICS tools and training. Be sure to check out these resources for continued professional growth in using ICS.

Sources for ICS Tools and Training

Emergency Management Services International, Inc. (EMSI)

Emergency Management Services International (EMSI) is a premier provider of the National Incident Management System Incident Command System (NIMS ICS) position-specific courses. EMSI's suite of ICS position training includes: Incident Commander, Operations Section Chief, Planning Section Chief, Logistics Section Chief, Resources Unit Leader, Situation Unit Leader, Safety Officer, and many others. In addition, EMSI provides Area Command training, Incident Management Team training; exercise development, execution and evaluation; and deployable ICS command and control expertise. Training and exercises can be tailored to specific organizational needs.

EMSI also develops operational handbooks, field operations guides, and ICS position-specific job aids; and can provide Incident Command Post support kits, including signage, posters, and large all-risk ICS-215s as well as other ICS material.

Web site: www.emsi-ics-services.com
E-mail: info@emsi-ics-services.com
Phone: (540) 423-9004

US Coast Guard

The US Coast Guard has been at the forefront of adopting the Incident Command System to manage its all-risk all-hazard missions starting in the early 1990s. You can find NIMS ICS forms and job aids (e.g., Meeting and Briefing Agendas) on the following Coast Guard Web site:

uscg.mil/mycg/portal/ep/home.do

Once you are at the Web site go to the link titled "Library."

National Interagency Fire Center (NIFC)

The National Interagency Fire Center (NIFC) in Boise, Idaho, is the nation's support center for wildland firefighting. Several federal and state agencies through NIFC, work together to coordinate and support wildland fire and disaster operations. These agencies include the Bureau of Indian Affairs, Bureau of Land Management, Forest Service, Fish and Wildlife Service, National Park Service, National Association of State Foresters, National Weather Service, and Office of Aircraft Services.

You can order T-cards, T-card holders, ICS forms, large ICS-215s and other ICS material from:

National Interagency Fire Center
Great Basin Cache Supply Office
3833 S. Development Ave.
Boise, ID 83705

Phone: (208) 387-5104
FAX: (208) 387-5573

National Wildfire Coordinating Group (NWCG)

The National Wildfire Coordinating Group (NWCG) is made up of the USDA Forest Service; four Department of Interior agencies: Bureau of Land Management, National Park Service, Bureau of Indian Affairs, and the Fish and Wildlife Service; and State forestry agencies through the National Association of State Foresters. The purpose of NWCG is to coordinate fire management programs, including ICS. Its goal is to provide more effective execution of each agency's fire management program. The group provides a formalized system to agree upon standards of training, equipment, qualifications, and other operational functions. You can get a catalog for ICS forms and other ICS material from www.nwcg.gov

Emergency Management Institute (EMI)

Emergency Management Institute (EMI) provides emergency management training to enhance the capabilities of federal, state, local, and tribal government officials, volunteer organizations, and the public and private sectors to minimize the impact of disasters on the American public. For information on the NIMS ICS courses that EMI offers, go to: www.training.fema.gov.

The Center for New Media & Resiliency

The Center for New Media and Resiliency is a non-profit organization formed to provide knowledge leadership to strengthen the integration of New Media into national resiliency and to increase safety, security and resiliency of citizens and communities through understanding of New Media and other emerging communications technologies in public safety, emergency preparedness & response and homeland security.

The Center supports and conducts non-partisan scientific research, education and engages in knowledge outreach activities all of which are designed to increase national resiliency and reduce time-to-competency.

www.thecenterfornewmedia.org

PO Box 1505
Castle Rock, Colorado 80104
Phone: (303) 671-0224

Sample Incident Action Plan (IAP)

This is an Incident Action Plan (IAP) for an incident involving a forest fire and release of oil near a waterway. The source of the incident is an overturned tank truck on a small road. The fire involves about 160 acres of forest, and is a threat to a national forest that is exacerbated by the season—it's during a hot and dry summer. The oil released from the overturned truck totals about 6,000 gallons and threatens to pollute the Animas River. In the scenario presented, there is a Unified Command comprised of federal and state agencies responding to the oil spill, and a separate federal agency dealing with the fire in the national forest. This IAP would represent the second operational period and covers 12 hours (0700-1900) on July 16. An Incident Command Post (ICP) has been established in Animas and at this point, the Operations Section Chief (OSC) has established two functional groups: one assigned to contain and remove the oil and another group assigned to keep the public and commercial vessels out of the oil-impacted areas. A division has been established to conduct fire suppression operations. The OSC has also established a staging area to provide immediate surge support in the event that more tactical resources are required during the operational period. A site safety plan has been developed for responder safety with emphasis on protection related to exposure to the oil. The site safety plan is referenced, but is not included with the IAP.

1. Incident Name: Animas	2. Operational Period to be covered by IAP (Date/Time): 16 July 0700-1900	COVER SHEET

3. Approved by Incident Commander(s):

ORG	NAME
USCG	N. Heath
ANF	M. Meza
State	D. Bartholomew

INCIDENT ACTION PLAN

The items checked below are included in this Incident Action Plan:

- [X] ICS 202 (Response Objectives)
- [X] ICS 203 (Organization List)
- [X] ICS 204 (Assignment Lists)
- [X] ICS 205 (Communications Plan)
- [X] ICS 206 (Medical Plan)
- [ ] ICS 220 (Air Operations Summary)
- [X] Map/Chart
- [ ] Weather Forecast/Tides/Currents

Other Attachments

- [ ] _____
- [ ] _____
- [ ] _____

4. Prepared by: J. Gafkjen (PSC)

Date/Time: 16 July 0400

1. Incident Name **Animas**	2. Operational Period (Date/Time) From: 16 July 0700 To: 16 July 1900	INCIDENT OBJECTIVES ICS-202

3. Objective(s)

Conduct operations in accordance with the site safety plan

Contain fire south of La Plata Road and east of County Road 501

Enforce a safety zone from Opir Road Bridge to La Plata Road Bridge

Contain and remove spilled product to protect the environment

4. Operational Period Command Emphasis (Safety Message, Priorities, Key Decisions/Directions)

Fire is burning in steep canyons with heavy brush. Be aware of rolling rocks and debris.

Weather: Temperatures will range from a low of 75 degrees to a high of 95 degrees. Humidity will be at 45 percent. Winds will be from the west at 7 to 12 knots, gusting to 20 knots. River current is 3.5 knots.

5. Prepared by: (Planning Section Chief) **J. Gafkjen (PSC)**	Date/Time **16 July 0430**

ORGANIZATION ASSIGNMENT LIST		Food Unit	
		Medical Unit	M. Lindaman
1. Incident Name **Animas**			
2. Date 16 July	3. Time 0300	**9. Operations Section**	
		Chief	L. Hewett
4. Operational Period 16 July 0700 to 16 July 1900		Deputy	
Position	Name	Staging La Plata	S. Kitchen
5. Incident Commander and Staff		Staging	
Incident Commander	N. Heath (USCG)	Staging	
Incident Commander	M. Meza (ANF)	**a. Branch I - Divisions/Groups**	
Incident Commander	D. Bartholomew (State)	Branch Director	
Safety Officer	A. Gresham	Deputy	
Public Information Officer	K. Roberts	Safety Group	M. Yale
Liaison Officer		Division A	S. van Valkenburg
6. Agency Representatives		Recovery Group	P. Gill
		b. Branch II - Divisions/Groups	
		Branch Director	
		Deputy	
		Division/Group	
7. Planning Section		Division/Group	
Chief	J. Gafkjen	Division/Group	
Deputy		**c. Branch III - Divisions/Groups**	
Resources Unit	R. Alworth	Branch Director	
Situation Unit	J. Strickland	Deputy	
Documentation Unit		Division/Group	
Demobilization Unit		Division/Group	
Technical Specialists		Division/Group	
Human Resources		**d. Air Operations Branch**	
Training		Air Ops Branch Director	
		Air Attack Supervisor	
		Air Support Supervisor	
8. Logistics Section		Helicopter Coordinator	
Chief	D. Cruz	Air Tanker Coordinator	
Deputy		**10. Finance Section**	
Supply Unit		Chief	
Facilities Unit		Deputy	
Ground Support Unit		Time Unit	
Communications Unit	J. Bell	Procurement Unit	
ICS-203	Prepared by Resources Unit		R. Alworth

1. BRANCH	2. DIVISION/~~GROUP~~ A	**ASSIGNMENT LIST**
3. INCIDENT NAME Animas	4. OPERATIONAL PERIOD DATE 16 July TIME 0700 - 1900	

5. OPERATIONS PERSONNEL

OPERATIONS CHIEF L. Hewett   DIVISION/~~GROUP~~ SUPERVISOR S. van Valkenburg

BRANCH DIRECTOR _____   AIR TACTICAL GROUP SUPERVISOR _____

6. RESOURCES ASSIGNED THIS PERIOD

STRIKE TEAM/TASK FORCE RESOURCE DESIGNATOR	EMT	LEADER	NUMBER PERSONS	TRANS NEEDED	DROP OFF POINT/TIME	PICK UP POINT/TIME
Butte #2		J. McDonald	12		Div A 0630	Div A 1900
E-2176		R. Routolo	4		Div A 0630	Div A 1900
Butte #7		K. Smith	20		Div A 0630	Div A 1900
D-12		R. Pond	2		Div A 0630	Div A 1900
E-120		K. Erickson	4		Div A 0630	Div A 1900
D-20		L. Slein	2		Div A 0630	Div A 1900

7. ASSIGNMENT

Conduct direct attack on the west side of the fire. Cut a dozer line on the north side of the fire and burn from west to east along La Plata Road. Establish a water source 1 mile east of the intersection of La Plata Road and County Road 501.

8. SPECIAL INSTRUCTIONS

Personnel are to be alert for spot fires north of La Plata Road. La Plata Staging Area is located just NW of intersection of La Plata Road and County Road 501. Watch traffic along La Plata Road -- numerous sightseers -- order Highway Patrol if necessary.

9. DIVISION/GROUP COMMUNICATIONS SUMMARY

FUNCTION		FREQ.	SYSTEM	CHAN.	FUNCTION		FREQ.	SYSTEM	CHAN.
COMMAND	LOCAL				SUPPORT	LOCAL			
	REPEAT					REPEAT			
DIV./GROUP TACTICAL		154.090		2	GROUND TO AIR				

ICS-204	PREPARED BY (RESOURCES UNIT LEADER) R. Alworth	APPROVED BY (PLANNING SECT. CH.) J. Gafkjen	DATE 16 July	TIME 0330

1. BRANCH	2. ~~DIVISION~~/GROUP **Recovery**	**ASSIGNMENT LIST**
3. INCIDENT NAME **Animas**	4. OPERATIONAL PERIOD DATE 16 July TIME 0700 - 1900	

5. OPERATIONS PERSONNEL

OPERATIONS CHIEF L. Hewett   ~~DIVISION~~/GROUP SUPERVISOR P. Gill

BRANCH DIRECTOR _____ AIR TACTICAL GROUP SUPERVISOR _____

6. RESOURCES ASSIGNED THIS PERIOD

STRIKE TEAM/TASK FORCE RESOURCE DESIGNATOR	EMT	LEADER	NUMBER PERSONS	TRANS NEEDED	DROP OFF POINT/TIME	PICK UP POINT/TIME
Vacuum Truck 12		K. Andrews	2			
Work Boat (24ft)		L. Hare	3			
Vacuum Truck 9		P. Richard	2			
Clean up Crew #2		S. Spitzer	25			
Skimmer Package		R. Campbell	4			
Clean-up Crew #4		E. Parsons	25			

7. ASSIGNMENT

Prevent the oil from migrating to the Animas River. Use a combination of dikes and containment boom where necessary to stop oil movement and to minimize damage to environmentally sensitive areas. Conduct removal operations using vacuum trucks and skimmers. Ensure that liquid and solid waste are separated for disposal.

8. SPECIAL INSTRUCTIONS

Conduct clean-up operations to minimize damage to riparian vegetation and bald eagle nest. Recovery Group personnel will meet at the end of County Road 501.
All personnel to be verified current in HAZWOPER training, medical surveillance, and respirator clearance and fit testing by the Safety Officer before work commences.

9. DIVISION/GROUP COMMUNICATIONS SUMMARY

FUNCTION		FREQ.	SYSTEM	CHAN.	FUNCTION		FREQ.	SYSTEM	CHAN.
COMMAND	LOCAL				SUPPORT	LOCAL			
	REPEAT					REPEAT			
DIV./GROUP TACTICAL				23A	GROUND TO AIR				

ICS-204	PREPARED BY (RESOURCES UNIT LEADER) **R. Alworth**	APPROVED BY (PLANNING SECT. CH.) **J. Gafkjen**	DATE 16 July	TIME 0330

1. BRANCH	2. ~~DIVISION~~/GROUP **Safety Zone**	**ASSIGNMENT LIST**
3. INCIDENT NAME **Animas**	4. OPERATIONAL PERIOD DATE 16 July TIME 0700 - 1900	

5. OPERATIONS PERSONNEL

OPERATIONS CHIEF L. Hewett   ~~DIVISION~~/GROUP SUPERVISOR M. Yale

BRANCH DIRECTOR _____ AIR TACTICAL GROUP SUPERVISOR _____

6. RESOURCES ASSIGNED THIS PERIOD

STRIKE TEAM/TASK FORCE RESOURCE DESIGNATOR	EMT	LEADER	NUMBER PERSONS	TRANS NEEDED	DROP OFF POINT/TIME	PICK UP POINT/TIME
State Police Boat 684		D. Schuster	3			
USCG Boat 47124		E. Doyle	4			

7. ASSIGNMENT

Enforce 24/7 safety zone from Opir Road Bridge to La Plata Road Bridge. Coordinate escort of commercial vessels through the safety zone. Notify the OSC when escorts commence and are completed.

8. SPECIAL INSTRUCTIONS

9. DIVISION/GROUP COMMUNICATIONS SUMMARY

FUNCTION		FREQ.	SYSTEM	CHAN.	FUNCTION	FREQ.	SYSTEM	CHAN.
COMMAND	LOCAL				SUPPORT	LOCAL		
	REPEAT					REPEAT		
DIV./GROUP TACTICAL				81A	GROUND TO AIR			

ICS-204	PREPARED BY (RESOURCES UNIT LEADER) **R. Alworth**	APPROVED BY (PLANNING SECT. CH.) **J. Gafkjen**	DATE 16 July	TIME 0330

1. BRANCH	2. ~~DIVISION/GROUP~~ La Plata Staging	**ASSIGNMENT LIST**
3. INCIDENT NAME Animas		4. OPERATIONAL PERIOD DATE 16 July TIME 0700 - 1900

5. OPERATIONS PERSONNEL

OPERATIONS CHIEF __L. Hewett__ ~~DIVISION/GROUP SUPERVISOR~~ S. Kitchen (STAM)

BRANCH DIRECTOR _____ AIR TACTICAL GROUP SUPERVISOR _____

6. RESOURCES ASSIGNED THIS PERIOD

STRIKE TEAM/TASK FORCE RESOURCE DESIGNATOR	EMT	LEADER	NUMBER PERSONS	TRANS NEEDED	DROP OFF POINT/TIME	PICK UP POINT/TIME
3,000 ft Boom (12")						
Work Boat (24 ft)		N. Jesse	4			
E-345		C. Knight	4			
Water Tender		S. Klein	2			
County EMS Team		R. Ward	2			
E-789		T. Warwick	4			

7. ASSIGNMENT

All resources remain in staging and be prepared to respond within 10 minutes once notified of assignment by the Staging Area Manager, via the Operations Section Chief.

8. SPECIAL INSTRUCTIONS

Be aware of moving vehicle hazards and refueling that may occur periodically.

9. DIVISION/GROUP COMMUNICATIONS SUMMARY

FUNCTION		FREQ.	SYSTEM	CHAN.	FUNCTION		FREQ.	SYSTEM	CHAN.
COMMAND	LOCAL				SUPPORT	LOCAL			
	REPEAT					REPEAT			
DIV./GROUP TACTICAL		900 MHz		3	GROUND TO AIR				

ICS-204	PREPARED BY (RESOURCES UNIT LEADER) R. Alworth	APPROVED BY (PLANNING SECT. CH.) J. Gafkjen	DATE 16 July	TIME 0330

INCIDENT RADIO COMMUNICATION PLAN

1. INCIDENT NAME	2. DATE/TIME PREPARED	3. OPERATIONAL PERIOD/TIME
Animas	16 July 0300	16 July 0700 to 16 July 1900

4. Basic Radio Channel Utilization

Radio Type/Cache	Channel	Function	Frequency/Tone	Assignment	Remarks
VHF RADIO USCG	81A	TACTICAL	N/A	SAFETY ZONE	COAST GUARD MARINE SAFETY PRIMARY WORKING FREQUENCY
UHF RADIO	CH. 3	OPERATIONS	900 MHZ.	STAGING AREA MGR.	LA PLATA ROAD STAGING AREA
VHF RADIO CONTRACTOR	23A	TACTICAL	N/A	RECOVERY GROUP	CONNECT ALL CONTRACTOR'S ASSETS
PHONE	407-782-8973	COMMAND	N/A	COMMAND/COMMAND AND GENERAL STAFF	UNIFIED COMMAND
VHF RADIO STATE	34A	LOGISTICS	N/A	LOGISTICS/OPERATIONS	N/A
VALLEY VIEW FIRE NET	CH. 2	TACTICAL	154.090	DIV A	N/A

5. PREPARED BY (COMMUNICATIONS UNIT)   J. Bell

ICS-205

MEDICAL PLAN	1. INCIDENT NAME Animas	2. DATE PREPARED 16 July	3. TIME PREPARED 0230	4. OPERATIONAL PERIOD 16 July 0700-1900

5. INCIDENT MEDICAL AID STATION

MEDICAL AID STATIONS	LOCATION	PARAMEDICS YES	NO
La Plata - EMT Capability only	La Plata Road Staging		X

6. TRANSPORTATION

A. AMBULANCE SERVICES

NAME	ADDRESS	PHONE	PARAMEDICS YES	NO
Animas Memorial Hsp	232 Main Avenue, Clay, CA	999-760-2312	X	

B. INCIDENT AMBULANCES

NAME	LOCATION	PARAMEDICS YES	NO
Same			

7. HOSPITALS

NAME	ADDRESS	TRAVEL TIME AIR	GROUND	PHONE	HELIPAD YES	NO	BURN CENTER YES	NO
Animas M. H.	232 Main Ave, Clay, CA	10	20	999-760-2312	X		X	

8. MEDICAL EMERGENCY PROCEDURES

Transport injured immediately to aid station at La Plata Staging and advise supervisor as soon as possible. To access a medical helicopter contact Animas Memorial Hospital at the above number.

ICS-206	9. PREPARED BY (Medical Unit Leader) M. Lindaman	10. REVIEWED BY (Safety Officer) A. Gresham

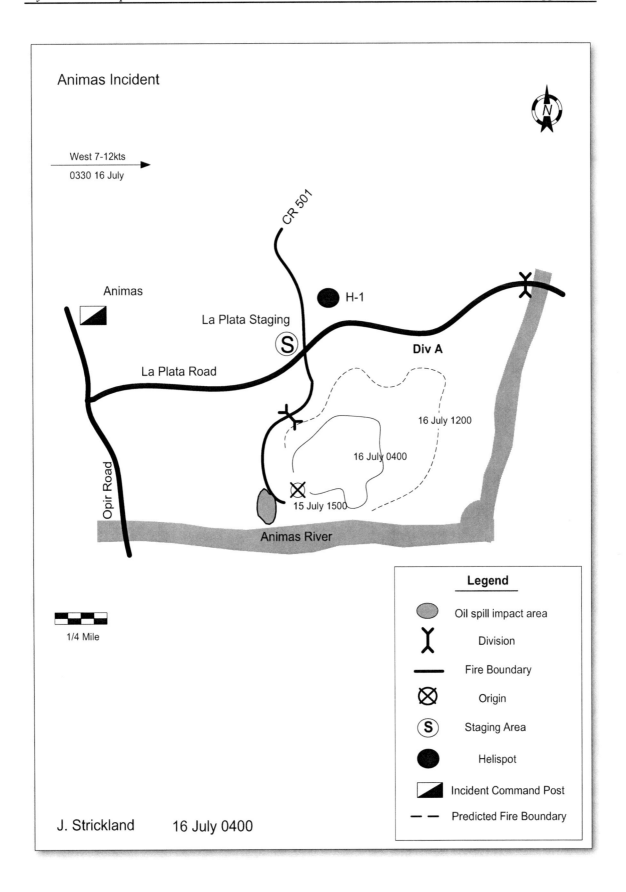

Position-Specific Operational Planning "P"s

The Incident Command System Planning Process is designed to methodically "walk" an incident management team through a series of activities or steps that result in the development of an Incident Action Plan, which in turn is used to manage response operations for the upcoming operational period.

For the planning process to work, designated Incident Command System positions must fulfill their individual responsibilities during each step. This Appendix contains seven position-specific Operational Planning "P"s:

- Incident Commander/Unified Command
- Safety Officer
- Operations Section Chief
- Planning Section Chief
- Resources Unit Leader
- Situation Unit leader
- Logistics Section Chief

Copies of the Planning Ps can be found at: www.emsi-ics-services.com

Incident Commander/Unified Command Operational Planning "P"

Preparing for the Tactics Meeting
- Meet one-on-one with Command & General Staff members for follow up on assignments.
- Prepare further guidance and clarification as needed
- Receive operations briefing

Preparing for the Planning Meeting
- Agree on who will present UC's response emphasis and motivation remarks
- Review task assignments, objectives, decisions & directions
- Receive operations briefing

Planning Meeting
- Provide opening remarks
- Review response plan as presented to ensure that Command's directions and objectives have been properly addressed
- Provide further guidance and resolve issues
- Give tacit approval of the proposed Plan
- Agree when written plan will be ready for review & approval

Command & General Staff Meeting
- Meet and brief Command & General Staff on IC/UC direction, objectives & priorities
- Assign work tasks
- Resolve problems & clarify staff roles and responsibilities

IAP Preparation & Approval
- Review IAP for completion and make changes as necessary
- Approve Plan

IC/UC Develop/Update Objectives Meeting
- Establish priorities
- Identify constraints & limitations
- Develop incident objectives
- Identify necessary IMT SOPs
- Agree on operating policy, procedures and guidelines
- Identify staff assignments
- Agree on division of UC workload

Operations Briefing
- Provide overall guidance and clarification
- Provide leadership presence and motivational remarks
- Emphasize response philosophy

Initial UC Meeting
- Finalize UC structure
- Determine overall response organization
- Identify and select support facilities
- Clarify UC roles and responsibilities
- Determine Operational period
- Select OSC & Deputy OSC
- Make key decisions

Execute Plan & Assess Progress
- Monitor ongoing operations
- Review progress of assigned tasks
- Receive periodic situation briefings
- Review work progress
- Identify changes that need to be made during current and future operations
- Prepare for UC Update Objectives Meeting

Initial Response & Assessment
- Ensure that an appropriate initial response is deployed
- Provide direction as needed
- Monitor initial response operations

Incident Brief ICS-201
- Determine ICS-201 briefing timeframe & receive briefing
- Clarify/request additional information
- Determine incident complexity
- Provide interim direction
- Initiate change of command
- Determine UC players
- Ensure interagency notifications
- Brief superiors

Planning P cycle stages (in order): Incident/Event → Notification → Initial Response & Assessment → Incident Brief ICS-201 → Initial UC Meeting → IC/UC Develop/Update Objectives Meeting → Command & General Staff Meeting/Briefing → Preparing for the Tactics Meeting → Tactics Meeting → Preparing for the Planning Meeting → Planning Meeting → IAP Prep & Approval → Operations Briefing → New Ops Period Begins → Execute Plan & Assess Progress

Safety Officer Operational Planning "P"

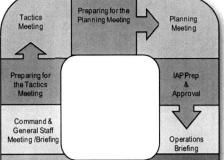

Preparing for the Tactics Meeting
- Obtain briefings from field personnel
- Work with Ops & Plans & develop risk analysis for draft 215 using a 215A
- Make notes on safety gear needed
- Identify safety support staff needed
- Identify locations for safety equipment and locations: 1st aid, eye wash, etc.

Tactics Meeting
- Continue to work with Ops & Plans & develop risk analysis for draft 215 using a 215A
- Make notes on safety gear needed
- Identify safety support staff needed
- Identify locations for safety equipment and locations: 1st aid, eye wash, personnel decon stations, etc.

Preparing for the Planning Meeting
- Contact field personnel
- Meet with Logistics to ensure ordering of proper safety gear
- Prepare Safety briefing prior to planning meeting

Planning Meeting
- Maintain listening mode
- Report on:
  - # injuries, near misses
  - Preventative/corrective actions
  - Top 3 hazards & safeguards
- Place emphasis on Safety Plan
- Report on status of any tasking assigned by IC/UC.

IAP Preparation & Approval
- Complete Safety Plan
- Complete Safety Message
- Add "General Safety Message" to ICS Form 202
- Ensure Safety Organization reflected in ICS Form 203
- Add Safety instructions in section 8 of ICS Form 204
- Review Medical Plan (206)
- Review Comms Plan (205)
- Place safe travel message & comms information inside Demobilization Plan

Command & General Staff Meeting
- Maintain listening mode. Provide input only when called upon or when a serious safety issue surfaces
- Review safety implications of Operational Periods (fatigue)
- When objectives are discussed, ensure there is a safety objective if necessary
- Begin identifying Safety Staff support for meeting IC/UC objectives

IC/UC Develop/Update Objectives Meeting
- If present, maintain a listening mode. Provide input only when called upon or when a serious safety issue surfaces
- Review safety implications of Operational Periods (e.g., fatigue)
- When objectives are discussed, ensure there is a safety objective if necessary
- Begin identifying Safety Staff support for meeting IC/UC objectives
- Identify availability of safety staff support from agencies/organizations represented in the Unified Command
- Identify technical specialists needed

Operations Briefing
- Report on top 3 hazards & safeguards
- Refer to important safety precautions in 204s
- Inform Div/Group supervisors of safety personnel in the field

Execute Plan & Assess Progress
- Continue to receive periodic updates from Assistant Safety Officers in the field
- Tour the field via air, ground, water to assess progress
- Tour the ICP to gauge crew performance & safety
- Review outstanding UC tasks and complete them
- Prepare for UC Update Objectives Meeting

Initial Response & Assessment
- Conduct full site characterization & risk assessment
- Identify hazards, evaluate exposures & implement controls to safeguard responders and public
- Establish control areas, exclusion zones, decontamination zones, support zones, safe refuge areas, evacuation distances and assembly areas

Operations Section Chief
Operational Planning "P"

Tactics Meeting
- Brief on current operations
- Divide incident into manageable units
- Develop work map
- Develop strategy/tactics to deploy
- Complete ICS-215
- Identify resource needs
- Identify contingencies
- Develop operations org chart
- Continue on-scene operations

Preparing for the Planning Meeting
- Make sure the ICS-215 is complete
- Continue to update work progress
- Continue on-scene operations

Planning Meeting
- Brief on planned strategy/tactics [ICS-215 & work map/chart]
- Identify how incident will be subdivided into mgmt/work units
- Identify resource needs & reporting locations
- Identify any contingencies as needed
- Identify organizational requirements

Preparing for the Tactics Meeting
- Develop draft strategies & tactics for each assigned objective, including alternative and/or contingency strategies
- Outline work assignments and required resources using ICS-215
- Develop/outline OPS Section organization for next operational period

IAP Preparation & Approval
- Provide information for IAP for Air Operations ICS-220
- Ensure ICS-204 taskings are clear
- Communicate incident status changes
- Continue on-scene operations

Command & General Staff Meeting
- Receive direction from IC/UC
- Clarify objectives & priorities
- Clarify organizational issues
- Identify any limitations & restrictions
- Reach agreement on IC/UC focus and direction
- Discuss interagency issues
- Prepare for tactics meeting
- Continue on-scene operations

Operations Briefing
- Provide operations briefing to Ops Sec Personnel
- Ensure support to operations in place
- Deploy next operating period resources

Incident Brief ICS-201
- Using ICS-201, brief on current operations
- Clarify issues & concerns
- Discuss planned operations & direction

Execute Plan & Assess Progress
- Monitor ongoing operations & make tactical adjustments
- Measure/ensure progress against stated objectives
- Debrief resources coming off shift
- Prepare to brief UC/Planning on accomplishments

Initial Response & Assessment
- Assess situation
- Develop ICS-201
- - Develop initial tactics & priorities
- - Develop sketch map
- - Summarize actions
- - Develop resource summary
- - List current organization
- Continue to update response using ICS-201

Planning P stages (center diagram):
- Tactics Meeting
- Preparing for the Planning Meeting
- Planning Meeting
- Preparing for the Tactics Meeting
- IAP Prep & Approval
- Command & General Staff Meeting / Briefing
- Operations Briefing
- IC/UC Develop/Update Objectives Meeting
- Execute Plan & Assess Progress
- New Ops Period Begins
- Initial UC Meeting
- Incident Brief ICS-201
- Initial Response & Assessment
- Notification
- Incident/Event

Planning Section Chief
Operational Planning "P"

Preparing for the Tactics Meeting
- Meet with Operations to determine strategies, tactics & resource requirements
- Complete ICS-215
- Notify meeting participants of scheduled meeting
- Set up meeting room

Tactics Meeting
- Facilitate meeting
- Provide Situation Briefing
- Review proposed strategy, tactics & resource requirements
- Identify resource shortfalls
- Assure the strategy & tactics comply with IC/UC objectives
- Mitigate Logistics and Safety issues

Preparing for the Planning Meeting
- Clean up ICS-215 & make hard copies for attendees
- Notify participants of meeting location & time
- Set up meeting room

Planning Meeting
- Facilitate meeting
- Provide Situation Briefing
- Confirm availability of resources
- Verify support for the proposed plan
- Document decisions & assigned actions

Command & General Staff Meeting
- Set up meeting room
- Facilitate meeting
- Provide Situation Briefing
- Receive work tasks & assignments
- Resolve conflicts & clarify roles & responsibilities

IAP Preparation & Approval
- Develop components of the IAP
- Review completed IAP for correctness
- Provide IAP to IC/UC for review and approval
- Make copies of IAP for distribution

IC/UC Develop/Update Objectives Meeting
- Set up meeting room
- Facilitate meeting
- Provide recorder to document decisions
- Distribute and post decisions

Operations Briefing
- Set up briefing area
- Provide situation briefing
- Distribute copies of IAP
- Facilitate briefing
- Make adjustments to IAP, if necessary

Initial UC Meeting
- Set up meeting room
- Facilitate Meeting
- Provide recorder to document discussion points

Execute Plan & Assess Progress
- Monitor progress of implementing the IAP
- Measure/ensure progress against stated objectives
- Maintain Situation and Resource status
- Debrief resources coming off shift
- Maintain interaction with Command and General Staff

Incident Brief ICS-201
- Facilitate ICS-201 brief
- Obtain ICS-201 & distribute to RESL & SITL
- Document results of ICS-201 briefing

Initial Response
- Check-in
- Receive IC/UC Briefing
- Activate Planning Section
- Organize & brief subordinates
- Acquire work materials

Resources Unit Leader Operational Planning "P"

Preparing for the Tactics Meeting

- Follow up on any open action items that are your responsibility
- Identify short and long-term staffing requirements
- Identify & request work space, equipment, & supplies
- Submit an ICS-213 for any required staffing needs
- Continue to update resource status display & be prepared to support the Tactics Meeting
- Ensure that the PSC is briefed on the status of the Resources Unit

Tactics Meeting

- Display current resource status on the incident
- Working with the OSC enter resource "have" & "need" information on the ICS-215
- Act as scribe & make any changes to the ICS-215 as required
- Consider potential locations for check-in & re-evaluate staffing requirements
- Begin to discuss resource needs needs with the LSC

Preparing for the Planning Meeting

- Maintain resource status displays
- Coordinate with OSC & LSC on offsite resource availability
- Submit ICS-213 for resources identified on the ICS-215
- Provide input to the SITL for inclusion in the ICS-209
- Identify any excess resources

Planning Meeting

- Make any changes needed to the ICS-215
- Confirm the availability of resources to meet the plan
- Request additional resources for any identified shortfalls
- Begin to set up resource status display for the upcoming operational period

IAP Preparation & Approval

- Develop the Organization List, ICS-203 & the Assignment Lists, ICS-204s for the IAP
- Coordinate with the OSC, COML, SOF, THSP in the development of the ICS 204s & 204a
- Coordinate resource assignments with the OSC
- Complete the resource status display for the upcoming Operational Period
- Assemble the IAP & submit to the PSC
- Coordinate with the DOCL development of an IAP distribution list

Command & General Staff Meeting

- Continue to collect resource information
- Establish an ICS-207

IC/UC Develop/Update Objectives Meeting

- Start a resource status display
- Monitor check-in process & ensure ICS-211s are being properly filled in
- Discuss with the PSC/LSC the resource ordering process
- In coordination with the SITL start a field verification process to account for initial response resources

Operations Briefing

- Ensure that any last minute IAP changes that will effect the status of resources is documented & displays updated
- Answer any resource status questions
- Document any "pen & ink" changes to the ICS-203 and ICS-204

Initial UC Meeting

- Continue to update resource information needed for the initial UC Meeting. Sources of initial information include:
  Command/dispatch centers
  Initial IC/OSC
  Field Observers
  Command & General Staff
  Agency Representatives
  Staging Area Manager(s)

Execute Plan & Assess Progress

- Monitor the check-in process & performance of the Resources Unit staff & make adjustments as necessary
- Continually update resource status displays
- Ensure that the PSC is fully aware of resource status
- Produce any special reporting requirements (e.g., ICS-209)
- Interact with all "customers" to ensure that the Resources Unit is providing satisfactory service
- Ensure that the Resources Unit staff is briefed on current & future activities

Incident Brief ICS-201

- Review your ICS tools: IMH, Job Aid, checklists, etc.
- Review your in-briefing checklist & formulate additional questions as the situation dictates (obtain copy of pages 3 & 4 of the ICS-201)
- Based on the in-briefing determine:
  Initial level of RESL staff
  Initial work location needs
- Establish & staff check-in locations
- Advise the PSC when you are operationally ready

Initial Response & Assessment

- Ensure readiness of your personal response kit
- Evaluate potential to being assigned to the incident
- Begin situational awareness

Situation Unit Leader
Operational Planning "P"

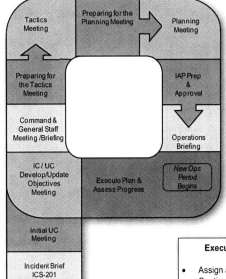

Preparing for the Tactics Meeting
- Follow up on any open action items that are your responsibility
- Identify short and long-term staffing requirements
- Identify & request work space, equipment, & supplies
- Submit an ICS-213 for any required staffing needs
- Formalize off-site reporting requirements
- Ensure that the PSC is briefed on the status of the Situation Unit

Tactics Meeting
- Prepare and deliver an up-to-date detailed situation briefing
- Provide any current modeling predictions
- Consult with any THSP and evaluate adequacy of THSP to support planned operations
- Consider potential locations for displays and re-evaluate staffing requirements
- Determine any mapping requirements for the IAP (e.g., Division Specific Map)

Preparing for the Planning Meeting
- Prepare briefing, displays, and handouts
- Coordinate with OSC, INTL and others who may provide a briefing
- Validate modeling predictions
- Coordinate with other IMT members on ICS-209 requirements

Planning Meeting
- Deliver an up-to-date detailed situation briefing
- Provide any current modeling predictions
- As needed, update the Meeting Schedule
- Resolve any unanswered questions as a result of the briefing

IAP Preparation & Approval
- Develop IAP support maps
- Provide weather, tides, currents etc. information to the PSC for inclusion in the IAP
- Prepare for the Operations Briefing
- Coordinate with OSC, INTL on who is delivering what parts of the briefing

Command & General Staff Meeting
- Prepare and deliver an up-to-date situation briefing
- Provide any initial modeling predictions
- Begin to formalize any threshold reporting requirements
- Document and post any decisions regarding Meeting Schedule

IC/UC Develop/Update Objectives Meeting
- Present the most up-to-date detailed situation briefing
- Begin to identify essential elements of information and threshold reporting requirements
- Identify off-site reporting requirements

Operations Briefing
- Deliver an up-to-date detailed situation briefing
- Provide any current modeling predictions
- Discuss if there will be FOBS in the field and their role
- Discuss any end-of-shift briefing requirements

Execute Plan & Assess Progress
- Assign and monitor the FOBS field activity
- Continually update displays
- Update and disseminate modeling predictions
- As required, prepare to deliver any special briefings (e.g., political, stakeholder)
- Prepare the situation briefing for the next Objectives Meeting
- Continue to evaluate Unit's performance and make adjustments as necessary
- Ensure that the PSC is up-to-date on incident situation (situational awareness)
- Produce any special reporting requirements (e.g., ICS-209)
- Interact with all "customers" to ensure that the Situation Unit is providing satisfactory service

Initial UC Meeting
- Present the most up-to-date information needed for the Initial UC meeting. Sources include:
  - Command centers
  - Dispatch centers
  - Media
  - Radio traffic
  - Command & General Staff

Incident Brief ICS-201
- Review your ICS tools: IMH, Job Aid, checklists, etc.
- Review your in-briefing checklist & formulate additional questions as the situation dictates (obtain copy of pages 1 & 2 of the ICS-201)
- Based on the in-briefing determine:
  - Initial level of SITL staff
  - Initial work location needs
- Advise the PSC when you are operationally ready

Initial Response & Assessment
- Ensure readiness of your personal response kit
- Evaluate potential to being assigned to the incident
- Begin situational awareness

Logistics Section Chief
Operational Planning "P"

Preparing for the Tactics Meeting
- Survey availability of tactical resources
- Report on status of resources already in the pipeline
- Summarize support capabilities, facilities, comms, etc.
- Identify resource ordering process

Tactics Meeting
- Review proposed tactics
- Identify resource needs and reporting locations from ICS215 and 215a
- Discuss availability of needed resources
- Identify resource shortfalls
- Identify resource support requirements

Preparing for the Planning Meeting
- Meet with Log Units to determine status and availability of required resources
- Order necessary resources
- Order support for resources
- Update OSC on resources unavailable to meet reporting requirements
- Suggest alternatives if necessary

Planning Meeting
- Confirm availability of required resources and timelines
- Determine additional resources necessary to support objectives
- Identify any contingencies as needed
- Verify support for upcoming plan
- Provide estimates of future service and support requirements

IAP Preparation & Approval
- Provide information for IAP [ICS-205, 206 and Traffic Plan]

Operations Briefing
- Provide logistics information briefing to Operations Section personnel
- Review Medical & Comm Plan
- Traffic Plan
- Other logistical information to support field operations

Command & General Staff Meeting
- Receive direction from IC/UC
- Clarify objectives & priorities
- Clarify organizational issues
- Identify any limitations & restrictions
- Reach agreement on IC/UC focus and direction
- Discuss interagency issues
- Agree on resource approval, requesting, and ordering process
- Identify Log Section assignments
- Identify support facilities
- Prepare for tactics meeting

Incident Brief ICS-201
- Obtain situation overview
- Anticipated Log Section activities
- Indication of required support

Initial Response
- Check-in
- Receive IC/UC briefing
- Assess situation
- Activate Logistics Section
- Organize & brief subordinates
- Acquire work materials
- Begin transition actions
    - -transportation
    - -medical
    - -resources
    - -communications
    - -facilities
    - -resource requesting
    - -safety issues
    - -environmental issues
    - -food/shelter

Execute Plan & Assess Progress
- Track resources effectiveness and make adjustments as needed
- Monitor ongoing logistical support & make logistical adjustments
- Meet with Unit personnel to monitor performance
- Maintain interaction with Command and General Staff

Planning "P" flow (bottom to top): Incident/Event → Notification → Initial Response & Assessment → Incident Brief ICS-201 → Initial UC Meeting → IC/UC Develop/Update Objectives Meeting → (loop) Command & General Staff Meeting/Briefing → Preparing for the Tactics Meeting → Tactics Meeting → Preparing for the Planning Meeting → Planning Meeting → IAP Prep & Approval → Operations Briefing → Execute Plan & Assess Progress → New Ops Period Begins

Managing Risks Using the ICS Planning Process

Managing risks during an emergency response operation can often be very difficult. Competing priorities, complex operations, and diverse challenges across disciplines can be daunting for the new Safety Officer. It is one thing to recognize the hazards inherent in an operation; another thing is to be able to characterize those hazards so that you focus on the ones that are most urgent and severe—this can be a major challenge.

The information in this Appendix offers one perspective on managing risks using the Incident Command System (ICS) processes. The suggestions incorporated here are consistent with the practices in the occupational safety and health discipline, and with the tenets of the National Incident Management System (NIMS). This appendix provides an overview of a process to characterize and quantify risks. The authors assume that the responder reading this section has some safety management training, understands the applicable safety and health regulations, and is intimately familiar with the NIMS ICS planning process.

An Overview

To better understand what risk management entails in emergency management, we have to understand two things—*hazards* and *risks*.

> Hazard: A condition that may cause an adverse effect on the response.

A hazard could be:

- Personnel hazards – unsafe behavior, inadequate knowledge due to lack of training, inexperience.
- Environmental hazards – hot weather, rain, sleet, steep terrain, mud, working on water, changing wind direction or strength.
- Equipment hazards – a crane failure due to overloading or inadequate maintenance, pump failure supplying firefighting water to the line; failure to detect an explosive atmosphere due to a faulty instrument.
- Systems Hazards – inadequate preparations, leadership shortfalls at the tactical level, communication system inadequacy, supply shortages (too little available, wrong materials), or other systems supported function.

All of these *hazards* can evolve into injury, illness, property damage, environmental damage, or other adverse impacts. Recognizing the hazards and then working to mitigate or address each hazard can range from straightforward to complex as each hazard presents a different potential for an adverse effect. Let's look at an example of what we mean here: suppose you are responding to a single vehicle accident on a remote stretch of road. The vehicle involved is a truck carrying water treatment supplies. The truck is overturned and is off the road. The driver is uninjured, and you are able to block approaching traffic from either direction successfully. The hazards remaining are with the truck and its cargo at this point. The driver tells you that he is carrying water treatment supplies including some chlorine in solid form, as well as some biocide in liquid form. You recognize that chlorine is often associated with strong oxidizers, and that means you have identified at least one hazard. The biocide is unknown at this point. To further develop the extent of the hazards you need to find out more about the material, how much is on the truck, its shipping containers, and whether the containers have been ruptured. Depending on the answers to your concerns, this could be a minor hazard, easily segregated as a small quantity of hazardous material; or a significant hazard, with hundreds of pounds of oxidizer in a truck with flammable materials.

Intuitively, we understand that all of the hazards we identify on an incident do not have the same potential for adverse impact. So how do we sort them out? That question is what leads us to risk.

> Risk: An expression of the potential severity and likelihood of a hazard.

Hazards can vary tremendously by incident. For example, a law enforcement response team managing an incident involving criminals with hostages in a bank presents a different risk profile than a response team managing a cargo vessel on fire off the coast of New York. In both cases, lives may be at risk, but the consequences to the responders, victims, and the environment will be different. The decisions made in each response by the incident management team relative to the risks will directly influence the outcome. The hazards that are present are managed through the skills of the responders, available resources, and the strategies and tactics that they employ. The effectiveness of the response effort depends, to a very large degree, on the ability of the incident management team to successfully manage the hazards inherent in the response. This is where you as the Safety Officer (SOF) are critical.

While it is certainly true that safety is every person's responsibility on an incident, there are four primary ICS positions on the incident management team that work together to identify hazards and manage risks: Incident Commander (IC), Operations Section Chief, Planning Section Chief, and you the Safety Officer. During large or complex responses, risk management is done throughout the ICS planning process.

Although the Incident Commander owns the function of risk management at all times, it falls on you as the Safety Officer to influence the proper management of risks. This is because as the SOF you have the implicit duty to ensure a safe response. As the SOF, you need to be focused on the larger hazard assessment and management of safety concerns beyond the urgency of the tactics being deployed and planning process supporting the operations. What we mean here is that you must step back and always maintain the big picture from a safety perspective. You must also prioritize the risks to enable the least amount of adverse exposure to the responders, victims, and communities.

Determining Risk Level (understanding risk severity by assigning each identified hazard a risk level)

Risk Management requires:

- An understanding of the hazards
- The *frequency* of the hazards associated with an activity
- The potential *severity* of the consequences if the hazard is not managed well

Once you leave the Command and General Staff Meeting you want to get with the Operations Section Chief (OSC) and Planning Section Chief (PSC) to help them as they develop the strategies and tactics that will be used to meet the incident objectives. By being involved at the start of the tactical planning conducted by the OSC and PSC, you can hear what strategies and tactics are being considered for the next operational period. From these planned operations, you can identify the hazards associated with each strategy, and the potential failures that could result in injury, illness, equipment damage, or any other adverse result.

Once the hazards are identified, you will be able to determine the risk level. To help you accomplish your assessment you can refer to *Figure C.1*, which is intended to provide a basic model for assessing risk and assigning severity. The concept incorporated into this table is that by categorizing the potential hazards and their associated risks, we can prioritize our attention and resources. As the Safety Officer you can use this method to determine resources that you need and the level of effort required to properly manage hazards. Higher risk hazards require many more resources to than lower risk hazards.

Level of Risk Matrix

LIKELIHOOD					
(5) Frequent Likely to occur many times	Moderate	Moderate	High	High	High
(4) Occasional Likely to occur sometimes	Low	Moderate	Moderate	High	High
(3) Remote Unlikely, but possible to occur	Low	Low	Moderate	Moderate	High
(2) Improbable Very unlikely to occur	Low	Low	Low	Moderate	Moderate
(1) Extremely improbable Almost inconceivable that the event will occur	Low	Low	Low	Low	Moderate
	(1) Negligible Little consequence	**(2) Minor** Nuisance to operating limitations. Use of Emergency procedures. Minor Incident, first aid treatment.	**(3) Major** Significant reductions in safety margins. Reduction in ability to deal with adverse operating conditions. Impact to community or customer. Loss of efficiency. Serious incident. Injury to persons.	**(4) Hazardous** Large reductions in safety margins. Operations unable to perform mission. Serious impact to community or customer. Serious injury or death to a number of people.	**(5) Catastrophic** Equipment or property destroyed. Multiple deaths.

SEVERITY →

Figure C.1. Matrix on level of risk associated with severity and likelihood.

Let's look at the Level of Risk Matrix and discuss how it works. First, there are a few definitions that we need to agree upon:

Likelihood of failure: A function of the probability (odds) of failure and the frequency of the task. In the example of the truck accident we discussed earlier; we could assess the likelihood of a release of chlorine based on the condition of the truck, the quantity of the chlorine material on board, and the type of packaging that the chlorine was shipped in. With some basic information available from the first responders, we can determine the likelihood of a release.

- Severity of the hazard: The extent of injury, illness, or impact on operations that could result if the hazard is allowed to become a consequence.

The likelihood and the severity of a hazard together become a level of risk.

Risk Categories and their Requirements

Low Risk: In this category of risk, the minimum level of safety controls should result in a notation on the ICS Form 204, Division Assignment addressing each identified hazard and the procedures, equipment, and expertise necessary to manage the risk present. The mitigation for low-risk activities is typically enhanced awareness, appropriate protective equipment, and normal protective measures.

Examples of Low-Risk activities might include:

- Food service personnel working in kitchen or serving food in a Camp or Base
- Law enforcement personnel directing traffic on a secondary road in a low traffic area
- Heavy equipment operations removing debris where there are no ground workers, the terrain is level and the ground is firm

Moderate Risk: In this category of risk, the ICS Form 204 will be annotated with specific and detailed safety controls. To facilitate the development of these controls, a written hazard assessment and description of mitigations should be developed. Safety professionals use tools like a Job Safety Analysis (JSA) for this purpose. A JSA is used for a specific job task, addressing what the task is, what the steps are in executing the task, the hazards in each step, and the processes, equipment and training required to manage the risks (other tools and processes used by safety professionals include a Job Hazard Analysis or an Activity Hazard Analysis see figure C.6).

By using a JSA process, the SOF can develop detailed and focused safety measures for the ICS Form 204 Assignment Lists for each functional Group or Division. The team members will receive a detailed safety briefing at the beginning of the Operational Period based on the safety measures and job assignments on the ICS-204.

Examples of moderate risk activities might include:

- Fall hazards from heights above 4 feet (1.3 meters)
- Confined space entry into permitted confined spaces
- Lockout/tag-out (energy management hazards on equipment and facilities)
- Work in excavations < 4 feet (1.3 meters) in depth
- Entry into nonhabitable single story structures (damaged due to flood, wind, earthquake, or other event)
- Working in areas where the infrastructure is damaged and wildlife has been disturbed (e.g., floods, hurricanes, earthquakes, wildland fires, or other natural disaster response)
- Critical crane lift operations (where lifts exceed 75% of the crane's rated capacity, or multiple cranes are used in a single lift)
- Work on steep slopes (>35 degrees)

Figure C.2 is the ICS-204 that we used in our flood scenario in the Safety Officer Chapter, Chapter 5. Block 8 of the form is where you want to ensure that you list the safety controls that you had identified through your analysis.

1. BRANCH	2. ~~DIVISION~~/GROUP Search	**ASSIGNMENT LIST**
3. INCIDENT NAME **MERIDIAN FLOOD**	4. OPERATIONAL PERIOD (Date and Time) 16 Nov 0600 to 16 Nov 1800	

5. OPERATIONS PERSONNEL

OPERATIONS CHIEF___L. Hewett___ ~~DIVISION~~/GROUP SUPERVISOR ___P. Robert___

BRANCH DIRECTOR _____ AIR TACTICAL GROUP SUPERVISOR _____

6. RESOURCES ASSIGNED THIS PERIOD

STRIKE TEAM/TASK FORCE RESOURCE DESIGNATOR	EMT	LEADER	NUMBER PERSONS	TRANS NEEDED	DROP OFF POINT/TIME	PICK UP POINT/TIME
MESA #3		B. Riggs	21		0530	
44120		V. Kammer	4		0530	
Ambulance #1		S. Miller	2		0530	
Helicopter 12		T. Troutman	2			

7. ASSIGNMENT

Conduct house-to-house search for injured persons. Mark each dwelling searched with a red "X" on the front door. Evacuate injured persons to triage center.

8. SPECIAL INSTRUCTIONS

Work in teams of two or more at all times. Snakes are suspected to be in the work area, wear snake gators. Personal floatation devices should be worn. Daylight operations only. Conduct regular communications checks. Send hourly updates on the group's progress to the Situation Unit.

9. DIVISION/GROUP COMMUNICATIONS SUMMARY

FUNCTION		FREQ.	SYSTEM	CHAN.	FUNCTION		FREQ.	SYSTEM	CHAN.
COMMAND	LOCAL REPEAT	CDF 1	King	1	SUPPORT	LOCAL REPEAT			
DIV./GROUP TACTICAL		157.4505	King	3	GROUND TO AIR				

PREPARED BY (RESOURCES UNIT LEADER) A. Worth	APPROVED BY (PLANNING SECT. CH.) J. Gafkjen	DATE 16 Nov	TIME 0400

Figure C.2. ICS-204 with safety instructions.

High Risk: In this category of risk, extensive efforts to manage the hazards are necessary. A detailed Safety Plan (or Standard Operating Procedure) is required, with additional mandated oversight by a number of competent and experienced Assistant Safety Officers with detailed knowledge of the tasks and operations.

Response personnel working in these operations often have specialized training, with robust redundant safety processes. The preshift briefings must be extremely detailed, and oversight of all operations is intense. High-risk hazards may include:

- Chemical, biological, or radiation hazard response activities.
- Demolition of multistory structure activities.
- Firefighting on vessels, aircraft, or large vehicles or facilities involving chemicals or highly volatile fuels.
- Response to high-level radiation source releases.
- Incidents involving unexploded ordnance recovery or destruction entailing munitions larger than 0.50 caliber (130 mm).
- Activities involving highly unstable multistory structures (e.g., earthquake, flood damage).
- Activities undertaken in extreme weather (e.g., hot, cold, high winds, heavy seas).
- Special Weapons and Tactics law enforcement operations.

Putting it all together using the ICS Planning Process

We are going to use the next few pages to walk you through what occurs in preparing for the Tactics Meeting and during the Tactics Meeting from a safety perspective. We will do it in *steps* and use our flood scenario from Chapter 5.

Scenario: In a flood response, search teams will go house-to-house looking for survivors. They will work in teams of three, and will proceed in a small 14-foot aluminum boat. They will work in daylight hours (September in Mississippi). They will carry a radio for communications. One of the team members is an emergency medical technician. They are equipped with food and water for the 12-hour operational period, and their boat will search an area containing about 50 homes and small businesses in an area under 12 feet of water. They are instructed to search each flooded structure and call for helicopter or boat evacuation of survivors when any are found (depending on the survivors' condition). They are to locate deceased victims for subsequent recovery.

Step 1: The OSC and PSC will complete a draft of the ICS-215 as shown in Figure C.3 (for our discussion, we are only going to look at the work assignment of the Search Group).

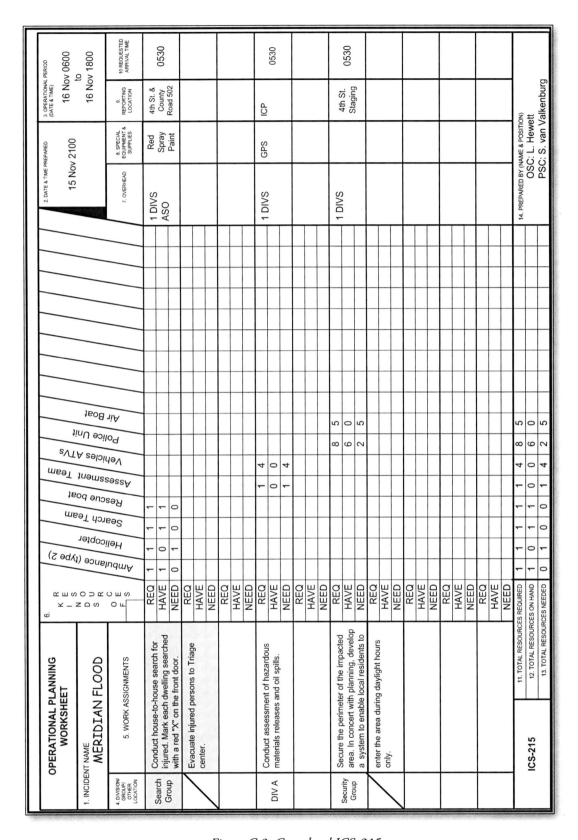

Figure C.3. Completed ICS-215

Step 2: As the SOF, you are identifying potential hazards with the work assignment that the OSC is planning for the Search Group. Some of the hazards that we identified for the Search Group include:

- Drowning
- Animals/pets that are agitated
- Survivors unwilling to be rescued and leave their property
- Criminal activities (e.g., looting) in the flooded areas
- Unlit spaces in structures (e.g., no power)
- Compromised utility infrastructure (e.g., gas, electrical, sewage)
- Structural hazards (e.g., flooded buildings that may collapse or move in the flood water)
- Fire
- Heat stress
- Navigation hazards (e.g., obstacles that may damage the small boat)

Each of these hazards presents a different level of risk for the responder. Drowning is a hazard, but only if the individuals leave the boat and have no personal floatation devices (e.g., boat hits submerged object and sinks). Gas leaks are a hazard, but not if the gas supply is properly secured to the area allowing only residual pressure in the lines.

Step 3: Using the ICS-215A, Incident Action Plan Safety Analysis you list the hazards that you identified across the top of the form as shown in *Figure C.4*. We only listed two hazards associated with the Search Group - drowning and snakes. What the ICS-215A does not do is show the *level of risk* that those two hazards pose to the Search Group. For that we will use the level of Risk Matrix table we discussed earlier.

INCIDENT ACTION PLAN SAFETY ANALYSIS		1. Incident Name Meridian Flood						2. Date 15 Nov
Division or Group	**Potential Hazards**							
	Type of Hazard: Drowning	Type of Hazard: Snakes	Type of Hazard: Blood Pathogen	Type of Hazard:	Type of Hazard:	Type of Hazard:	Type of Hazard:	Type of Hazard:
Search Group	X	X						
Medical group			X					

Figure C.4. The upper left side of the ICS-215A, Incident Action Plan Safety Analysis showing potential hazards.

Step 4: Let's look at the first hazard: drowning. Use our matrix in *Figure C.5* to find our level of risk. First, start with the horizontal axis - severity rating. Note the description of the Hazardous Severity, which includes death as an outcome. Drowning can result in death so this is the column that we will use. Second, on the vertical axis, likelihood, note that drowning is a remote but possible outcome or unlikely but possible. Correlating these two categories (severity rating and likelihood) we see that the level of risk that we would assign to this hazard is *Moderate*.

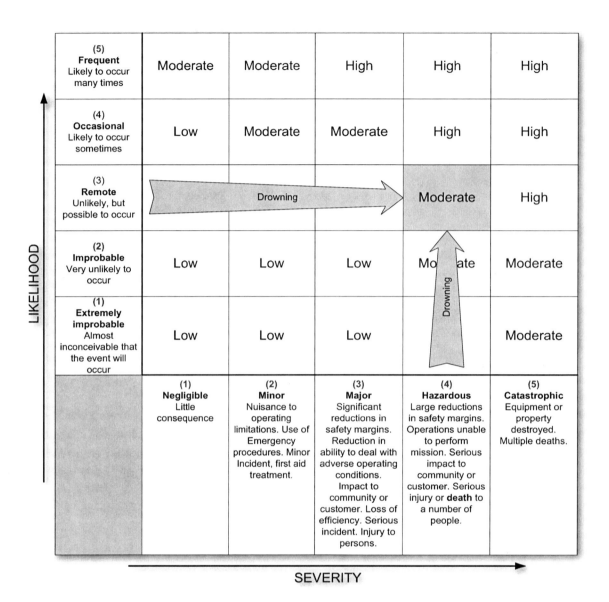

Figure C.5. Finding level of risk use severity and likelihood.

A moderate risk level suggests a written hazard assessment and management process. A Job Safety Analysis (JSA) should be completed for the moderate risk hazards. This will facilitate the development of notations on the ICS-204 Assignment List. The JSA will result in specific safety controls including such things as the type of personal floatation devices necessary for each member of the Search Group, specific requirements to address heat stress or hypothermia, daylight operations to avoid accidents that might put a responder in the water, and the use of the buddy system to assure there is always someone else around during higher risk activities. *Figure C.6 is an example of a Job Safety Analysis.*

Job Safety Analysis	JOB TITLE (and number if applicable): Search Group (waterborne)	PAGE __ OF __ JSA NO.	DATE: 16 Nov	REV0
	TITLE OF PERSON WHO DOES JOB: Search Group	SUPERVISOR: P. Robert	ANALYSIS BY: N. Knapp (ASO)	
COMPANY/ORGANIZATION: Search Group	LOCATION: Meridian Flood	DEPARTMENT: Operations	REVIEWED BY: J. Gafkjen (PSC)	
REQUIRED AND/OR RECOMMENDED PERSONAL PROTECTIVE EQUIPMENT: Type III PFD, self-inflating, low profile. Long sleeve shirt and pants, knee and elbow guards for entry, helmets for entry, lights, mace/pepper spray for animal deterrence. Carry extra water.			APPROVED BY: T. Burke (SOF)	
SEQUENCE OF BASIC JOB STEPS	POTENTIAL HAZARDS	RECOMMENDED ACTION OR PROCEDURE		
NOTE: This JSA includes all relevant quality, safety, and environmental factors associated with this job.				
Mobilize team and gear to the boat ramp (4-person team with boat driver, EMT, two structural entry and recovery specialists)	• Vehicle accident • Leaving behind important resources, supplies, or equipment.	• Qualified and experienced driver for towing and launching of boat. • Use checklists for equipment		
Launch 14' skiff with 35 hp outboard	• Slips/falls on ramp • Vehicle backing down ramp with trailer	• Qualified boat operator to launch and operate boat. • Lugged footwear for personnel on ramp • Launch in daylight only • Wear personal floatation device during launching of boat and boat operations.		
Search structures within the flooded area	• Slip/falls in unstable and wet structure • Material falling • Animals (domestic) disturbed and protecting territory • Animals (e.g., snakes)	• Properly trained search team members only allowed to enter a structure. • Wear designated protective equipment for all entries. Buddy system for all entries • Flashlights and headlamps for all entry personnel • Animal control to be called for aggressive animals • Team members to carry radios • Daylight operations only		
Demobilization	• Risk during boat recovery • Fatigue during drive to Camp	• Recover boat during daylight hours • Limit distance from launch to camp to < 10 miles		

Figure C.6. Completed Job Safety Analysis.

In Summary

The tools in this appendix will allow you as the incident Safety Officer to determine the hazards and the level of risk that the hazards pose to the responders. This will then allow for best use of response team time, effort, and focus. The end result will be a safer, better-managed response.

The process we have shown you is complementary to the ICS Planning Process. The ICS forms 215 and 215A used in the planning process can support a hazard/risk assessment process. If the hazards are determined to be Moderate or High risk, the Job Safety Analysis tool described above is one suggested method to make hazard recognition and mitigation a more transparent process. The JSA also provides for documentation of the hazards and the selected mitigations. It simply needs to get done in a structured, well thought out manner.

More complex responses with high-risk scenarios should have a dedicated Safety Plan. This Safety Plan is often based upon the Standard Operating Procedures that are prepared in advance by agency and organization contingency planning efforts. These contingency plans can be modified quickly for a specific response operation. This saves time and allows response teams to prepare for various contingencies through training and exercises.

In either case, the JSA process or a Safety Plan will identify controls and measures to manage higher risk hazards. These measures can then be incorporated into the ICS-204 Assignment List—which will be used to brief the teams at the Operations Briefing at the start of each Operational Period.

Example ICS-214 Unit Log

1. Incident Name MERIDIAN FLOOD	2. Operational Period (Date/Time) From: 11/15 1200 To: 11/16 0600	UNIT LOG ICS 214
3. Unit Name/Designators Planning Section	4. Unit Leader (Name and ICS Position) J. Gafkjen--Planning Section Chief	

5. Activity Log (Continue on Reverse)

TIME	MAJOR EVENTS
1230	Received an in-brief from the IC
1310	Ordered Resources and Situation Unit Leaders
1420	IC has shifted the operational period from 12 to 24 hours. Night operations are deemed too dangerous.
1700	Ordered a structural engineer technical specialist to support development of building survey plan.
1740	Met with State Historic Preservation Officer to discuss location of historic sites that are within the recovery area.

6. Prepared by: J. Gafkjen   Date/Time 16 Nov 0600

UNIT LOG   ICS 214

Joint Information Center Media Analysis Worksheet

Date: ____/____/____

Media Outlet Name: _____

Current Release: _____

Daily Broadcast Synopsis: _____
(If recorded please mark Y or N after time)

Daily Coverage Synopsis: _____
_____
_____
_____
_____

Issues: _____
_____
_____

Inaccuracies: _____
_____
_____

Viewpoints: _____
_____
_____

Fixes: _____
_____
_____

Who Replied: _____

Example Incident Briefing Form, ICS-201

The ICS-201, Incident Briefing Form is designed for the initial Incident Commander to quickly record the incident situation, initial actions he or she has taken, the organization, resources that are on-scene and those that have been ordered. It's an excellent document to brief incoming personnel from and to use to conduct a transfer of command.

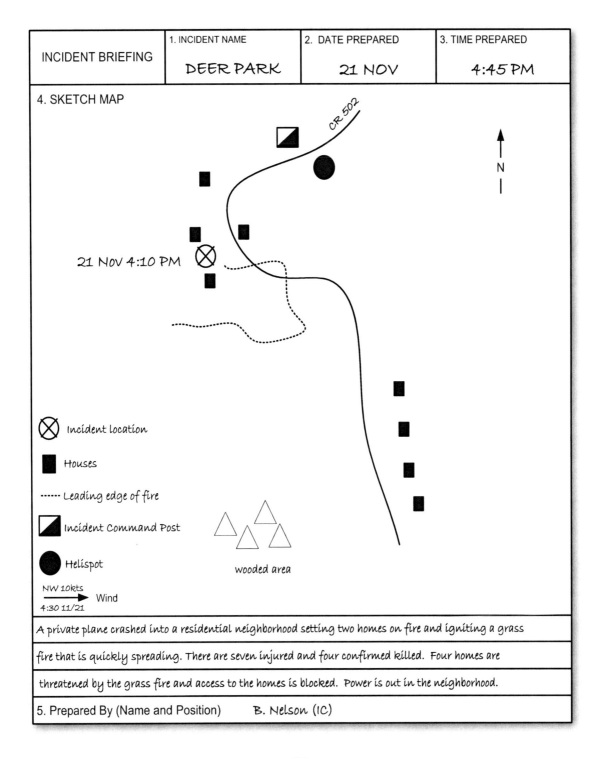

6. SUMMARY OF CURRENT ACTIONS

Objectives:

(1) Ensure response operations are conducted in accordance with safe work practices
(2) Remove, triage, and transport the injured
(3) Evacuate nearby residents in the path of the grass fire
(4) Protect the remaining homes from fire damage
(5) Establish perimeter control and secure the incident area
(6) Contain the fire west of CR 502

Current Actions:

Medical Group - Stabilize injured and transport to Durango Hospital

Fire Group - Conduct fire suppression operations and protect remaining structures. Hold the fire west of CR 502 and north of the wooded area.

Law Enforcement Group - Evacuate residents in the path of the fire. Establish traffic control point at the entrance of CR 502.

ICS-201

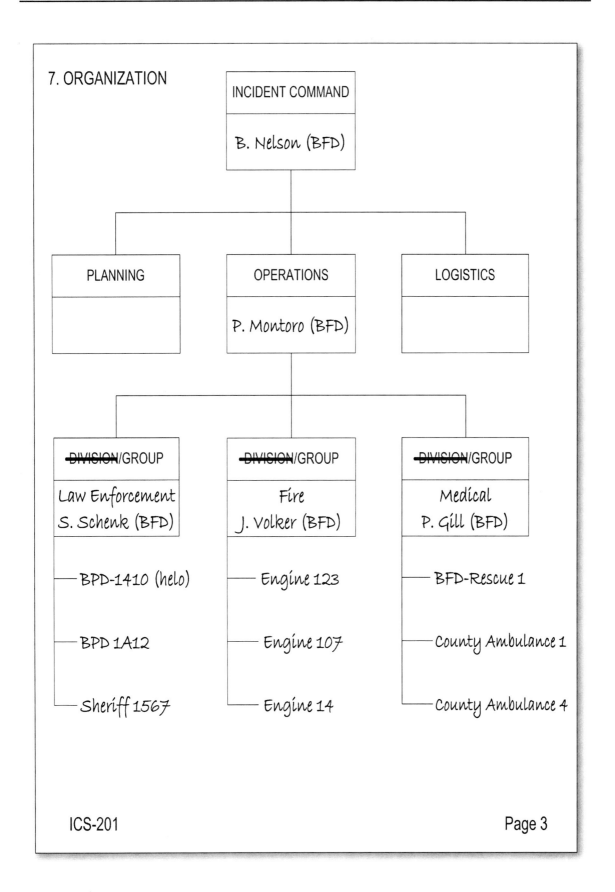

8. RESOURCES SUMMARY				
RESOURCES ORDERED	RESOURCES IDENTIFICATION	ETA	ON-SCENE (X)	LOCATION ASSIGNMENT
	B. Nelson		X	IC/ICP
	P. Montoro		X	OSC/ICP
Helicopter	BPD-1410		X	Law Enforcement Gp.
Fire Engine	Engine 123		X	Fire Group
Fire Engine	Engine 107		X	Fire Group
	S. Schenk		X	LE Group Supervisor
LE Unit	Sheriff 1567		X	Law Enforcement Gp.
LE Unit	BPD 1A12		X	Law Enforcement Gp.
	P. Gill		X	Medical Gp. Sup.
Ambulance	Ambulance #1		X	Medical Group
Ambulance	Ambulance #4		X	Medical Group
	J. Volker		X	Fire Gp. Supervisor
Fire Engine	Engine 14		X	Fire Group
Rescue Unit	BFD Rescue 1		X	Medical Group

ICS-201   Page 4

Joint Information Center Query Record

Person Calling: _____

Date/Time of Call: _____

Organization: _____

Phone Number: _____

Fax: _____

Address: _____

Inquiry: _____

_____

_____

_____

_____

_____

Deadline: _____

Person taking call: _____

Reply made by: _____

Date/Time: _____

Reply: _____

_____

_____

Example Operations Briefing Checklist

To help demonstrate the use of the Operations Briefing Checklist let's use our Meridian Flood scenario where the Florida River has overrun its banks and the town of Meridian is inundated with flood waters. The Operations Section Chief established a Search Group (to conduct house-to-house searches), Division A (to conduct assessment of hazardous material releases and oil spills), and a Security Group (to secure the perimeter of the impacted area). In addition to the Groups and Divisions, there is a Deputy Operations Section Chief that will be overseeing the field operations while the Operations Section Chief will work in the Incident Command Post to participate in the planning process.

Meridian Flood Operations Organization

(For operational period 16 November 0600 to 16 November 1800)

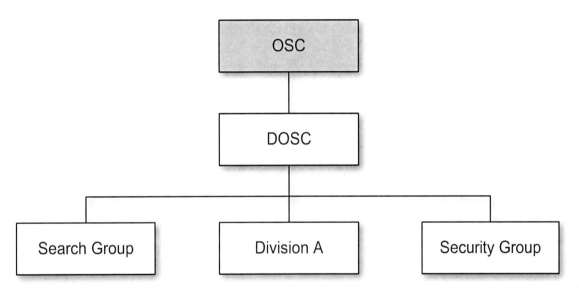

The checklist below is tailored to the Meridian Flood incident, but it should give you an idea of how you want to structure and conduct your own Operations Briefing. The briefing should be broken into two parts: Part One is to first go over *general issues* that apply to everyone, and Part Two is to brief on *specific issues* for each individual Division, Group, or Staging Area Manager.

Use the following as a guide when reading the Meridian Flood briefing outline below:

- **Bold words:** These are the areas you want to cover during the briefing.
- <u>Nonitalicized underlined words</u>: These words are provided for guidance to help you focus your comments. The comments are generic and apply to any incident.
- *Italicized words:* These are examples of what the Operations Section Chief might say during the Operations Briefing to his or her staff using the Meridian Flood incident as background.

Meridian Flood

Operations Briefing Checklist

(Operational period beginning 16 November 0600)

General Briefing Issues

- **Chain of command** - Make sure that everyone understands the chain of command within the Operations Section.
    - Supervisors report to the Deputy Operations Section Chief who will report to me (the Operations Section Chief). I will be in Incident Command Post.
- **Resource requesting process** - Tell your Branch Directors, Supervisors, and Staging Managers the process for how you want resources to be requested when they are out in the field.
    - All resource requests will go through the Deputy Operations Section Chief (DOSC).
- **Situation status** - Report all significant changes in field conditions or occurrences through the chain of command.
    - The DOSC needs to know when you encounter problems out there that are going to impact your ability to accomplish your assignment or when an unanticipated situation develops that threatens the safety of responder.
- **Resources status**- Notify your supervisor of any additional resource needs, and if the resources that are assigned to you are not appropriate for the task.
    - We need to know if the resources that are assigned to you are not adequate for what you are facing. Let the DOSC know so that we can get the right kind and type of resources.
- **Resupply issues**- Ensure that you go through your chain of command for any resupply issues.
    - As soon as you know that you will require critical resupply such as fuel let the DOSC know. I do not want operations to be impacted because we did not provide enough notice to get critical supplies to you.
- **Real-time work progress reporting**- When briefing your personnel let them know who they should report work progress to and how often.
    - All supervisors are to report work progress to the Deputy Operations Section Chief every 3 hours.
- **Managing critical/sensitive information**- <u>Here you want to let your field personnel know of any restrictions on how you want them to report certain information.</u>
    - Should any of you discover a deceased person you will notify the DOSC by cell phone or face-to-face. At no time, will you transmit that information by radio.

- **Accidents/injuries-** <u>Request medical assistance and then notify supervisor.</u>
    - Call 911 immediately for any medical assistance and then notify your direct supervisor.
- **Environmental/property damage issues-** <u>Report any damage, either found or caused by our resources to your supervisor.</u>
- **End of shift debriefing process-** <u>Make sure you debrief with your relief, and after returning to ICP, provide full debriefing to SITL. Areas to cover in your debriefing include:</u>
    - Work accomplished- what specifically has been accomplished in your area of responsibility (e.g., number of homes searched, boundaries of the flood waters)
    - Performance issues- anything hindering performance (e.g., lack of training, personnel issues, equipment not performing as expected)
    - Proper resource mix- do you need additional or different resources to complete your assignments.
- **ICS-214 (Unit Log)-** <u>Fill out for each operational period.</u>
    - I do not want to hear from the Planning Section at the end of each shift that your ICS-214s are not being turned in.

Specific Issues

Once you have completed your general comments, take a few minutes and make sure to go through each of the ICS-204s with the Division/Group Supervisors and Staging Area Managers. The idea here is to ensure that each supervisor or manager understands their assignment. If you and Planning have done your job right, this should not take too long. The more dynamic the incident the more chance there is for confusion so make sure everyone understands their role. Consider using the incident map/chart if it helps you explain what you want done or to clarify questions. At a minimum the briefing should cover the following on the ICS-204.

- Work assignment
- Assigned resources
- Special instructions
- Safety considerations
- Radio / telephone communications

This is the time that the supervisors and managers will ask for clarification so make sure that you respond to their questions. If you know that one of the supervisors is working in a challenging area of the response and that it will require more discussion, talk to them after the briefing so everyone else can go to work.

Planning Section Chief Support Kit Checklist

Ensure that the following materials are available to the Planning Section Chief (PSC) during an incident, if not already provided in a unit or section specific support kit.

- ☐ ICS Forms Catalog
- ☐ Incident Management Handbook
- ☐ Local charts and maps
- ☐ Rulers
- ☐ Mylar sheets
- ☐ Felt-tip pens
- ☐ Dry-erase markers (wide and thin line)
- ☐ Pencils (lead and grease)
- ☐ Paper, sticky notes
- ☐ Scissors
- ☐ Notebooks (some waterproof)
- ☐ ICS-213 (general message forms)
- ☐ White-out
- ☐ Masking tape
- ☐ Easel chart
- ☐ In-boxes or large envelopes
- ☐ Stapler
- ☐ Push pins
- ☐ 3- or 6-part folders
- ☐ 2-hole punch
- ☐ Preprinted meeting agendas
- ☐ ICS forms on CD or flash drives etc
- ☐ Laptop computer with printer and software
- ☐ T-cards
- ☐ Position manuals for Planning Section positions
- ☐ ICS-215 (operational planning worksheet)
- ☐ Other IAP forms
- ☐ Intercom headsets for PSC and PSC Unit Leaders
- ☐ Copy enlarger/poster printer

Example Demobilization Plan
31 August 2008
For the Yaz Northern Incident

I. GENERAL INFORMATION

The response is rapidly transitioning from the emergency response phase to a planned recovery effort. The demobilization of incident resources must be conducted in a manner that is safe and efficient and not to interfere with ongoing operations. Every Staff Officer and Section Chief are to ensure that they maintain the appropriate level of staff to support the planned recovery phase. The following will be incorporated into the demobilization effort:

A. Responders that were operating within the Stauffer Chemical Facility will be offered the opportunity to undergo critical incident stress management.

B. Decontamination of personnel, personnel clothing, and equipment will undertaken under the direction of the Safety Officer.

C. All responders who are traveling by vehicle for more than 3 hours must have a minimum of 4-hours rest prior to being released. Any exceptions to the 4-hour rest rule must have the approval of the Unified Command.

D. Driving between the hours of 2200 and 0600 will be limited to airport transport to facilitate demobilization. Point-to-point driving for returning responders will be limited to 12 hours with sufficient breaks outside of 2200-0600 rest hours.

II. RESPONSIBILITIES

A. Planning Section Chief

i. Ensure that the demobilization process and expectations receive wide distribution and that there is an orderly release of resources

ii. Ensure that all agency/industry specific requirements regarding the demobilization of the agency's/industry's resources are followed. Any deviations must have the approval of the agency/industry Incident Commander

iii. Review the demobilization plan prepared by the Demobilization Unit Leader. Review Command and General Staff comments and make changes as appropriate prior to presenting the Plan to the Unified Command for approval

B. Operations Section Chief

i. Identify any excess personnel and equipment available for demobilization and provide list to the Planning Section Chief

ii. Identify and decontaminate all tactical resources that require decontamination. Coordinate the decontamination effort with the Safety Officer and Logistics Section Chief

    iii. Where possible, release resources that have preestablished shared transportation together to facilitate demobilization

C. Logistics Section Chief

    i. Coordinate all personnel and equipment transportation needs to designated location to meet travel needs

    ii. Ensure that the Supply and Communications Units are prepared to accept and document the return of all equipment that was checked out through them

    iii. Ensure you coordinate all vehicle inspections with the Finance/Administration Section Chief

    iv. Provide courtesy vehicle safety inspections for all noncontract vehicles

D. Finance/Administration Section Chief

    i. Ensure that all personnel and equipment time reports are complete and accurate

    ii. Ensure that any injury and/or equipment claims are well documented and complete

    iii. Adjust Equipment and Time Recorder's schedule to meet demobilization needs

III. RELEASE PRIORITIES

A. The following are the Release Priorities:

    i. Local government response resources

    ii. Federal government response resources

    iii. Industry resources

    iv. Release priorities may be adjusted to better serve the changing incident situation. Ensure that you obtain concurrence from agency that is providing the resource.

IV. RELEASE PROCEDURES

A. Section Chiefs and Command Staff:

    i. Have the authority to approve the tentative release list of resources to the Demobilization Unit Leader

    ii. Submit tentative release list of surplus resources to the Demobilization UnitLeader a minimum of 24 hours prior to the resource's anticipated departure

B. Demobilization Unit Leader:
   i. Prepare the Demobilization Checkout Form, ICS-221, when the tentative release list is approved by the Unified Command
   ii. Ensure that it is noted on the ICS-221 that resources requiring decontamination were decontaminated
   iii. Ensure that resources requiring critical incident stress debriefing are noted on the ICS-221
   iv. Effectively communicate with all staff members in order to identify any changes in the transportation needs of personnel. Ensure timely notification of anyone who will be impacted by changes in established transportation times
   v. Note on the ICS-221 any travel and arrival notification procedures that were established between the resource provider and the resource

C. Excess resources being demobilized are to follow the directions outlined on their respective Demobilization Checkout Form to ensure that all required signatures are obtained. Signatures may include the following Units:
   i. SPUL
   ii. COML
   iii. TIME
   iv. DOCL

V. PHONE DIRECTORY

Any time during the demobilization process that there is concern over the status of a released resources contact the Demobilization Unit Leader 999-555-3491. Other important contact points include:

- Hiatusport County Emergency Operations Center 999-555-4632
- Coast Guard Sector Hiatusport 999-555-8965
- Delaware State Emergency Operations Center 999-444-4021

VI. APPROVAL

Prepared by:
_____   _____
Demobilization Unit Leader   Date

Reviewed by:
_____   _____
Planning Section Chief   Date

Reviewed by:
_____   _____
Logistics Section Chief   Date

Reviewed by:
_____   _____
Finance/Admin Section Chief   Date

Reviewed by:
_____   _____
Operations Section Chief   Date

Approved:
_____   _____
Command   Date

Meeting Rules and Agendas

Ground Rules:

GROUND RULES

CELL PHONES AND PAGERS ON VIBRATE OR TURNED OFF

RADIOS OFF

STICK TO THE AGENDA

NO TEXT MESSAGING

NO SIDEBAR CONVERSATIONS

Initial Unified Command Meeting Agenda

INITIAL UNIFIED COMMAND MEETING AGENDA

1. PSC BRINGS MEETING TO ORDER, COVERS GROUND RULES, AND REVIEWS AGENDA.

2. VALIDATE MAKEUP OF NEWLY FORMED UC.

3. IDENTIFY JURISDICTIONAL BOUNDARIES AND FOCUS.

4. ESTABLISH AND DOCUMENT LIMITATIONS AND CONSTRAINTS.

5. ESTABLISH AND AGREE ON RESPONSE PRIORITIES.

6. DESIGNATE THE BEST-QUALIFIED OPERATIONS SECTION CHIEF AND DEPUTY OSC.

7. AGREE ON OTHER STAFF ASSIGNMENTS AS NEEDED.

8. AGREE ON INCIDENT SUPPORT FACILITIES AND THEIR LOCATION.

9. AGREE ON OVERALL ORGANIZATIONAL INTEGRATION AMONG ASSISTING AGENCIES.

10. AGREE ON RESOURCE-ORDERING AND COST-SHARING PROCEDURES TO FOLLOW.

11. AGREE ON OPERATIONAL PERIOD AND WORK SHIFTS.

12. AGREE ON SENSITIVE INFORMATION, INTELLIGENCE, AND OPERATIONAL SECURITY MATTERS.

13. DESIGNATE LEAD ORGANIZATION FOR INFORMATION, SAFETY, INTELLIGENCE/INVESTIGATION, AND LIAISON OFFICERS.

Unified Command Develop/Update Objectives Meeting

UNIFIED COMMAND DEVELOP/UPDATE OBJECTIVES MEETING

AGENDA

1. PSC BRINGS MEETING TO ORDER, CONDUCTS ROLE CALL, COVERS GROUND RULES, AND REVIEWS AGENDA

2. DEVELOP OR REVIEW/SELECT OBJECTIVES

3. DEVELOP TASKS FOR COMMAND AND GENERAL STAFF TO ACCOMPLISH

4. REVALIDATE PREVIOUS DECISIONS, PRIORITIES, AND PROCEDURES

5. REVIEW ANY OPEN ACTIONS FROM PREVIOUS MEETINGS

6. PREPARE FOR THE COMMAND AND GENERAL STAFF MEETING

Command and General Staff Meeting

COMMAND & GENERAL STAFF MEETING AGENDA

1. PSC BRINGS MEETING TO ORDER, CONDUCTS ROLE CALL, COVERS GROUND RULES, AND REVIEWS AGENDA

2. SITL CONDUCTS SITUATION BRIEFING

3. IC/UC:
   - PROVIDE COMMENTS
   - REVIEW RESPONSE POLICIES, PROCEDURES, AND GUIDELINES
   - REVIEW DIRECTION AND DECISIONS
   - DISCUSS INCIDENT OBJECTIVES AND PRIORITIES
   - ASSIGN FUNCTIONAL TASKS TO COMMAND AND GENERAL STAFF MEMBERS

4. PSC FACILITATES OPEN DISCUSSION TO CLARIFY PRIORITIES, OBJECTIVES, ASSIGNMENTS, ISSUES, CONCERNS, AND OPEN ACTIONS/TASKS

5. IC/UC CLOSING COMMENTS

Tactics Meeting

TACTICS MEETING AGENDA

1. PSC BRINGS MEETING TO ORDER, CONDUCTS ROLE CALL, COVERS GROUND RULES, AND REVIEWS AGENDA

2. SITL REVIEWS THE CURRENT AND PROJECTED INCIDENT SITUATION

3. PSC REVIEWS INCIDENT OPERATIONAL OBJECTIVES AND ENSURES ACCOUNTABILITY FOR EACH

4. OSC REVIEWS THE OPERATIONS WORK ANALYSIS MATRIX (STRATEGY AND TACTICS)

5. OSC REVIEWS AND/OR COMPLETES A DRAFT ICS-215, WHICH ADDRESSES WORK ASSIGNMENTS, RESOURCE COMMITMENTS, CONTINGENCIES, AND NEEDED SUPPORT FACILITIES (E.G., STAGING AREAS)

6. OSC REVIEWS AND/OR COMPLETES OPERATIONS SECTION ORGANIZATION CHART

7. SOF IDENTIFIES AND RESOLVES ANY CRITICAL SAFETY ISSUES
   LSC DISCUSSES AND RESOLVES ANY LOGISTICS ISSUES
   PSC VALIDATES CONNECTIVITY OF TACTICS AND OPERATIONAL OBJECTIVES

Planning Meeting

PLANNING MEETING AGENDA

1. PSC BRINGS MEETING TO ORDER, CONDUCTS ROLE CALL, COVERS GROUND RULES, AND REVIEWS AGENDA

2. SITL PROVIDES BRIEFING ON THE CURRENT SITUATION, RESOURCES AT RISK, WEATHER/SEA FORECAST, AND INCIDENT PROJECTIONS

3. PSC REVIEWS COMMAND'S INCIDENT OBJECTIVES, PRIORITIES, DECISIONS, AND DIRECTION

4. OSC REVIEWS THE OPERATIONS STRATEGIES AND TACTICS

5. OSC PROVIDES BRIEFING ON CURRENT OPERATIONS FOLLOWED BY AN OVERVIEW OF THE PROPOSED PLAN, INCLUDING STRATEGY, TACTICS/WORK ASSIGNMENTS, RESOURCE COMMITMENT, CONTINGENCIES, OPERATIONS SECTION ORGANIZATION CHART, AND NEEDED SUPPORT FACILITIES (E.G., STAGING AREAS)

6. PSC REVIEWS PROPOSED PLAN TO ENSURE THAT COMMAND'S DIRECTION, PRIORITIES, AND OPERATIONAL OBJECTIVES ARE MET

7. PSC REVIEWS AND VALIDATES RESPONSIBILITY FOR ANY "OPEN ACTIONS/TASKS" AND MANAGEMENT OBJECTIVES

8. PSC CALLS ON ALL COMMAND AND GENERAL STAFF MEMBERS TO SOLICIT THEIR FINAL INPUT AND COMMITMENT TO THE PROPOSED PLAN

9. PSC REQUESTS COMMAND'S TENTATIVE APPROVAL OF THE PLAN AS PRESENTED

10. PSC ISSUES ASSIGNMENTS TO APPROPRIATE INCIDENT MANAGEMENT TEAM MEMBERS FOR DEVELOPING THE IAP SUPPORT DOCUMENTATION ALONG WITH DEADLINES

Operations Briefing

OPERATIONS BRIEFING AGENDA

1. PSC OPENS BRIEFING, COVERS GROUND RULES, AGENDA, AND TAKES ROLE CALL OF COMMAND AND GENERAL STAFF AND OPERATIONS PERSONNEL REQUIRED TO ATTEND

2. PSC REVIEWS IC/UC OBJECTIVES AND CHANGES TO THE IAP (E.G., ANY PEN AND INK CHANGES)

3. IC/UC PROVIDE REMARKS

4. SITL CONDUCTS SITUATION BRIEFING

5. OSC DISCUSSES CURRENT RESPONSE ACTIONS AND ACCOMPLISHMENTS

6. OSC BRIEFS OPERATIONS SECTION PERSONNEL

7. LSC COVERS TRANSPORT, COMMUNICATIONS, AND SUPPLY UPDATES

8. FSC COVERS FISCAL AND TIMEKEEPING ISSUES

9. SOF COVERS SAFETY ISSUES, PIO COVERS PUBLIC AFFAIRS AND PUBLIC INFORMATION ISSUES, LOFR COVERS INTERAGENCY ISSUES, AND INTELLIGENCE/INVESTIGATIONS OFFICER COVERS INTELLIGENCE/INVESTIGATIONS ISSUES

10. PSC SOLICITS FINAL COMMENTS AND ADJOURNS BRIEFING

Instructions To The Field Observers (FOBS)

General:

- Establish contact with the field supervisor(s) whose area(s) you are working in
- Discuss with the field supervisor your information-reporting requirements
- Do not go into any areas where there is not adequate communications (you must be able to have communications with someone on the incident)
- Ensure that you have read and initialed the site safety plan and adhere to the Plan's requirements
- Ensure that you have the contact information for the field supervisors whose area you will be operating in
- Ensure that all equipment is in working order before going into the field (e.g., communications equipment (both radio and cell phone), safety equipment, GPS, digital camera, binoculars)
- Ensure that you have a copy of the base map and/or other more detailed maps to use as common references when reporting information back to the Situation Unit
- Make sure that all nonexpendable equipment is returned
- Ensure that you have the right clothing for predicted weather conditions
- Have on hand adequate water and food for the estimated time you will be in the field
- Make sure that you have coordinated your transportation requirements with logistics
- Use common map references (latitude and longitude) when communicating back to the Situation Unit

Information to Collect (list is not specific to any incident and not inclusive)

- Safety hazards (Safety Officer)
    - Power lines (lines down are lying across access roads)
    - Hazardous materials
    - Unique weather conditions (ice, fog)
    - Topography (steep slope, narrow canyons)
    - Water conditions (swift current, extreme tides)
- Discrepancies in resource deployment based on the IAP (RESL)
- Transportation (GSUL)
    - Condition of roads within the incident area (e.g., bridge limited to 5,000 lbs, traffic choke points)

- Work Accomplished
  - Measurement of fire line production
  - Amount of boom deployed and location
  - Status of mitigation activities (e.g., chlorine release secured, hole in levee wall 50% filled)
- Impacts of the incident on:
  - Transportation infrastructure
  - Wildlife
  - Commercial and private property
  - Historic properties
  - Cultural sites
  - Hindrance (e.g., private property)
- Amount and location of shoreline contaminated
- Impact of the response efforts on the environment (e.g., improper disposal of contaminated debris)
- Any suspicious activities
- Any spontaneous special interest group activities (e.g., they may be in harm's way)
- Validate prediction modeling (e.g., hazardous materials, fire, oil spill)
- Conduct weather observations (requires weather kit)
- Any established or potential sites for support facilities (e.g., helispots)

Report in times

- You are to regularly report in to the Situation Unit and provide updates (e.g., every hour)

Best Briefing Practices

The following practices should be of assistance if you're preparing and/or presenting a briefing.

As the Briefer:

- Plan ahead by arranging your source and display material in a logical sequence or use provided format that is expected and is easily understood
- At the start of your briefing ensure that you professionally introduce yourself. Your briefing should include:
    - The time and date the briefing material covers
    - Title of the briefing (e.g., Planning Meeting)
    - Incident situation, area impacted, any new support facilities established
    - Impact to infrastructure, modes of transportation (e.g., road closures)
    - Number of injured and fatalities
    - Success of mitigation efforts (e.g., 50% of buildings searched, 8 miles of shoreline boomed)
    - Major considerations (e.g., weather, tides and currents, high priority activities, political sensitivities)
    - Forecast, predictions, trajectories
- Understand the target audience for the briefing and tailor the briefing to meet the information requirement
    - If audience is mixed agency/organizations, avoid acronyms
- Anticipate potential questions in advance and have the answers ready. If you don't know the answer to a question, say you don't know and make note of the question for prompt follow-up. The Documentation Unit should be capturing open issues.
- Check the presentation area for lighting, display area, seating, and size for the anticipated audience
- Review preparations with the Planning Section Chief for advice and guidance
- Contact the key presenter (e.g., OSC, INTL) informally prior to getting together to make sure that there is a clear understanding of who will be briefing what material so that the briefing is coordinated
- Determine in advance if material of a sensitive nature is to be discussed, and if so, limit attendance according to presenters direction
- Speak in a strong well-modulated voice and avoid distracting mannerisms
- Ensure you know how the Planning Section Chief wants to work questions and answers (Q & A). Will Q&A be allowed during the briefing or following?

- Use presentation technology (e.g., PowerPoint) as appropriate and only if you have mastered it. You do not want the briefer or the technology to distract from the presentation.
- At the end summarize key points as necessary

Supporting Material:

- All display and handout material must have a date and time shown along with the person's name who prepared and or approved the material
- If you're using audiovisual material have spare bulbs, cords, handouts, and other material that might fail or not be sufficient for extra attendees
- If you're using wall displays that might be hard to read, then provide duplicates in smaller sizes for key attendees

Beyond Initial Response　　Appendix N

Facilities Needs Assessment Worksheet

FACILITY NEEDS ASSESSMENT WORKSHEET																				
1. INCIDENT NAME: MERIDIAN FLOOD												2. DATE & TIME PREPARED: 15 Nov 2100				3. OPERATIONAL PERIOD (DATE & TIME): 16 Nov 0600 to 16 Nov 1800			
4. DIVISION/GROUP/OTHER LOCATION	5. WORK ASSIGNMENTS	6. KIND OF RESOURCES	# of Personnel	Workspace (Sq. ft.)	Work stations	Speaker phones	Wall Space (sq ft)	Conf./Work Tables	Chairs	Clock	Easels	Power Outlets	Telephones	Video Projectors	Meeting Room (sq. ft.)	Printers	Fax	Copy Machine	Chart Pro Plotter	7. OVERHEAD	8. SPECIAL EQUIPMENT & SUPPLIES	9. REPORTING LOCATION	10. REQUESTED ARRIVAL TIME
ICP	Unified Command	REQ	6	400	1	1	60	1	8	1	1	3										
		HAVE																				
		NEED																				
	Liaison Officer and Agency Reps.	REQ	10	600	2		60	4	6	1	1	4	2					1				
		HAVE																				
		NEED																				
	Safety Officer	REQ	4	400	1		60	2	2	1	1	2	1									
		HAVE																				
		NEED																				
	Public Information Officer	REQ	2	200	1		60	2	4	1	1	2	2									
		HAVE																				
		NEED																				
	Intelligence / Investigation	REQ	2	200	1		60	2	4	1	1	2	2									
		HAVE																				
		NEED																				
	Planning Section	REQ	18	2K	8		120	10	16	1	4	20	8	1	800	4	2	1				
		HAVE																				
		NEED																				
	Operations Section	REQ	2	300	2				2	1												
		HAVE																				
		NEED																				
	Logistics Section	REQ	12	1K	6		120	12	14	1	2	8	10			2	2	1				
		HAVE																				
		NEED																				
	Finance/Admin Section	REQ	6	400	2		60	3	6	1	1	4	3									
		HAVE																				
		NEED																				
	Common Areas	REQ		500																			
		HAVE																				
		NEED																				
11. TOTAL RESOURCES REQUIRED			62	6K	24	2	600	36	62	9	12	45	28	1	800	6	2	3	1			
Facilities Needs Assessment Worksheet																			12. PREPARED BY (NAME & POSITION): LSC: T. Steves FACL: R. Pond			

N-1

Incident Command Post Check-off Sheet

The checklist below is provided to help you establish an Incident Command Post (ICP) that provides the incident command team with a proper facility and equipment to perform their job. The list contains information on ICP site selection, setup, and operating and equipment requirements. The list is not inclusive, but it will get you started in the right direction.

Site Selection Criteria

- ☐ Determine organization size and the space requirements of each function
- ☐ Is the proposed Incident Command Post (ICP) facility in a secure area?
- ☐ Is it located in proximity to the Incident?
- ☐ Is location convenient for Agency/Organization Executives to access?
- ☐ Is there adequate secure parking?
- ☐ Is there appropriate work space separation?
- ☐ Is there adequate meeting/briefing room space?
- ☐ Are there additional telephone lines available and will the facility accommodate them?
- ☐ Are you able to control public access?
- ☐ Is it near a helicopter (pad/landing zone)?
- ☐ Is it in a quiet area away from major distractions such as airports and railroads?
- ☐ Is it in close proximity to billeting and feeding facilities such as other agency operations centers and Emergency Operations Centers?
- ☐ Do you know the rental or lease cost of the facility?
- ☐ Is there adequate wall space for required displays?
- ☐ Is it located out of harm's way?
- ☐ Would it be able to accommodate the potential need for a separate Joint Information Center?
- ☐ Is there additional space available for co-locating the incident base?

Setup and Operating Requirements

- ☐ Develop a sketch map of the facility
- ☐ Develop clear directions and a map along with reference points for location of the facility
- ☐ Establish a check-in desk with a check-in recorder and ICS-211 forms
- ☐ Assign work space and identify each functional area (Planning, Operations, Logistics, Finance/Administration, Incident/Unified Command, etc.)
- ☐ Ensure that the check-in recorder knows the location of all functional areas
- ☐ Provide security for the facility and the parking area
- ☐ Establish facility and services contract and agreement to include daily maintenance
- ☐ Procure required furniture, equipment, and supplies for the ICP
- ☐ Install communications system
- ☐ Conduct facility and grounds safety, security evaluation, and mitigate problems as needed

- ☐ Develop and post an emergency evacuation plan and brief staff
- ☐ If necessary, augment sanitation facilities
- ☐ If necessary, negotiate facility use agreement

Equipment Requirements

- ☐ Fax machines – incoming and outgoing
- ☐ Professional-quality copy machine
- ☐ Video projector and projection screen
- ☐ Easels and flip charts
- ☐ Wall clocks
- ☐ Television with necessary connections to be compatible with audiovisual equipment
- ☐ Computers and printers, radio display processor for displays, digital and video camera
- ☐ ICS position vests
- ☐ Maps and charts as needed
- ☐ Dry-erase boards
- ☐ T-card racks to support Resources Unit Leader (resource status)
- ☐ Administrative support kits for Planning Section Chief, Logistics Section Chief, and Finance/Admin Section Chief

Incident Command Post Moving Plan

for the

Meridian Flood Incident

November 29, 20XX

Contents

Objectives	1
Directions	2
Map..	3
Parking	4
Meals	5
Command Post Layout	6
Command Post Phone Numbers	7

Objective

The purpose of this Plan is to ensure a timely and effective movement of the Incident Command Post from the Bayfield Operations Center to the Bayfield Town Hall. All personnel are requested to bring all equipment necessary to conduct work required by their ICS positions (printers, laptop computers, etc.).

Driving Directions from the
Bayfield Operations Center to the
Bayfield Town Hall

1. From Bayfield Operations Center parking lot, turn RIGHT onto Bucks Highway. Proceed approximately one mile.

2. Make a RIGHT onto East Mills Street.

3. Turn LEFT onto South East Street.

4. South East Street will turn into East Court.

5. The Incident Command Post is located at the end of the Court in the Bayfield Town Hall.

Map to the Incident Command Post

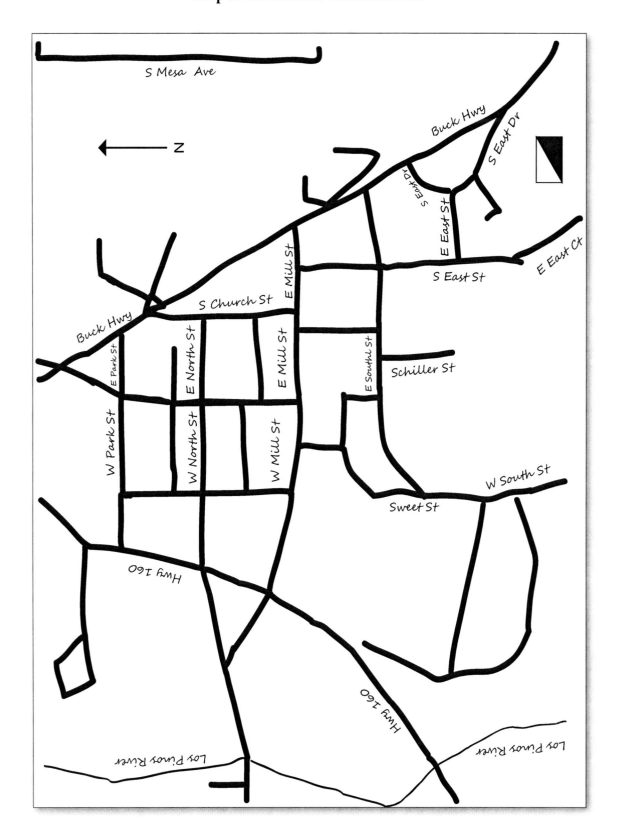

Parking

1. Parking is free to all Incident Command Post personnel.

2. Personnel are to park behind the building and check in. Check-in has been established just in front of the building's front entrance.

3. After checking in, you will receive a parking pass to be displayed in your vehicle windshield and the Check-in Recorder will direct you to your area of the Incident Command Post.

Meals

1. A working breakfast, lunch, and beverage service will be provided.

2. These services will be provided for Incident Command Post personnel ONLY.

Bayfield Community Center
Incident Command Post Layout

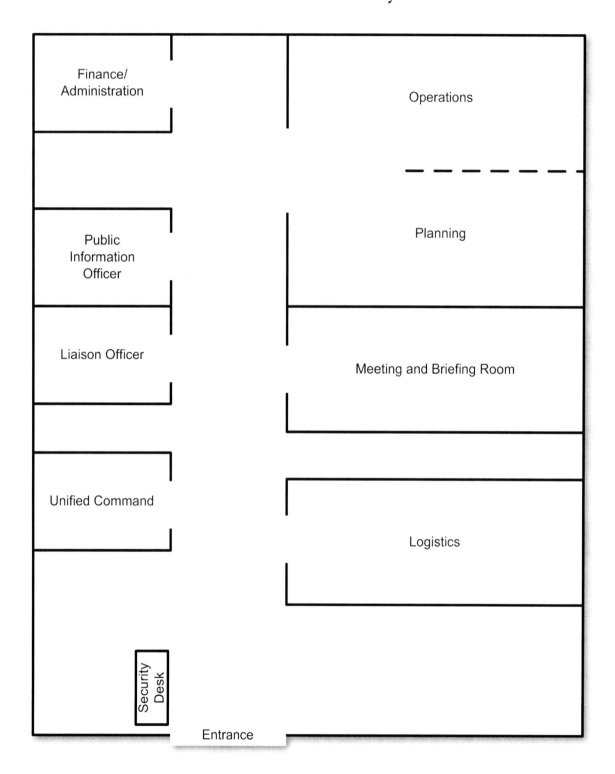

Phone Numbers for the Meridian Flood Incident Command Post

Command Table	808-555-3433
	808-555-3445
Operations Table	808-555-3434
	808-555-3439
	808-555-3449
	808-555-3450
Planning Table	808-555-3440
	808-555-3437
	808-555-3443
	808-555-3444
Logistics Table	808-555-3432
	808-555-3453
	808-555-3454
Finance/Admin Table	808-555-3435
	808-555-3451
	808-555-3452
	808-555-3455
Fax – Incoming	808-555-4476
Fax – Outgoing	808-555-3456
Command Staff Room	808-555-3431
Public Information Room	808-555-3436
	808-555-3438
	808-555-3441
	808-555-3442
Resources Unit	808-555-3446
	808-555-3447
	808-555-3448

Area Command Operational Cycle (Command Activities)

Meeting with ICs/UCs

- Receive IC/UC situational brief (copies of ICS-201s or IAPs)
- Discuss AC interim operating procedures
- Clarify AC roles and expectations
- Provide policy and direction
- Provide ground rules or procedures for on-scene IC/UC to follow
- Reach agreement with IC/UC on division of responsibilities (e.g., media, stakeholder meetings)

AC/UC Meeting

- Reach agreement on what are the critical resources
- Discuss limitations & constraints
- Prioritize incidents and critical resource allocation
- Develop overall priorities, objectives, and strategies
- Finalize the AC operating procedures (e.g., core hours of operation, meeting schedule)
- Identify any specific tasks for the AC staff

AC Staff Meeting

- Present AC decisions, directions, priorities and objectives
- Present AC operating procedures
- Discuss overall response emphasis including any limitations and constraints
- Present functional work assignments

Check-in AC Briefing & Establish ACP

- Provide guidance to AC staff on scope of assignment
- Convey agency executives' expectations, policy guidance, authorities, etc.
- Convey AC decision on staffing and support needs
- Assign tasks, as necessary (e.g., prepare for meeting with the ICs)
- Declare AC operational and notify the ICs, EOCs, JFO, command centers, and Agency Executives

AC Approve Operating Guide

- Review and approve Operating Guide
- Distribute the Operating Guide

Brief Operating Guide

- Ensure that the IC/UC, EOC, JFO and Agency Executives are familiar with the Operating Guide
- Provide leadership presence and motivational remarks
- Provide clarification and guidance

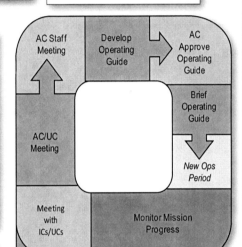

Activate AC organization

- Identify & agree on initial AC Staffing requirements
- Agree on who will fill key AC positions
- Agree on AC support needs
- Develop initial operating procedures

Monitor Mission Progress

- Maintain close liaison with IC/UC, EOC, JFO, and Agency Executives
- Attend functional meetings and briefings
- Evaluate overall effectiveness of the AC organization
- Resolve problems as they occur
- Follow up on staff work assignments/open actions
- Evaluate staff effectiveness and order additional staff as required
- Provide guidance and clarification

Executive Briefing

- Obtain Briefing
- Clarify scope of effort
- Discuss makeup of Unified AC
- Agree on critical information reporting
- Discuss limitations & constraints
- Define social, political, economic and environmental issues
- Identify cost constraints
- Identify on-scene commanders
- Discuss Area Command Post location

DELEGATION OF AUTHORITY/
DELINEATION OF RESPONSIBILITIES
(AREA COMMAND)

Date: 12 October

From: Sandy Kitchen, Director, Department of Public Safety

To: Thomas Yale, Chief, St. Louis Fire Department
Subject: Delegation of Authority

You are hereby delegated the authority to represent the Department of Public Safety as an Area Commander on the Unified Area Command organization that is being established in response to the earthquake that struck the city of St. Louis this morning. You have full authority and responsibility for managing the Department's activities and committing resources and funds necessary to support the Department's response within the framework of law, agency policy, and broad direction provided in this Delegation of Authority.

The St. Louis Area Command that you will be serving on is responsible for the following incident management teams that are operating throughout the city:

- Arch Incident
- University City Incident
- Kirkwood Incident
- Clifton Heights Incident

My specific expectations are that you will:

- Provide for Department personnel and public safety.
- Manage the incident cost-effectively.
- Serve as the day-to-day representative for the Department with the various incident management teams, providing strategic direction, support, advice, and coordination.
- Ensure that I am briefed at least twice daily or anytime a significant change in the status of the response effort occurs.
- Submit an ICS-209, Incident Status Summary no later than 2:00 PM each day.
- Maintain close coordination with the city Emergency Operations Center and FEMA Joint Field Office.

Specific circumstances in which you are to report to me immediately include:

- Injury (that requires hospitalization) or death of a Department of Public Safety responder
- Disagreement among the Area Commanders that cannot be resolved and is impacting the ability of Area Command to perform its responsibilities
- Any issues of a sensitive nature such as claims, litigation, or political impacts
- Negative public perception
- Serious shortfalls of specialized resources
- Coordination issues with city, county, state or federal entities
- Major changes in mass casualty status

This delegation is effective at 1:00 PM 12 October, and will remain in effect until command is transferred to a relieving Area Commander or the need for an Area Command organization is no longer necessary.

Sandy Kitchen
Director, St. Louis Department of Public Safety

Saint Louis Area Command
Area Command Operating Guide

—

Saint Louis Fire Department
Saint Louis Police Department
Saint Louis Department of Health

Saint Louis Area Command

October 12, 20XX

I. AREA COMMAND PRIORITIES
- Save lives and reduce suffering
- Provide essential lifesaving services
- Restore local infrastructure
- Maintain a common operating picture
- Ensure effective communications with all responding agencies
- Ensure Area Command is providing timely and effective support to the on-scene Incident Commanders
- Instill good resource management practices

II. AREA COMMAND OBJECTIVES
- Establish a critical resource management system – develop a resource allocation process that includes:
    - Assessment of on-scene incident resource needs to determine which resources are critical
    - Criteria for determining the priority for the allocation of identified critical resources
    - Process for obtaining the critical resources
    - Process for tracking critical resources
    - Assignment of the resources to a particular incident
    - A critical resource demobilization process
- Ensure the health, safety, and welfare of the responders and the public at the incident and Area Command level
- Ensure effective coordination/communications occur among the Area Command, the Incidents, Agency Executive(s), Emergency Operations Centers (EOC), Joint Field Office (JFO) and other response entities
- Ensure coordinated effort at the incident and Area Command level in the planning for and restoration of critical infrastructure
- Implement an aggressive media strategy in coordination with the incidents and other response entities for keeping the public informed
- Minimize adverse impact on the incidents through proactive Area Command support of incident level needs and activities
- Ensure an effective strategy is developed and implemented to provide stakeholders and elected officials with timely and accurate information
- Establish and maintain an effective Area Command organization that is capable of supporting the Incident Commanders and coordinating entities

III. AREA COMMAND MANAGEMENT PHILOSOPHIES

- Equal authority for all Area Commanders. One speaks for all
- Operate on a strategic basis only; do not become involved in tactical operations
- Act as facilitators, strategic priority setters
- Apply unbiased decision making in the allocation of critical and/or specialized resources
- Keep the Agency Executives informed and involved
- Maintain and facilitate a common operating picture among response entities (e.g., EOC, JFO)
- Minimize report demands on the Incident Commanders/Unified Commands
- Always strive to maintain a value added service to the Incident Commanders/Unified Commands and other coordinating bodies
- Continue to evaluate Area Command's usefulness and support to the Incidents
- Set priorities for assignment of critical resources to incidents, based on applicable criteria
- Whenever feasible, try to reduce workload associated with Incident Commanders/Unified Commands external influences (e.g., political stakeholders)
- Treat all Incident Commanders/Unified Commands as equals
- Provide the opportunity for Incident Commanders/Unified Commands to advocate their particular critical resources needs
- Keep the Area Command organization efficient, sized right, safe, and cost effective

IV. SAFETY MESSAGE

Area Command and incident management teams must take all measures necessary to ensure that the overall response is conducted in a safe and secure manner. We must respond aggressively but always maintain our focus on the safety and welfare of the responders and the public. The AC Safety Officer will be available to assist the incident Safety Officers when necessary. In case there is a serious safety situation, Area Command must be immediately informed of any accident or injury requiring follow-up medical treatment. Let's all work together to maintain a safe and secure work environment.

V. KEY DECISIONS – Area Command will:

- Coordinate all Regional and National Media upon establishing an Area Command Joint Information Center (JIC)
- Coordinate all VIP visits and briefings both at the Area Command Post and the Incident Command Posts
- Coordinate the use of aircraft among the incidents
- Manage the ordering, assignment, and demobilization of critical resources
- Prepare and distribute a situation summary (ICS-209) for off-site reporting
- Operate 24/7 on 12 hour shifts throughout the emergency response phase of the incidents
- Provide legal advice for the Incident Commanders/Unified Commands

VI. AREA COMMAND RESPONSE EMPHASIS

Using good response management practices Area Command must ensure that both the Incidents and the Area Command organization are working in harmony and that all entities are operating both effectively and efficiently as well as maintaining strong stakeholder coordination, public confidence, and political support.

VII. AREA COMMAND ORGANIZATION CHART

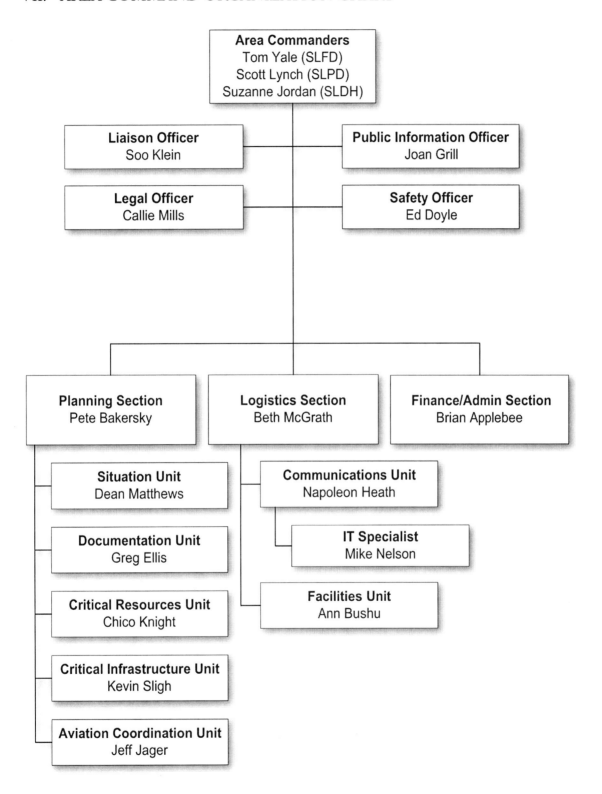

AREA COMMAND DAILY ACTIVITIES FOR OCTOBER 12, 20XX

Time	Activity	Participants
0600	Area Command Team Meeting	AC Staff
0730	AC/IMT IC Conference Call	AC Staff
0930	AC/Agency Executive Conference Call	AC, AACP, AACL
1100	AC Public Information Call	AC, PIO
1300	ICS-209s Submitted by the IMTs	SITL
1500	Incident Action Plans Submitted by the IMTs	AACP
1600	AC/IMT IC Conference Call	AC Staff
1800	AC Team Debrief	AC Staff

VIII. AREA COMMAND DAILY CONFERENCE CALL AGENDA

    A. Morning AC/IMT IC Conference Call

Summary by incident:

- Situational update
- Major activities
- Evacuations
- Injuries/accidents
- Resource needs next 24-hours
- Constraints and limitations
- Number of trapped victims
- Victims rescued
- Numbers receiving medical treatment

    B. Evening AC/IMT IC Conference Call

IX. CRITICAL RESOURCE ALLOCATION

Not all critical resource requests could be filled for the upcoming operational period. There is a significant drawdown on these specialized resources due to the far-reaching affects of the earthquake. Because of the critical resource shortfalls, incidents will have to use resources of lesser capability (e.g., Type II ambulance versus Type I). Area Command is working closely with the state Emergency Operations Center to identify additional resources to meet the requirements of the incident management teams and will keep the incidents informed. Each incident can be expected to receive the number and kind of critical of resources indicated below.

Resources	Arch	University City	Kirkwood	Clifton Heights
US&R	4	6	2	1
Ambulances	7	4	5	3

ABOUT THE AUTHORS

Michael de Bettencourt — is a Health, Safety, and Environment Program Director for a defense services contractor and is a retired US Coast Guard officer. He has provided emergency management program assessment services and conducts ICS training for industry and public sector clients. He has worked as a responder and response manager for oil spills, chemical releases, shipboard and facility fires, floods, large-scale law enforcement operations, and many maritime incidents. He served as Executive Officer on the US Coast Guard's National Strike Force Pacific Strike Team. He has lectured on ICS at many conferences and has trained hundreds of responders in the US, Europe, Asia, and New Zealand. Mr. de Bettencourt has a BS from the University of the State of New York, and an MS from the University of Houston.

Tim Deal — is the Federal Preparedness Coordinator for the Federal Emergency Management Agency (FEMA) in Region VIII. Prior to coming to FEMA he worked two years as a vice president in the emergency management training field. He is a retired US Coast Guard officer and has extensive emergency response experience with natural disasters, oil spills, and chemical spills, and he served as Operations Officer on the US Coast Guard's National Strike Force Pacific Strike Team. Tim was in charge of developing the US Coast Guard's ICS implementation strategies in Washington DC between 1999 and 2003. He has response experience in several ICS positions, including Planning Section Chief, Operations Section Chief, and Deputy Incident Commander and has trained hundreds of responders from government and private sector organizations. Mr. Deal has a BS from Humboldt State University and an MPP from the University of California at Berkeley.

Vickie Deal — works for a federal government agency in Denver, CO. Prior to that, she was an Operational Planner for the Federal Emergency Management Agency (FEMA) in Region VIII. She is also a retired US Coast Guard officer and has a variety of operational experience in emergency response management for oil spills, hazardous materials releases, as well as homeland security and port operations. Vickie Deal has been an Operations Section Chief and Deputy Incident Commander. She also served as a response program manager for the US Coast Guard in Washington DC and developed the ICS program implementation guidelines for the Coast Guard in the 1990s. She has a BS from the US Coast Guard Academy, an MS from Central Michigan University, and an MPA from the George Washington University.

Gary Merrick — is an Associate with Booz Allen Hamilton and is a retired US Coast Guard officer. In the Coast Guard he served as Commanding Officer of Marine Safety Office Charleston, SC. Gary Merrick also had a tour as Executive Officer on the US Coast Guard National Strike Force's Gulf Strike Team. He has been a Planning Section Chief, Operations Section Chief, and Incident Commander for oil spills, hazardous materials releases, security functions, and many other marine incidents. He has trained hundreds of responders from many state and federal agencies, and has served as a curriculum developer and instructor at the Coast Guard's prestigious training center in Virginia. He is a graduate of the US Coast Guard Academy, and received his MA from the George Washington University.

Chuck Mills — is the President of Emergency Management Services International, Inc. (EMSI). He has more than 40 years of experience in emergency management and provides emergency management services to government and nongovernmental organizations. He was previously Program Manager for the National Association of Search and Rescue (NASAR). In this capacity, he administered programs for the Office of Foreign Disaster Assistance (OFDA), the US Public Health Service (USPHS), the Federal Emergency Management Agency (FEMA), Veterans Administration (VA), and various state and local government organizations. He was instrumental in the development of both national and international disaster response teams and actively participated as a member on these teams. Mr. Mills was also an OFDA Disaster Management Specialist working on the development of their International Disaster Response System. He served 32 years in the US Forest Service, specializing in emergency management. During this period he was the federal representative for the development and implementation of the National Interagency Incident Management System Incident Command System. He has also served as a National Incident Commander, providing management oversight for numerous complex emergencies using the ICS. He is recognized worldwide as a subject matter expert on preparedness and disaster response operations.

INDEX

A

Air Operations Branch 1-22
Air Operations Branch Director 1-22
Area Command 10-1, 11-2, 11-6–11-8
Assigned Resource 1-5, 7-23
Assistants 1-4, 1-17
Available Resource 1-5, 7-23

B

Base 1-8, 1-11
Branch Director 1-22, 6-16, 6-17
Branches 1-14, 1-22, 1-26, 7-62

C

Camp 1-12
Check-in 1-8, 7-23
Command 1-16
Command and General Staff Meeting 2-9, 3-18–3-19
Command Staff 1-17, 5-1
Communications Unit 1-31, 8-11
Compensation/Claims Unit 1-32, 9-7
Cost Unit 1-32, 9-7

D

Divisions 1-20–1-22, 1-25, 6-12
Documentation Unit 1-28, 7-8

E

Emergency Operations Center 11-2–11-4
Emerging Communications 12-1. *See also* New Media
Execute Plan and Assess Progress 2-15, 3-23, 6-24, 7-17

F

Facilities 1-9
Facilities Unit 1-30, 8-6
Finance/Administration Section Chief 1-31–1-32, 9-1
Food Unit 1-31, 8-11

G

General Staff 1-18–1-19, 3-1, 3-14
Ground Support Unit 1-30, 8-6, 8-9
Groups 1-21–1-22, 6-12

H

Helibase 1-8, 1-10, 7-62
Helispot 1-10, 7-62

I

ICS Planning Process 1-34, 1-35, 2-15, 2-19
ICS symbols 7-61
Incident Action Plan 1-34, 1-36, 2-5, 2-11–2-14, A-1
Incident Base 1-8, 1-11, 7-28
Incident Commander 1-16, 3-1, 3-11–3-15, 4-1
Incident Command Post 1-9, O-1, P-1
Incident map 1-13, 7-64, A-11
Initial Unified Command Meeting 2-7, 2-8, 4-6–4-7, 7-13, K-2
Intelligence/Investigations Officer 1-16, 1-18, 5-41, 5-43, 5-44

J

Joint Field Office 11-2, 11-8
Joint Information Center 1-9, 1-12, 5-34, 5-37

K

Kind 1-6

L

Liaison Officer 1-18, 5-24–5-31
Logistics Section Chief 1-29, 8-1

M

Management Functions 1-15
Medical Unit 1-31, 8-10, 8-18

N

New Media 12-2. *See also* Emerging Communications

O

Objectives 3-4, 3-17
Operational Period 2-3, 2-15–2-18
Operations Briefing 2-15, 2-19, 6-22–6-23, 7-16
Operations Section 1-23, 1-26, 6-1, 6-11–6-14, 6-25–6-26
Operations Section Chief 1-19, 6-1
Operations Section Organization 1-20, 6-12
Organizational Elements 1-2, 1-3
Out-of-Service 1-6, 7-23

P

Planning Meeting 2-11
Planning Process. *See* ICS Planning Process
Position Titles 1-3, 1-4
Public Information Officer 1-16, 1-17, 5-1, 5-31

R

Resource Check-in 1-8
Resource Status 1-5
Resources Unit 1-28, 7-7, 7-23
Resources Unit Leader 7-23
Resources Unit Organization 7-27

S

Safety Officer 1-17, 5-1
Shifts 2-19. *See also* Work Shifts
Single Resource 1-7
Site Safety Plan 5-19
Situation Unit 1-28, 7-7
Situation Unit Leader 7-57
Span-of-Control 1-2, 1-14
Staging Areas 1-11, 6-13
Strategies 7-14, 7-15
Strike Team 1-7, 1-8, 6-13
Supply Unit 1-30, 8-6, 8-8
Symbols. *See* ICS Symbols

T

Task Force 1-7, 1-8, 1-24, 6-14
Technical Specialist 1-29, 7-5
Terminology 1-2, 1-5, 1-13
Time Unit 1-32, 9-5–9-6
Type 1-6

U

Unified Command 4-1

W

Work Shifts 2-19. *See also* Shifts